PRODUCT DEVELOPMENT AND DESIGN FOR MANUFACTURING

QUALITY AND RELIABILITY

A Series Edited by

EDWARD G. SCHILLING

Coordinating Editor
Center for Quality and Applied Statistics
Rochester Institute of Technology
Rochester, New York

ADDITIONAL VOLUMES IN PREPARATION

PRODUCT DEVELOPMENT AND DESIGN FOR MANUFACTURING

A Collaborative Approach to Producibility and Reliability

Second Edition, Revised and Expanded

John W. Priest
University of Texas at Arlington
Arlington, Texas

José M. Sánchez
Instituto Tecnologico y de Estudios Superiores de Monterrey
Monterrey, Mexico

MARCEL DEKKER, INC.   NEW YORK · BASEL

The first edition was published as *Engineering Design for Producibility and Reliability*, John W. Priest, 1988.

ISBN: 0-8247-9935-6

This book is printed on acid-free paper.

Headquarters
Marcel Dekker, Inc.
270 Madison Avenue, New York, NY 10016
tel: 212-696-9000; fax: 212-685-4540

Eastern Hemisphere Distribution
Marcel Dekker AG
Hutgasse 4, Postfach 812, CH-4001 Basel, Switzerland
tel: 41-61-261-8482; fax: 41-61-261-8896

World Wide Web
http://www.dekker.com

The publisher offers discounts on this book when ordered in bulk quantities. For more information, write to Special Sales/Professional Marketing at the headquarters address above.

Dedicated to our families:

Pat, Audrey, and Russell Priest
Mary, Paul, Carlos, and Christy Sanchez

ABOUT THE SERIES

The genesis of modern methods of quality and reliability will be found in a sample memo dated May 16, 1924, in which Walter A. Shewhart proposed the control chart for the analysis of inspection data. This led to a broadening of the concept of inspection from emphasis on detection and correction of defective material to control of quality through analysis and prevention of quality problems. Subsequent concern for product performance in the hands of the user stimulated development of the systems and techniques of reliability. Emphasis on the consumer as the ultimate judge of quality serves as the catalyst to bring about the integration of the methodology of quality with that of reliability. Thus, the innovations that came out of the control chart spawned a philosophy of control of quality and reliability that has come to include not only the methodology of the statistical sciences and engineering, but also the use of appropriate management methods together with various motivational procedures in a concerted effort dedicated to quality improvement.

This series is intended to provide a vehicle to foster interaction of the elements of the modern approach to quality, including statistical applications, quality and reliability engineering, management, and motivational aspects. It is a forum in which the subject matter of these various areas can be brought together to allow for effective integration of appropriate techniques. This will promote the true benefit of each, which can be achieved only through their interaction. In this sense, the whole of quality and reliability is greater than the sum of its parts, as each element augments the others.

The contributors to this series have been encouraged to discuss fundamental concepts as well as methodology, technology, and procedures at the leading edge of the discipline. Thus, new concepts are placed in proper perspective in these evolving disciplines. The series is intended for those in manufacturing, engineering, and marketing and management, as well as the consuming public, all of whom have an interest and stake in the products and services that are the lifeblood of the economic system.

The modern approach to quality and reliability concerns excellence: excellence when the product is designed, excellence when the product is made, excellence as the product is used, and excellence throughout its lifetime. But excellence does not result without effort, and products and services of superior quality and reliability require an appropriate combination of statistical, engineering, management, and motivational effort. This effort can be directed for maximum benefit only in light of timely knowledge of approaches and methods that have been developed and are available in these areas of expertise. Within the volumes of this series, the reader will find the means to create, control, correct, and improve quality and reliability in ways that are cost effective, that enhance productivity, and that create a motivational atmosphere that is harmonious and constructive. It is dedicated to that end and to the readers whose study of quality and reliability will lead to greater understanding of their products, their processes, their workplaces, and themselves.

Edward G. Schilling

PREFACE

The objective of this book is to illustrate the strategies and "best practices" for ensuring a competitive advantage. Successful product development has become a necessary but difficult collaborative task that requires effective and timely communication between various disciplines. Everyone involved in product development should have a basic knowledge of product development processes and the design fundamentals of producibility and reliability. A key benefit of this book is that it serves as a single informational source for the relationships between the many disciplines and methodologies. It is intended for college students and professionals involved with the design, development, manufacturing and support processes.

Technological change, global markets and the importance of knowledge are fundamentally changing product development. One unique aspect of this text is the large number and wide variety of topics presented. We believe that successful product development requires a systematic application of many methods and techniques that are tailored to the particular product and market environment. No single method is emphasized. Some of the many methods discussed include collaborative development, technical risk management, producibility measurement, mistake proofing, Boothroyd and Dewhurst DFMA, Taguchi methods, Six Sigma quality, rapid prototyping, testability, self-diagnostics, self-maintenance, environmental design, Isakawa, thermal analysis, and others. The relationships between these methods and the topics of the Internet, electronic commerce and supply chain are discussed.

Another unique aspect is the inclusion of the software design process. This is especially important in today's market, as many products are either software themselves or contain software. Software development requires different strategies and processes than hardware development. Today's design team needs to recognize these important differences.

Major industry studies of product development that the authors participated in provided the foundation for this book. These major studies include:

- Producibility Systems Guidelines for Successful Companies (1999 BMP Producibility Task Force Report)
- Transitioning from Development to Production DOD 4245.7M (i.e., often called the Willoughby templates)

- Best Practices for Transitioning from Development to Production (NAVSO P-6071)
- Producibility Measurement Guidelines (NAVSO P-3679)

This applied industry approach is helpful to both university students and practicing professionals. Much of this text was developed as a joint industry, government, and university project. Years with Texas Instruments, General Motors, and the U.S. Navy were also major influences when writing this text. Special thanks are given to Willis Willoughby, Jr., Ernie Renner, Douglas Patterson, Bernie List, Ed Turner, Gail Haddock and Lisa Burnell.

The book describes both what needs to be done (the best practices) and how to do them (product development steps or tasks). Brief examples start each chapter to give the reader a feel for the material that will be covered. Many of the examples are brief summaries of published articles used to highlight key design aspects, and expose the reader to current issues in design. Readers are encouraged to read the entire reference for greater understanding. Each chapter includes important definitions, best practices, an explanation of why the topic is important, and the steps for applying the methodology. These discussions will provide the reader with a working knowledge of critical techniques and terminology that will help in communicating with other engineering disciplines.

This book is divided into three major parts:

Part I introduces the reader to how change is affecting product development, the value of knowledge, and an overview of the design process.

Part II reviews the major stages in the design and development process. These chapters examine the processes of requirement definition, conceptual design, trade-off analysis, detailed design, simulation, life cycle costing, test and evaluation, manufacturing, operational use, repairability, product safety, liability, supply chain, logistics, and the environment.

Part III reviews producibility and the specialized techniques that have been proven in industry for designing a highly producible and reliable product. These include simplification, standardization, producibility methods, testability, and reliability design methods.

As mentioned earlier, this text is the result of joint industry, government and university projects involving many contributors. Some of the many contributors were:

BMP Producibility Measurement Task Forces both 1991 and 1999

All task force members but especially Robert Hawiszczak, Raytheon; Erich Hausner, TRW; Michael Barbieri, Lockheed Martin Tactical Aircraft; Scott McLeod, Dr. Mikel Harry, Motorola, and finally Amy Scanlon and Roy Witt, BMP.

Texas Instruments, Inc. (now Raytheon)

Paul Andro, Reliability and Design; Dennis Appel, Safety; Chris Beczak, Producibility; Lou Boudreau, Reliability; Bob Brancattelli, Documentation; Jim Brennan, Life-cycle cost; Phil Broughton, Maintainability; Bob Hawiszczak, Producibility; Oscar Holley, Engineering design; Chuck Hummel, Human engineering; Mike Kennedy, Manufacturing; Leland Langston, Engineering design; Skip Leibensperger, Software quality; Jim Marischen, Testability; Bill McGuire, Production planning; Jim Polozeck, System design; Henry Powers, Failure modes and effects; John Pylant, Thermal and packaging; Herb Rodden, System design; Jamie Rogers, Planning; Doug Tillett, Failure reporting; Merle Whatley, Requirements definition and conceptual design; Larry Franks and Jim Shiflett, Editors. **AT&T**: Key Cheaney, Software; **Rockwell International:** Ed Turner, Manufacturing and group technology; **Hewlett Packard:** Robin Mason, Software; **Naval Material Command**: Doug Patterson, DOD directive; **The University of Texas at Arlington**: Gail Haddock, Software; **Best Manufacturing Program**: Ernie Renner.

This project includes the research work, write-ups and suggestions of many students. The purpose was to insure that the information was effective for college students and recent college graduates. Special thanks are given to those who provided major materials including Richard McKenna (7 steps for assembly), Tracey Lackey (software testability rules), Marcelo Sabino (software translation), Heather Stubbings (software interfaces), Eloisa Acha (artwork), and the many other students who helped. Additional thanks are given to the many authors cited in the book.

Without the technical expertise of those listed above, this book would not have been possible. We especially appreciate the support of our families: Pat, Audrey, and Russell; Mary, Paul, Carlos, and Christy. Finally, we would like to acknowledge the editorial assistance and the technical support of Ann von der Heide and Ginny Belyeu, who lived through the book's many revisions.

John W. Priest, Ph.D. P.E.
jpriest@uta.edu

José Manuel Sánchez, Ph.D.
msanchez@campus.ruv.itesm.mx

CONTENTS

Part II   Stages of Product Development

Chapter 1

PRODUCT DEVELOPMENT IN THE CHANGING GLOBAL WORLD

The Times They Are A-Changing

We are entering a new era of fundamental change in products, services, and how they are delivered. Satellite phones, the Internet, intelligent machines, biotechnology, electronic commerce and many other new technologies are changing the world that we live in. As customers, we want customized products with more performance and options at a lower cost. At the same time, the very resources that organizations need to remain competitive; knowledge, people, equipment, facilities, capital, and energy are scarce or more costly. Manufacturing industries and business organizations must be able to react quickly to prevailing market conditions and maximize the utilization of resources. Product development requires better strategies and methods that are flexible, fluid, and promote simplicity. The key is a systematic application of best practices that focus on reducing technical risk in a changing environment. The only thing that will be constant in the future is change itself.

Current Issues

- New Business Models and Practices
- Global Business Perspective
- Trends Affecting Product Development
- Best Practices for Product Development

LIGHT SPEED CHANGE

Attributed to racecar drivers Mario Andretti and Richard Petty, "if things seem under control, you're not going fast enough." Today's world of business is like the speed-oriented, technology driven world of car racing. Things are changing so quickly due to advances in technology, that to keep pace and stay competitive requires a company to be able to adapt to the changes at lightning-fast speed. Time is a scarce resource. The future promises to be even more of a challenge. What's coming is a world of intelligent machines, telecomunications, nano-bots, and biotech that is so advanced, we will probably look back on these first days of commercial technology as very slow and mellow (Peters, 2000). Due to the completely new playing field, some of the more traditional approaches that a company might use to remain competitive are no longer appropriate. Can a company even succeed today when systems are so much more complex and when technology changes so rapidly? The answer is yes, if we focus product development efforts not only on a product's function, project schedules and deadlines, and cost, but also in other life cycle issues such as customization, technical risk, simplicity, producibility, quality, innovation and service.

Today's customer wants value, a product that is easy to purchase and meets specific needs. Value is expressed as the relative worth or perceived importance of a product to the customer. It can be measured by a series of critical marketing parameters such as innovation, styling, performance, cost, quality, reliability, service, and availability. Customers define value in relation to their personal expectations for the product or service offered, which means every company must meet an infinite number of expectations. The more you exceed those expectations, the more value you have delivered.

Finding ways to create innovative solutions provides the greatest opportunity for distancing yourself from the competition. Innovation, however, does not come cheap. "At one company, it can take as many as 250 raw ideas to yield one major marketable product." (Peters, 1997) A company that is committed to innovation should have a high tolerance for many failures. In Search for Excellence, Tom Peters makes numerous references to the fact that it is possible to finish projects under budget and on time (Peters, 1989). He notes the success of the "skunk" work at Lockheed. Many factors that contributed to its success, including reliance upon the best technical expertise available and off-the-shelf solutions.

A company has to be willing to put itself at risk everyday through innovation. Small, incremental changes of the past won't work in today's marketplace. Somewhere there is a competitor who is committed to change and developing innovative solutions. New technologies, markets and aggressive schedules require a company to take some risks. A key goal of product development to is to identify these technical risks early in the development process and implement methods to minimize their potential occurrence and effect. Best practices provide the framework and systematic process for success.

1.1 NEW BUSINESS MODELS AND PRACTICES

New technologies are driving change in every aspect of our lives. The future will be even more chaotic with more advances at a faster rate. For many companies the barriers to competitive entry will be lower than ever. Knowledge and time will be the scarcest resources. Successful companies will be those whose business model can respond to change more quickly and more effectively than their competitors.

Much has been written on the potential of technology to drastically change business models and manufacturing practices. Often technology has been used to replace an existing process. For example, CAD systems are used to replace drawing tables. Presentation software such as Powerpoint has replaced the overhead transparency. In such cases, technology enhanced a particular process and has been introduced without too much disruption, but the benefits are equally modest. Technology's true potential is realized when it is employed in innovative ways that change traditional business practices.

The "New Economy" is a name for the future business environment that will consist of new business models using the Internet to interact with customers and suppliers. It will be a technology driven, knowledge rich, collaborative interactivity. Computers and the appropriate software will enhance collaborative product development through decision support systems, engineering analyses, intelligent databases, etc. Collaboration is not new; however, computers will improve collaboration independent of time and place.

Technology will cause more new customized products to be introduced more rapidly resulting in shorter product lives. For example, every time a new microprocessor is introduced to the market, new laptops are designed with the new product. A typical company may have to introduce one or two new notebook computers per month, leaving their two-year-old laptops worthless for sale.

The tremendous increase in information and the knowledge for these new technologies is also increasing the role of design specialists or experts. Thus, design team members will be selected for specific areas of expertise such as knowledge in circuit design, programming skills, composite materials, or electronic assembly, financial markets, supply chain rather than for their generalized degrees in business, electrical, computer, mechanical, industrial, or manufacturing engineering. To find these many specialists and experts the company may have to go outside of the company. Developing products in this collaborative environment is like running a relay. Speed and innovation are meaningless when a handoff of the baton from one team member to another is dropped. It is essential that each member of the design team, management, vendors etc. is communicating and knowledgeable of the development process. The key is to insure effective collaboration between these many different specialists even though each has unique locations, objectives, methods, technologies and terminology.

Everyone and everything will be connected and the digital economy will change accordingly. In the future, businesses will be open 24 hours a day, 7 days a week. Areas where e-commerce will change the way we do business in the future include collaborative product development, real time buying and selling, customer service - one to one dialog, communities of key partners and vendors, employee communication, and knowledge management. Smart products with embedded microprocessors will communicate not only with the user but also with other products and service departments when needed.

1.2   GLOBAL BUSINESS PERSPECTIVE

The Internet, fax machine, higher bandwidth, and faster methods for travel/shipping have made globalization a reality. This provides new opportunities as well as threats. Consumers can receive worldwide information freely from around the world and purchase items offshore in many countries. Sales can come from anywhere and will be globally visible. Firms are now able to acquire resources, skills and capital from global sources that best satisfy their business objectives. Both the place and scale of all types of changes continue to accelerate as we move into the 21st century. Multiple technological revolutions will have a substantial impact on almost all consumer and business activities.

Global customers are becoming more demanding expecting greater product value although their definition of value continually expands. They want products that meet their specific needs. Products must be customized for these needs including culture, language, and environments. Thus, Global companies must become more flexible, efficient (i.e. lean), and focused on their "core competencies" outsourcing those items that they are not proficient in. This is also resulting in many partnerships and alliances between companies specialized in different aspects of the market.

In response, business models and product development strategies have to be more tailored to the particular situation; the challenge is to quickly adapt to any type of new customer requirements, or to move in a totally new direction while maintaining a consistent vision for the enterprise. Several marketing, economic, and technology driven issues are fueling this perspective. For example, we are observing shifts in the nature of business practices in order to:

- **Anticipate future market demands** – These rapid large-scale changes mean that companies can no longer wait for customers to tell them what is expected; if they wait, they'll be too late. There is simply not enough time to merely satisfy customer's current needs. Companies must anticipate the effects of future technologies on customer expectations and position themselves, well in advance, to satisfy them.
- **Manage global relationships** - The global business environment is increasingly competitive and continuously changing. Survival in

this environment demands agility and an absolute commitment to excellence. To be successful a manufacturing company needs to offer products with world-class quality. A company is said to be "world class" when it is capable of providing products or services that can successfully compete in an open global market. Extended enterprises, partnerships and alliances will be used in a multitude of forms to create synergy and a competitive advantage. Successful businesses will be truly global in their organizational structures and value-creating processes. Through an extensive network of relationships, they will conduct R&D, design and manufacturing activities in any world region and will gain access to all lucrative markets.

- **Reduce time to market** - In today's market, there is pressure to produce products with much shorter lead times. This is often called "speed" and is true for both new products and products already in production. New products must be designed and manufactured quickly in order to get out into the market before the competition. If a product is late, market share is lost. Lead-time from order to delivery of existing products, for many companies, is expected to shrink over 30% per year.

- **Excel in customer service.** World-class organizations put customers' needs at the top of the agenda. For such companies, the idea of being close to the customer is very important to compete against global competition. In a global economy, customer requests may come in many ways: e-mail, web forms, fax or the old-fashioned telephone call. Customer response management (CRM) will continue to become more important for success. The key is to do what it takes to meet customer service requirements.

1.3   TRENDS AFFECTING PRODUCT DEVELOPMENT

Companies are continually responding to change. A few of the most important trends affecting product development and product life cycle are:

- Rate of innovation
- Software tools, rapid prototyping, and virtual reality
- Mass customization and customized "on-demand" production
- Core competency, partnerships and outsourcing
- Internet and telecommunication
- Electronic commerce
- Flexibility and agility
- Global manufacturing
- Automation
- Environmental consciousness

1.3.1 Rate of Innovation

Innovation is a critical design factor in the success of most products and a core competency for many companies. Customers want exciting, unique, colorful, and interesting products now. The advances of new technologies are providing many opportunities for increasing the rate of innovation. Embedded processors for smart products, new materials, and virtual reality will all provide new and exciting products for the future. For many companies, software is becoming the focused area for innovation that differentiates their product from its competitors.

Significant pressures exist to develop new products much faster since the first innovative product is often the competitive winner in a "winner take all" market. Many product development projects are now started before some key technologies are even ready. This co-development method increases the level of technical risk. In the future there will be less and less time to develop innovative products. Time constraints will become an even greater factor in the development and production of new innovative products.

Automobile Innovation

As noted by Jerry Flint in Forbes Magazine, success in the automobile industry depends on innovation and passion. It's not enough to find out what buyers want. Companies have to be out ahead of them, producing things the consumer can't resist. Customers want looks, power, and gadgets. A car that's good enough but nothing special will sell, but there's no profit in it because people will buy it only with a fat discount or rebate. If automobile makers build some passion into their cars, customers will feel it and go out and buy them, in spite of sticker shock. (Flint, 1997)

1.3.2 Software Tools, Rapid Prototyping, and Virtual Reality

Most businesses are driven by three essential activities: design, manufacturing, and service. Competitive advantage is gained in the design process when software products or tools can improve or shorten product development and enhance service support. Available software tools such as computer aided engineering (CAE/CASE), expert systems, virtual reality, and prototyping tools can assist the design team in customer needs assessment, innovation, product layout, selection of components or materials, simulation testing and simulating the manufacturing process. Intelligent CAE systems are able to quickly access large knowledge bases and provide answers to questions that might arise during the design process. Technical developments allow companies to have visualization and simulation capabilities that are more complete than a conventional CAD system. Virtual reality utilizes interactive graphics software to create computer generated prototypes and simulations that are so close to reality that users believe they are participating in a real world situation or observing a real final product. Similarly, prototyping consists of building an experimental product rapidly and inexpensively for users to evaluate. By interacting with a prototype, users and business personnel can gain a better idea of final manufacturing and functional requirements. The prototype approved by final users can then be used as a template to develop the final product requirements and the corresponding manufacturing operations. Someday the design team will define the product's requirements and then the software tools will develop the possible design options, perform engineering analyses, simulate testing, build prototypes to evaluate, purchase parts, and send all of the necessary information for manufacturing the product to the factory.

Computer Testing for Verification

As published in the Dallas Morning News, "The latest generation Gulfstream V corporate jet will fly farther than any existing corporate jet. The key is its wing, which was designed using a radical new approach by a team at Northrop Grumman's Commercial Aircraft Division in Fort Worth, Texas. The wing was built in less time and with far fewer parts than a typical wing. For the first time at Northrop, a powerful three-dimensional computer-aided design system was used that let designers test the parts without building an expensive, full-scale metal model. (The giant Boeing 777 was also designed this way). No wind-tunnel testing was needed, either: The computer did it." (Dallas Morning News, 1996)

1.3.3   Mass Customization and Customized "On-Demand" Production

Customers want products that are tailored to meet their unique needs. Rather than buying a generic product and being satisfied, customers want to purchase a "customized" product. Mass customization is the ability to provide specific product or service solutions while still realizing the benefits of large-scale operations. One technique is to build a generic product and have someone else (e.g. retailer, customer) modify the product or service as needed. An increasingly more popular solution is called "build to order" or "on demand" production, which requires manufacture of a customer, specified product only after it is ordered. This requires manufacturing to build many more versions of similar products, i.e. flexibility.

Notebook Computer Customization

Customers can now order a notebook computer with customer selected features. The number of possible selections for each option is limited such, as 3 different processors or 2 different hard drives. Other selections are whether to include an item such as a DVD or CD-ROM drive. The order is immediately sent to the factory, which builds the notebook in the configuration ordered. Within a very short period of time, the product is delivered to the customer. To do this, some companies are changing their workstations to allow a single worker to assemble an entire "built to order" computer. IBM, Dell, Hewlett Packard and Compaq are all implementing build to order.

1.3.4   Core Competency, Partnerships and Outsourcing

The cost of staying competitive and developing new technologies is forcing most companies to focus internal development on just the few core competencies that the company considers strategic to being competitive. Core competencies can be in any area of design, marketing, manufacturing, or service. These are the critical areas that make the company unique or better than the competition. Sometimes companies will buy or merge with other companies that have needed core competency. Areas that are not part of the core (i.e. not critical) but are still needed are purchased or developed for the company by outside vendors and partners. The result of companies focusing on core competencies is a rising trend called outsourcing. Historically, companies have tried to manufacture almost all of a product's parts and software "in-house". This required a company to be good in many different areas of manufacturing and software development. For many companies, the cost of doing everything has become too high. Outsourcing parts, assemblies, and software modules can save money. Successful outsourcing companies have developed into world-class manufacturers with high efficiency and lower costs. A great example of this is in the field of electronics. Companies who develop consumer products usually specialize in the design process and protecting their brand name and then leave manufacturing and distribution to other companies. This increase in outsourcing places more importance on vendor selection and supply chain management

Purchasing core competency

CISCO had the second highest market capitalization in the world in early 2000. They built their empire on identifying what additional services/products that their customers wanted, and then purchasing companies that provided these services to build core competencies within the company. To build this empire they purchased 51 companies in 6.5 years and 21 companies in 1999 alone. CISCO successfully rebuilt itself by using acquisitions to reshape and expand its product line. (Thurm, Wall Street Journal, 2000).

1.3.5   Internet and Telecommunications

The Internet and telecommunications have provided the communication platform for the increased level of collaboration that is required in today's business world. Since the Internet has become an integral part of daily life, it is becoming a business need, not a technological option. It has created a new universal technology platform upon which to build a variety of new products or services.   The Internet's potential for reshaping the way products are developed is vast and rich, and it is just beginning to be tapped. By eliminating geographic barriers obstructing the global flow of information, the Internet and telecommunications are paving the way for increasing collaboration in new product development.

1.3.6   Electronic Commerce

Another benefit of the Internet is electronic commerce or E-commerce. E-commerce is the paperless exchange, via computer networks, of engineering and business information using e-mail and Electronic Data Interchange (EDI). Its introduction has dramatically changed the procurement process and will eventually impact all suppliers and vendors. Internet file transfer of design CAD data has become a part of the printed circuit board production cycle at many companies. The factory electronically receives the data and then immediately manufactures the printed circuit board to specification. Another way for companies to utilize electronic commerce is to solicit price quotes and bids from vendors. The EDI can also be used to combine electronic transactions with manufacturing inventory and planning information to have parts directly delivered to the factory floor. As technology and the Internet evolve, file transfer methods will continue to improve.

Areas that electronic commerce will affect include:

1. Marketing – customers and vendors provide up-to-date information for designers.
2. Purchasing and inventory management – lower prices based on paperless transactions and increased number of bids.
3. Re-order schedules – real time sales data and inventory information from the factory floor is provided to vendors.
4. Supply chain management – allow companies and vendors to work closely together as partners.
5. Design tasks – real time multidiscipline design group collaboration.

The key to e-business success revolves around these questions: What you can offer your customer? How do you improve your customer or client's market position? How do you increase value?

E-commerce/business/web may change product development forever. Companies will have to predict where this technology will take the global markets, product design, and manufacturing. The key to success will always be innovation. There are predictions that company-to-company Internet trade will hit $134 billion a year by the end of the decade. A web site, coupled with its custom software, can enable users to contact thousands of suppliers, who can also respond over the Internet saving time, money and a lot of paperwork. All three US car manufacturers are combining on an on-line purchasing system to leverage their size to win price cuts from suppliers. E-commerce will also provide 24-hour collaboration as team members, vendors and customers will communicate automatically between databases using software agents.

Finding Replacement Parts

When machinery at GE Lighting's factory in Cleveland broke down they needed custom replacement parts, fast. In the past, GE would have asked for bids from just four domestic suppliers. This required getting the paperwork and production line blueprints together and sent out to suppliers. This time they posted the specifications and "requests for quotes" on GE's Web site and drew seven other bidders. A Hungarian firm's replacement parts arrived quicker, and GE Lighting received a 20% savings (Woolley,2000).

1.3.7   Flexibility and Agility

The manufacturing requirements of reduced lead-time, fast changeovers for new products and mass customization have forced manufacturing to become more flexible. Manufacturing must now produce a wide variety of product configurations quickly without modifying existing lines or adding new production lines. Flexibility can include product mix, design changes, volume level, and delivery. As described by Stewar (1992), investment in flexibility is not cheap. Flexibility offers economies of scope with the ability to spread cost across many products. New highly flexible lines cost more to build (10% more at Toyota, 20% at Nissan), but a single model change pays more than the difference due to shorter change over-times and lower costs. Flexibility pays off when a factory can change its product line by reprogramming existing equipment rather than by replacing it.

The Flexible Process Manifold Room

In the early 1990's Merck, a major pharmaceutical company was a typical inefficient manufacturing company. Most of its competitors went to outsourcing their production. Merck did not. Instead they streamlined existing operations and implemented flexibility. They wanted to be able to quickly ramp up production of new products and change volume levels of existing products. To do this, they developed a giant room called the "process manifold room" where giant flexible tubes come out of the walls. (Wall Street Journal, 2000) These flexible tubes connect chemical reaction chambers, tanks, centrifuges, and dryers with one another. It was more expensive to build this way but it provides great flexibility. With the right combination of tubes it can build almost any of their drugs.

1.3.8   Global Manufacturing

Globalism is changing product development and manufacturing. Products are now being collaboratively developed in many different locations all over the world. For manufacturing, this has resulted in worldwide production facilities and the use of many overseas vendors. The same products may be manufactured in many different locations in several different countries. The different manufacturing facilities can vary in terms of process capabilities, labor skills, quality, cost, etc. Critical parts that are purchased from another country must then be shipped to the various facilities. This causes many communication and logistics problems. Effective communication is difficult due to the differences in languages, cultures, time zones, education, and government regulations. Logistic costs such as shipping and handling become an even larger portion of a product's cost. Scheduling, ordering and shipping of parts and products all over the globe also becomes much more complex.

Global High-tech Hubs

Guadalajara, Mexico for example is becoming a major manufacturing center for electronic products sold in the United States. Large world-class contract electronic manufacturers (CEM) are manufacturing products for companies such as Compaq, Cisco, Ericsson and others (i.e. outsourcing). Although their labor costs are higher than Asia, their closer location allows faster delivery and cheaper shipping costs. Seven of the ten largest electronics contract manufacturers in the world have now located major factories there.

1.3.9   Automation

Manufacturing improvement has often focused on technological improvements. Automation of manual tasks has been a major strategy for many years. The current automation phase is driven by the development of complex product technologies such as semiconductors and optoelectronics that require automation and new integrated manufacturing planning and control. This reason for automation is called "product necessity" (e.g., stringent product requirements force a company to use automation in order to manufacture the product). An example of where automation is a product driven necessity is the semiconductor industry, due to the design's decreasing circuit path sizes and increasing clean room requirements.

Another major use of automation is to improve quality. A major misconception is that automation itself results in higher quality, but this is not necessarily true. An advantage of automation is that it produces a highly consistent product. This consistency, when combined with an effective total quality control program, provides the ideal basis for quality improvement. By solving quality problems in an automated process, the highest levels of product quality can be obtained.

1.3.10   Environmental Consciousness

Companies are incorporating environmental consciousness into product and process development. Activities focus on minimizing process waste, minimizing energy use, developing manufacturing process alternatives, using environmentally friendly materials, and integrating pollution control and abatement into their facilities. In some cases companies may invest in a particular technology only to discover a few years later that changes in some environmental factors make the investment worthless. For example, car companies are somewhat uncertain about investing in technology to manufacture electric cars, because they are not sure about the future of federal regulations on this matter. Over 500 million computers with 1 billion pounds of lead based wastes will be disposed of in the next few years. Manufacturers may become financially responsible for the recovery of used computers. The future ramifications of the environment on product development could be huge.

1.4   BEST PRACTICES FOR PRODUCT DEVELOPMENT

This period of rapid changes caused by new technologies and globalization suggests a new focus on product development. Innovation and implementing new technologies require companies to take greater risks. The key is to identify, reduce and control these risks without limiting innovation, creativity, simplicity, and flexibility. Most of the problems found in current product development can be traced back to inadequate design or design management. The focus needs to be on the fundamentals of successful product development which we will call "best practices". They are developed from lessons learned that are examples of the best and worst methods that have been found on previous projects and at different companies. In general, a Best Practice for product development is valuable if it:

- Improves communication among all the members of the product development team.
- Provides "what to" and "how to" recommendations that have been proven to be successful in industry.
- Develops scientific recommendations for both current and future product development.
- Helps measure the progress of the product development process and the technical risk of new technologies/methods.

Using best industry practices have been shown to be a very effective method for product development. This approach identifies the "best" practices for a particular application form both inside and outside of the company. It hopefully ensures the design team to not repeat the same mistakes from previous projects. A nation wide initiative called the Best Manufacturing Practices (BMP) program, sponsored by the Office of Naval Research, began in 1985. The objective was to identify, research, and promote exceptional manufacturing practices in design, test, manufacturing, facilities, logistics, and management. The focus was on the technical risks highlighted in the Department of Defense's 4245-7.m Transition from Development to Production Manual. The primary steps are to identify best practices, document them, and then encourage industry, government, and academia to share information about them. This project has also sponsored the largest and most popular repository of lessons learned for producibility. As noted in their website, **www.bmpcoe.org**, it has changed American industry by sharing information with other companies, including competitors. Their unique, innovative, technology transfer program is committed to strengthening the U.S. industrial base.

A successful design strategy requires a multidisciplinary collaborative approach that focuses on using best practices. The strategy identifies tasks and activities that are the most likely to produce competitive products, promote simplicity and reduce technical risk. Of course which best practice to use and its application all depends on the company, business environment and the product itself.

This book provides the elements of a flexible framework for collaborative product development resulting in a competitive advantage. It includes an overview of product development strategies as well as a description of best design practices. These practices have been proven in industry to (a) shorten time-to-market; (b) reduce development, manufacturing, and service costs; (c) improve product quality; (d) reduce technical risk, and (e) strengthen market product acceptance.

The main purpose of the book is to document and illustrate design practices that improve product competitiveness, not to train any type of product support specialist. The book is intended for all engineers, managers, designers, and support personnel who need to be collaboratively involved in the product life cycle. The documented material describes tested methods that identify areas of technical risk where new product development can bring the largest benefits.

1.5   REVIEW QUESTIONS

1. Identify at least three global issues that affect current practices of product development.
2. Compare today's Internet business methods to those of the past.
3. Identify other future trends of technology and global markets and how they might affect product development.
4. Think about a particular product (i.e. a cellular phone, a notebook computer) you can purchase today. Are they about to be rendered obsolete by something new? Explain
5. Discuss how advances and new directions in technology over the last 50 years affect business organizations

6. Identify and describe three emerging technologies that are likely to impact the way manufacturing companies develop new products.

1.6   SUGGESTED READINGS

1. T. Allen, Managing the Flow of Technology, The MIT Press, Cambridge, MA. 1993.
2. D. Burrus, Techno Trends: How to Use Technology to Go Beyond Your Competition, Harper Business, 1999.
3. Davis, et.al, 2020 Vision: Transform Your Business Today to Success in Tomorrow's Economy. Fireside/Simon & Schuster, 1991.
4. Hayes and S. C. Wheelwright, Restoring Our Competitive Edge. John Wiley and Sons, New York, 1984.
5. R. McKenna, Real Time: Preparing for the Age of the Never Satisfied Customer, Harvard Business School Press, Boston, MA. 1999.

1.7   REFERENCES

1. Dallas Morning News, Dedicating the New Wing, Dallas Morning News, page 2D, September 11, 1996.
2. J. Flint, Why Do Cars Cost so Much, Forbes Magazine, March 10, 1997.
3. T. Peters and R. H. Watesman. In Search for Excellence, Warner Books Co., New York, 1989.
4. T. Peters, Foreword for Innovation, edited by Kanter, Kao, and Wiersema, HarperBusiness, 1997.
5. T. Peters, New Economy, ASAP, Forbes Publishing, 2000.
6. T. Stewar, Brace for Japan's Hot New Strategy, Fortune, p. 63, September 21, 1992.
7. S. Thurm, Under CISCO's System, Wall Street Journal, p. 1. March 1,2000. Wall Street Journal, How Merck Is Prepared to Cope, Wall Street Journal, p. A8, February 9, 2000.
8. S. Woolley, "Double click for resin," Forbes, March 10, p. 132. 2000.

Chapter 2

PRODUCT DEVELOPMENT PROCESS AND ORGANIZATION

Collaborative Multidiscipline Design

    Modern product development needs to collaboratively manage many conflicting and complex requirements in a rapidly changing environment. Best practices must focus on meeting today's requirements of reduced cycle time; higher manufacturing quality; greater design flexibility for expanded or optional features demanded by the customer; lower cost, and higher reliability. Thus, a collaborative multidiscipline team effort is required to achieve these goals. An effective infrastructure includes managerial support that collaboratively promotes early and constant involvement of all team members. Resources such as knowledge, computers, communication networks, networked databases, project management and risk management techniques, time and money are all needed to insure highest levels of quality.

Best Practices

- Collaborative Product Development
- Product Development Teams
- Concurrent Engineering
- Technical Risk Management

BRAINS, NOT BRAWN

More than ever, value is being leveraged from innovative solutions and designs coming from knowledgeable employees. This is causing a major shift away from thinking of a company's key assets as machinery, buildings, and land. The assets that really count are increasingly becoming a company's intellectual assets – the knowledge of its workers and networked databases. Skill levels of workers are being judged on their intellectual abilities (Tapscott, 2000). The company is then judged on their ability to leverage this intellectual ability into innovative ideas, products and services.

Most companies have similar technologies to their competition and most designs and services can be copied or duplicated. All personal computer companies buy their microprocessors, disk drives, and memory from the same small number of suppliers. The key then is to implement new innovative solutions faster than the competition. Many products, such as software, can be manufactured or distributed over the web in large volumes for small incremental costs. Knowledge is the competitive weapon. Effectively using this asset requires companies to change their methods of using and managing people and networking knowledge. Knowledge management and intellectual property provide competitive advantage.

One executive noted that a "great" employee is worth 1000 times more than an average employee because of the quality of their ideas. As noted by Don Tapscott (2000), "knowledge workers must be motivated, have trust in their fellow workers and company, and have a real sense of commitment, not just compliance, to achieving team goals. When people are offered challenging work with sufficient resources, great things can happen." In his book Managing For The Future, Peter Drucker stresses that knowledge will empower companies. His view is that, just as the industrial revolution stimulated tremendous change, so too will the upcoming knowledge revolution. It is improved productivity in this area that will enable companies to successfully compete. Product development requires an infrastructure that enables the generation and distribution of both expert knowledge and lessons learned.

The next generation of people between the ages of 2 and 22, known as the N-generation, is the largest generation ever and the first generation to come of age in the Digital Age. It will be the largest market in the near future. Managers know that they cannot just rely on their own baby boomer instincts to identify new opportunities and design new products. It is important for technology professionals, marketers, and business leaders to understand its culture, psychology, and values, and how its members might change the world. This generation's ease with digital tools will create pressure for radical changes in the ways existing companies do business. This generation has the potential to transform the nature of the enterprise and how wealth is created.

2.1   IMPORTANT DEFINITIONS

Quality for both hardware and software is a measure of how well a product satisfies a customer at a reasonable price. Dr. Juran, a famous pioneer in quality, defined quality as "fitness for use". A product is fit for use if the product satisfies a customer's needs and requirements. Any deviation from the customer's requirements is called the "cost of quality" whether it is caused by design or manufacturing. Product quality is measured by sales, customer satisfaction, customer feedback, and warranty costs. Quality depends on correct requirements, user interface, failure free, service and documentation.

Design quality is measured as how well the design meets all requirements of the customer and other groups that interact with the product. Design quality can be measured by how well the product's design performs as compared to its product requirements and to the competition.

Software quality is when the final product performs all functions in the manner intended under all required conditions. To achieve quality, software must contain a minimum of mistakes as well as being void of misconceptions. This includes problems in requirements, architecture, domain, design, coding, testing, and installation.

Manufacturing quality is often measured as the percentage of products that meets all specified design and manufacturing requirements during a specified period of time. This is also expressed as failures, yield or as a percentage of products with defects. Many experts believe that manufacturing quality should be measured as a process's variance or uniformity about some target parameter.

2.2   COLLABORATIVE PRODUCT DEVELOPMENT

A modern product development process needs to establish a corporate culture in which everyone involved can freely and effectively communicate to collect knowledge, detect and resolve problems or suggest areas for improvement. The key objective is to improve communication between the many involved people including management, designers, product support, vendors and customers. This practice enables a collaborative and multi-disciplined product development approach.

Collaborative product development can be defined as the process of people working in teams to pursue design innovation. In collaborative product development, information, ideas, and problem solving are actively shared among the team. It can be synchronous, where team members meet together either face-to-face or via audio or video conferencing tools that bring them together when they are located in different places. Collaboration can also be asynchronous, where product development personnel log onto a computer network at different times and locations leaving their contributions for others to see and discuss.

For getting started in collaborative product development, the physical infrastructure and technical support must be in place to allow collaboration via computers. Once a decision has been made to develop a new product collaboratively, it is necessary to examine the design objectives and the available manufacturing capabilities. What needs to be changed? Where can collaboration significantly enhance product development? How can the product development team members become actively involved? A few suggestions may help to get started in collaborative product development.

- On-line product development materials should be up to date, well organized and easy to use.
- Collaborative product development may be new for some team members so training is needed. A project leader should be appointed right at the beginning of the project. He will devote some time up front to explain how the new strategy works, how to get help, and what is expected from each member of the team.
- The product development project should be started with a short task or two to get each member of the team involved in collaboration and become familiar with the use of new technology.
- The project leader should lead and encourage participation and keep a close watch on participation, especially during the beginning of the project.

Current strategies to support collaborative product development include implementing new design methodologies (e.g. Product Development Teams, Concurrent Engineering, or Product Development for the Life Cycle) and development tools (e.g. CAD/CAM/CAE/CASE systems, Web-Enabled Product Development, and virtual reality). An important characteristic observed here is the ability to simultaneously share data or information (on-line or off-line) among the members of a product development team. Using networked computers and the internet provides this capability. Additional research work is still being conducted on communicating product design data; defining exchange formats and interface standards for communicating data, plant wide computer control, use of knowledge databases for decision-making, and the use of browsing techniques for continuous product improvement.

2.3   PRODUCT DEVELOPMENT TEAMS

One product development strategy is to organize the assets and resources of a company into integrated Product Development Teams or PDT, with complete responsibility for designing, producing and delivering valuable products to customers. These teams are accountable for delivering quality, performance, program profitability, and additional business. They manage all the assets and resources necessary to meet their obligations to totally satisfy their customers and meet business objectives. Every team member is problem solver. The team is made up of combinations of people from different disciplines or functional organizations. Vendors and customers are often included. This approach relies on teams of people with the right skills working together smoothly to meet business objectives. In the future, firms will compete more on the basis of what they *know*, than on what they do. The skills and knowledge embodied in the work force will become *the* key competitive asset. Three reasons why a Product Development Team approach is vital for a business organization come from understanding that:

- Without a product design that is compatible with manufacturing or service capabilities and life cycle requirements, a company can miss market windows and incur excess cost.
- The opportunity to speed up the design process, delivery and service with an integrated strategy is critical in a global economy.
- New technologies and tools, such as the Internet, enable communication and collaboration between personnel in different organizations, functional areas, disciplines, and locations.
- For many, 80-90% of product costs are external. There is no choice but to work collaboratively and as early as possible with external areas in the supply chain such as suppliers, vendors and partners.

No matter what role the company plays, it is essential to deal with these facts from the very beginning of the design process. A multifunctional approach is paramount to focusing on integrating product development with life cycle issues to collectively achieve flexibility, higher profit margins, speed and customer satisfaction.

Today's technologies offer an opportunity to resolve the issues that have stood in the way of streamlining the product development process. It is now possible to organize product development projects globally and work locally. Information technologies such as e-mail, the Internet, and videoconferencing allow great coordination of geographically dispersed workers. Collaborative work across thousands of miles has become a reality as designers work on a new product together even if they are located in different countries. One example is the cross-continent collaborative approach applied by Ford Motor Company in developing the 1994 Ford Mustang. Supported by communications networks and CAD/CAM systems, the company launched the new design in Dunton, England. Designers in Dearborn, Michigan, and Dunton collaboratively worked on the design with some input from designers in Japan and Australia. Once the design was finished, Ford engineers in Turin, Italy, developed a full-size physical model.

2.4   CONCURRENT ENGINEERING

The term Concurrent Engineering (CE) or Simultaneous Engineering (SE) is a watchword for world-class companies to speed up and improve their product development process. **CE is defined as "a systematic approach to the integrated, concurrent design of products and their related processes, including manufacture and support. This approach is intended to cause the developers, from the outset, to consider all elements of the product life cycle from conception through disposal, including quality, cost, schedule, and user requirements. " (Winner et.al., 1988).** Benefits are realized quickly by utilizing CE concepts in the form of reducing direct labor costs, life cycle time, inventory, scrap, rework and engineering changes. More intangible benefits include part number reductions, process simplification and process step reduction (Gould, 1990).

The major objective of concurrent engineering is to overlap the different phases of design to reduce the time needed to develop a product. It requires the simultaneous, interactive and inter-disciplinary involvement of design, manufacturing and field support engineers to assure design performance, product support responsiveness, and life cycle reliability products. This requires the front-end or early involvement of all disciplinary functions, which improve quality, reduce cost, and shorten cycle time. Product development teams use CE to break down the traditional functional barriers by integrating team members across different business entities within an organization.

Each team member is involved in all aspects of product development from the very beginning, and each member has a respected voice in this development process. The different groups of experts might be very knowledgeable on a particular subject; however, conflicts may appear that are unrecognized until product manufacturing or product utilization begins. In order to reduce or even eliminate these problems, the development and deployment of an effective CE approach requires:

- Flexible decision models to represent the process by which a product development team could simultaneously design, debate, negotiation, and resolve.
- Knowledge representation schemes and tools to support and implement the integration requirements imposed by CE .
- Tools that facilitate simultaneous collaborative communication .
- Quantitative and qualitative tools that measure the impact of decisions on all product parameters. Assessment tools could be used to provide the design team with the capability to judge the relative merits of various design criteria and to evaluate alternatives with respect to life cycle implications.

CE has proved to be a valuable tool for maintaining competitiveness in today's ever changing and expanding world market. Some Japanese companies take half the time that U.S. companies do to deliver major products, such as aircraft and automobiles (Evanczuk, 1990). This success is due to the fact that CE contributed significantly to the reduction in the product development cycle. In addition, "CE methods help ensure that a design is compatible with a company's established manufacturing resources and processes" (Evanczuk, 1990).

2.5   THE PRODUCT DEVELOPMENT PROCESS

For the purposes of discussion, the traditional design process can be divided into seven linked and often overlapping phases:

1. Requirements definition
2. Conceptual design
3. Detailed design
4. Test and evaluation
5. Manufacturing
6. Logistics, supply chain, and environment

In a large developmental program more phases may be required. An example of the many tasks that may be required in each of these phases is shown in Figure 2.1. This extensive list helps to illustrate the large number of disciplines that can be involved.

Requirement Definition

The first phase of the design process is to identify the overall needs of the user and define the business and design objectives for the product. Requirement definition is the process of identifying, defining, and documenting specific needs for the development of a new product, system or process. It is the first step in the product development cycle. The major objective for this step is to identify, consolidate, and document all the features that the system could have into a feasible, realistic, and complete specification of product requirements. During this early activity, a universe of potential ideas for the product is narrowed to practical requirements. Product requirements or specifications are the final output of this early phase of product development.

Conceptual Design

The conceptual design process is the identification of several design approaches (i.e., alternatives) that could meet the defined requirements, performance of trade-off analyses to identify the best design approach to be used, and to then develop design requirements based on the selected approach.

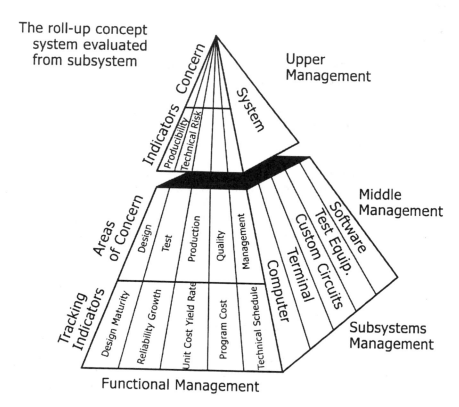

FIGURE 2.2 Technical risk management pyramid

It begins when a need for a new product is defined and continues until a detailed design approach has been selected that can successfully meet requirements.

In addition, design goals and requirements are allocated to the lowest levels needed for each member of the design team and then finalized during this process. Trade-off studies, analyses, mathematical models, simulations, and cost estimates are used to choose an optimum design approach and technology. Products of the conceptual design phase include guidelines, design requirements, program plans, and other documentation that will provide a baseline for the detailed design effort. All the "what if's" become "how's" and dreams become assignments. It is during this phase that the initial producibility, quality, and reliability design requirements are documented.

Detailed Design

Detailed design is the process of finalizing a product's design which meets the requirements and design approach defined in the early phases. Critical feedback takes place as the design team develops an initial design, conducts analyses, and uses feedback from the design analyses to improve the design. Design analysis uses scientific methods, usually mathematical, to examine design parameters and their interaction with the environment. This is a continuous process until the various analyses indicate that the design is ready for testing. Many design analyses are performed, such as stress analysis, failure modes, producibility, reliability, safety, etc. These require the support of other personnel having specialized knowledge of various disciplines. During this stage, the product development team may construct prototypes or laboratory working models of the design for testing and evaluation to verify analytic results.

The detail design stage therefore requires the most interaction of the many disciplines and design professionals. Communication and coordination becomes critical during the evaluation and analysis of all possible design parameters. This does not imply a design by committee approach but rather an approach in which the product development team is solely responsible for the design and uses the other disciplines for support.

Test and Evaluation

Test and evaluation is an integrated series of evaluations leading to the common goal of design improvement and qualification. When a complex system is first designed, the initial product design will probably not meet all requirements and will probably not be ready for production. Test and evaluation is a "designer's tool" for identifying and correcting problems, and reducing technical risks. A mature design is defined as one that has been tested, evaluated, and verified prior to production to meet "all" requirements including producibility. Unless the design's maturity is adequately verified through design reviews, design verifications, and testing, problems will occur because of unforeseen design deficiencies, manufacturing defects, and environmental conditions. A goal of every test and evaluation program should be to identify areas for design improvement, which improve producibility and reliability and reduce technical risks.

Manufacturing

The product development effort does not end when the product is ready for production. Problems found in production require the design team to perform analyses. Additional team efforts continually try to reduce manufacturing costs and improve quality throughout the product's useful life.

Logistics, Supply Chain, and Environment

The design teams role does not end when the product or service is sold. Products often require delivery, installation, service support, or environmentally friendly disposal. Logistics is a discipline that reduces life cycle installation and support costs by planning and controlling the flow and storage of material, parts, products, and information from product conception to product disposal. For many companies, logistic costs surpass all other direct costs. Supply chain is the "flow" and includes all of the companies with a collective interest in a product's success, from suppliers to manufacturers to distributors. It also includes all information flow, processes and transactions between vendors and customers. Packaging's purpose is to reduce shipping costs, increase shipping protection, provide necessary information, minimize the environmental impact and keep the product, delivery personnel, and customer safe. A product's integrity (i.e., reliability) may be compromised upon delivery unless the package is able to properly protect the product during distribution and storage. Operational problems may result in alternative maintenance techniques, customer comments, and other field environment issues that were not readily known during the design stage. Design and management personnel can evaluate how well the system works in the field to determine what problems justify corrective action

A rational balance between economic product development and environmental responsibility is a difficult task in product development. The difficulty of the problem includes the many environmental and governmental issues involved and the unknowns in directly relating design decisions to environmental results. The idea behind design for the environment (DFE) is to consider the complete product life cycle when designing a product.

2.6   PROGRAM ORGANIZATION

Organizational design is the process of organizing and developing managerial controls for a product design program. Nothing is accomplished without some degree of planning. A primary goal is to facilitate communication between people whose work is interrelated. A design team should have representation from many disciplines. This type of team, often referred to as a multidiscipline, multifunctional, or cross-functional team, may consist of representatives from management, systems engineering, electrical design, mechanical design, software development, manufacturing, industrial engineering, computer support services, reliability, marketing, purchasing, maintenance, and vendors. A key task to ensure success is to develop a program organization which is the process of allocating the resources needed to develop a product and the organizational leadership to successfully manage the program organize the team by establishing roles, encouraging risks, and rewarding innovation.

Many companies miss market penetration, revenue and margin opportunities because business and design objectives were not clearly understood, evaluated and enabled during key points in the development process. The design team defines customer and company needs based upon multiple perspectives and varied experiences within the company and its environment are in an excellent position to achieve successful product development and full customer satisfaction. The best companies proactively focus on the customer and consider life cycle performance in determining how the product is designed and how it will be delivered and serviced.

A design policy is a management's statement of its overall goals for the design process and includes proven product development methods and guidelines. The policy should be supported by design guidelines that attempt to improve the design process by implementing fundamental engineering and business principles and setting the right climate for good design practices. When company policy ignores the importance of setting the right climate for design, schedule and cost become the overriding design goals. This results in products that are designed in a manner in which emphasis is placed on short-term goals. Program management is responsible for the design team having all the proper tools, including a well-structured design program, design guidelines, resources, and facilities. Design policies should include a mechanism for setting schedules that are realistic and allow adequate time for design, analysis, and development testing.

2.6.1   Typical Program Organization and Major Techniques

The first major step in system design for management is to identify the resources necessary to develop the product. The new business model calls for an extended enterprise that provides access to capabilities and assets no single firm can afford to own alone. Such an enterprise leverages and focuses the capabilities of all participants to generate increased product value, as measured in capability, cost and quality. The application of this strategy extends to our customers, as well as to our suppliers. Risk and revenue sharing arrangements may be used. Design programs must be staffed with a proper mix of technical and management personnel with the necessary level of education and experience. The organization of these many functions requires thorough planning. A road map of activities must be made from the start of the program to its completion.

Work Breakdown Structure

Management develops a structure that defines responsibilities, resources, goals, and milestones. One of the tasks in this process is to define a Work Breakdown Structure (WBS) for the program. This structure is a hierarchical family tree (either in a listing or block pictorial format) that identifies and defines all task elements required for the program. A unique identification number is assigned to each task element. An effective structure identifies all reporting-level task elements of the program and provides a baseline for measuring financial, technical, and schedule progress on each individual element. Table 2.1 illustrates a portion of a typical work breakdown structure. The next step is to develop a definition (or dictionary) of briefly worded statements for each element, indicating the scope of effort for each element or task to be performed. The dictionary is used to communicate task element coverage within the project team, as well as to the customer. Table 2.1 also illustrates a portion of a dictionary.

The WBS provides a framework for detailed resource and financial planning of the program. Staffing requirements are then determined and the proper personnel are organized. Actions are started for developing various plans, standards, goals, and schedules. Determination of resource requirements and allocation of these resources is a major activity of program organization. Some of the useful techniques for program organization, planning, and control are described below.

Network Diagram

Design activities usually involve a multitude of smaller tasks that must be done to complete the project. Each of the smaller tasks requires specific time and resources. In considering each work element, two important questions must be answered:

1.   What other work must be completed before this work can start?
2.   What other work is dependent upon completion of this particular work?

The planning can quickly become mind-boggling. Network diagrams aid in visualizing the various connections between these work elements and the sequence in which they are required to be performed. The Program Evaluation and Review Technique (PERT) is one type of network diagram approach used on many programs, in which start and completion events are shown as blocks with interconnecting lines showing the dependency relationship and present activity times allowed. The amount of time required for each activity is estimated and entered on the dependency relationship line and represents activity time. With identifier numbers, the network information can be input to a computer program

TABLE 2.1 Work Breakdown Structure and Dictionary for a Notebook Computer Printed Circuit Board (AIAOS4)

Work breakdown structure

AI   Notebook computer design
AIA   Circuit board design
   AIAS   System engineering
   AIAO   Design engineering
       AIAOA   Circuit board assembly
            AIAOA1   Bare board
            AIAOA2   Assembly hardware
      AIAOS   Support equipment
            AIAOS1   Test equipment
            AIAOS2   Test fixtures
   AIAM   Manufacturing engineering

Work breakdown dictionary

AIA   **Printed circuit board:** the engineering, material, and computer resources required to design, analyze, and document. This is used in a notebook computer.
The major tasks include:
      Design of the printed circuit board
      Reduction of control logic to an integrated circuit and minimum number
           of parts
      Thermal stability/control system analysis
      Mechanical design and documentation

for plotting the network diagram and identifying critical paths for further planning consideration.

Schedule Diagrams

Schedule diagrams are prepared by listing task activities in a column on the left and a calendar-related time scale on a row across the top of the chart. A start-and-stop time span line (or bar) is placed in the row corresponding to each task activity to indicate the calendar relationship. These charts, with variations, are very useful planning and control tools but provide limited visibility of task interrelationships.

Task Resource Budgets and Schedules

Carefully prepared task estimates are necessary to establish project or program budgets. Several cycles of adjustment and review by program management are usually required to establish a realistic budget for production of a customer acceptable product at a management-acceptable profit margin. The budget and schedule for each task element are established after these results have been reviewed and agreed upon by management. Progress and performance are then expected to stay within the budget and schedule constraints and produce a product that meets requirements.

Technical Controls

Controlling the design process is not simple, since seldom does everything work as planned. A major method of technical control is documentation. Technical documentation includes specifications, block and interface diagrams, design guidelines, drawings, process capabilities, purchased part information, and technical files. Specifications are usually required for the system and address the performance, environmental, reliability, producibility, and quality requirements. From these specifications, the program prepares written requirements of lower level subsystems to guide the individual design activities. These documents must be sufficiently comprehensive to control the design task. No design activity should proceed without written task definitions and specifications. The design team must take the initiative to search out all-pertinent specifications related to the particular design task and understand the requirements and implications. Typical specifications for a program are as follows:

- The system specification is an approved document that states what the product is to be and what it is to do and the business model.
- The subsystem specification is system engineering generated or approved specification for a subsystem.
- The major vendor and unit specification are a lower specification generated by the design activity.
- Specifications, governmental regulations, and professional standards are imposed as applicable documents.
- The company or program imposes design guidelines, internal standards, and process specifications.

The system block diagram depicts in block form the functional and physical partitioning of the elements of the system and indicates the major I/O flow. A well-prepared block diagram is a very useful working document for communicating essential interface information and functional operations. Lower level block diagrams complement the system diagram by expanding the level of detail for the system-level blocks. For example, system interface drawing shows cable interconnections between system units. Detailed signal characteristics, cable type, connector type, and pin assignment information are defined in the system interface specification.

Design guidelines and rules are prepared to encourage the use of proven design practices and promote consistency of design for the specific development program. Typically, design guidelines include the following:

- The general design guidelines and rules are a summary of common electrical, mechanical, software, and etc. design requirements for the particular program at hand.
- Specialized guidelines and rules are prepared by groups, such as design, manufacturing, quality, reliability, packaging, repair, and safety. These contain requirements, helpful information, and best practices.
- A design policy must specify who should participate in the design analyses, accepted data gathering techniques, practices for common databases, and standards for reporting the results.

Cost and Schedule Controls and Assessment

Budgets and schedules are established at the start of the program for each element of the WBS. A forecast of resource expenditures are prepared by labor category, time frame, and entered into a cost and schedule status-reporting system. Periodic updates are made available for review of trends and corrective action. Gantt charts can then be used effectively to indicate the progress of each activity against a time scale. The end points of the task are clearly defined, but judgment is required to measure progress between the end points. In setting up the lower level activities, the task scope and time span should be broken into small enough units of time to provide good visibility of progress. Schedule control is then aided by weekly informal reviews and periodic formal reviews. Computer software programs are commercially available to help in these tasks.

Design Reviews and Audits

Design reviews are a crucial communication link between the designer and specialists from all the applicable disciplines. The purpose of design reviews is to evaluate technical progress, identify potential problems, and to provide suggestions for design improvements. It also provides an opportunity for the support areas, such as manufacturing, maintenance, test, and logistics to communicate with the project. The intention should be to evaluate and criticize the design, not the designer! It is not design by committee, but rather a systematic method to ensure that all aspects of the design are thoroughly evaluated prior to production. Best in class companies use weekly meetings with a strategic planning team made up of upper management from various areas. This leads to better design and better acceptance of the final design. Several types of reviews may occur in typical design and development programs.

Production Readiness and Design Release

At some point in product development, creative design must cease so that the product can be released to production. This point in the developmental phase is called design release. Scheduling a design release is closely related to the status of other design activities, such as design reviews, production design, test results, and configuration control. Care must be taken to prevent the release of a design that is incomplete, inaccurate, or premature. When this happens, problems result. Establishing release schedules by "back-planning" from manufacturing schedules often requires the design team to meet unrealistic dates. Deviating from standard procedures allows inferior-quality products to reach users. By using uniform practices and procedures concerning technical requirements and by evaluating current manufacturing capability, more realistic design release dates can be established.

The design should be validated in stages, using experienced personnel from technical and production disciplines to ensure that the design is producible, documentation is complete, and released on schedule. Proofing the design on manufacturing models and providing the results to the team ensure that the documentation maintains its integrity.

Before a design is ready to release for production, several actions must be completed:

- All engineering and business analyses, process capability studies, and vendor verification tests must be completed and corrective actions incorporated.
- Final technical reviews of hardware and software must be completed and corrective actions incorporated in the design.
- Manufacturing and service plans and procedures must be revised to reflect the latest design changes.

2.7   TECHNICAL RISK MANAGEMENT

Most product development problems are a result of "unexpected" events. Technical risk is a measure of the level of uncertainty for all of the technical aspects of the development process. Technical risk management identifies and tries to control this uncertainty found in product development. It is essential for identifying and resolving potential problems to ensure that the proposed system will work as intended and be reliable when it reaches the user. The steps are to:

1. Systematically identify areas of potential technical risk
2. Determine the level of risk for each area
3. Identify and incorporate solutions that eliminate or reduce the risk
4. Continue to monitor and measure progress on minimizing

Too many managers try to control complexity by eliminating risk. Innovative products and new technologies, however, require the need to take risks. The greatest risk is not taking some. The technological risk of developing new innovative products begins at the onset of the design process. The design team should identify all potential technical risks, monitor and control these risks as much as possible and finally have contingency plans when things go wrong. Management helps by developing systems for monitoring and evaluating technical progress throughout the development process. History provides many lessons for identifying what risks may occur in the future. Experienced designers and managers know there are conditions, requirements, and situations that almost always create problems. When problems do occur as they most certainly will, the design team should take contingency actions that are more likely to have positive outcomes. Companies that do not study and learn from history are doomed to repeat the same mistakes.

Technical risk assessment is a managerial planning and control system for quantifying design and technical progress during program development. Technical risk assessment can be performed to decide whether to start new designs, to evaluate alternative technologies, or to make or buy a particular technology. Through technical risk assessment, management can identify early a product with serious production or reliability problems. This systematic approach allows management to regularly evaluate program status and assign additional resources as problems are identified.

Identification and assessment of technical progress are essential for resolving problems, ensuring that the proposed system will work as intended, and be producible when it reaches the production phase. Early detection is critical because it is easier and cheaper to make changes early in a design. An effective technical risk management program should provide the following:

- System for identifying and measuring important technical risks
- Measurement and monitoring from the very beginning to track indicators that realistically demonstrate technical progress
- Instantaneous "real time" assessment of technical status with early problem identification
- Direction and trend measurement
- Contingency responses when problems occur
- Sufficient information for trade-off decisions and crisis identification

Technical risk management is similar to other control systems in its utilization of a "roll-up" or pyramidal approach, in which subsystems are evaluated to provide an indication of overall system status (Figure 2.2). The various tracking indicators are monitored for progress to provide program status. These indicators must be able to provide a direct indication of how the design is progressing. One problem with this, however, is that a steady improvement may

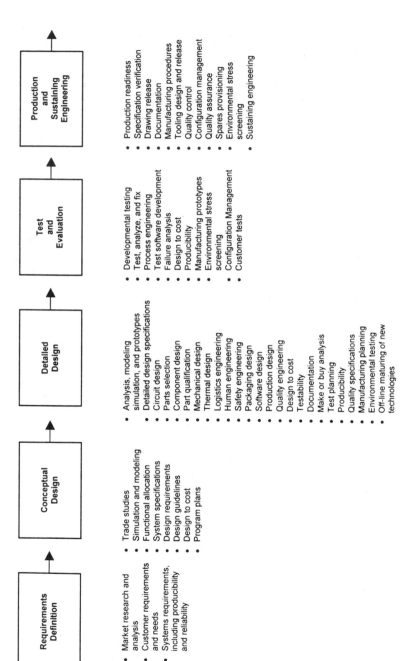

FIGURE 2.1 Engineering design process.

be reflected, but the actual results show little progress. Since most complex systems have several distinct subsystems, a third dimension is added to allow for each subsystem to be separately monitored. As with most planning tools, technical risk management requires a considerable amount of effort and cost to be properly incorporated into the program. The up-front cost, however, is well spent and should provide a superior product with reduced risk in the production phases.

The U.S. Department of Defense (DOD) has attempted to become a leader in developing methodologies for addressing the technical risks associated with the transition from development to product. The emphasis on technical risks is necessary to establish a balance with traditional administrative risk (i.e., cost and schedule). The team focuses on the technical issues of design, test, and production, which are the root cause for most cost and schedule overruns and performance shortfalls, rather than the overruns and shortfalls themselves. DOD Document 4245.7-M, "Transition From Development To Production," was promulgated in 1985 in a joint industry-government project to identify the most critical areas of risk affecting the transition processes in product development. This document emphasizes that the transition from development to production process is primarily a technical process that is normally misunderstood. The methodology stresses that to reduce technical risk; fundamental engineering and technical disciplines such as those described in this book must be integrated into the transition process. Later studies and reports identified the best practices for reducing technical risk. Information on this document and its "Best Practices" can be found at www.bmpcoe.org.

As noted earlier, the first step is to identify technical risk areas that should be evaluated. The critical areas of technical risk the DOD study identifies are shown in Table 2.2. These are often referred to as the "Willoughby Templates," and would be applicable for our notebook computer. The database consists of eight categories (funding, design, test, manufacturing, transition to production, facilities, logistics, management), which are subdivided into multiple, process-oriented templates (e.g., design reference, manufacturing strategy, piece part control). These templates relate to the areas in which past programs have experienced difficulty and/or where the potential for failure is most likely to occur if ignored. Currently 70 templates exist and new ones are being added as the program evolves and matures. Each template incorporates a series of approximately 500 expert questions that invoke best practices in its process area. It is highly flexible, and allows the user to tailor the system to specific aspects of their product by adding their own categories, templates. Although it was developed for the US Defense department, it is applicable to the commercial sector since it is process, not product, oriented. The tool enables the user to ask questions to avoid potential problems and effectively use limited resources. This helps the user make informed decisions and avoid surprises.

The Technical Risk Identification & Mitigation System (TRIMS) is another software tool available at www.bmpcoe.org. It is a risk management system based on technical measures rather than costs and schedules. Early

TABLE 2.2 Critical Areas of Technical Risk

DESIGN

Use Profile	Software Design
Design requirements	Computer-Aided Design (CAD)
Trade Studies	Design for Testing
Design Policy	Built-in test
Design Process	Configuration Control
Design Analysis	Design reviews
Parts and Materials Selection	Design Release Test

TEST

Integrated Test	Design Limit
Failure Reporting System	Life
Uniform Test Report	Test, Analyze, and Fix (TAAF)
Software Test	Field Feedback

PRODUCTION

Manufacturing Plan	Tool Planning
Qualify Mfg. Process	Special Test equipment
Piece Part Control	Computer-Aided Mfg. (CAM)
Subcontractor Control	Manufacturing Screening
Defect Control	

FACILITIES

Modernization
Factory Improvements

MANAGEMENT

Manufacturing Strategy	Technical Risk Assessment
Personnel Requirements	Production Breaks
Data Requirements	

identification enables corrective actions to be applied in a timely manner, and prevents problems from developing into cost and schedule overruns. TRIMS can be tailored to the user's needs. It identifies and ranks those program areas with the highest risk levels, provides the ability to conduct continuous risk assessments for preemptive corrective actions and tracks key project documentation and concept through production.

The document provides a "template" that describes what risk exists, how to identify the level of risk and outlines what can be done to reduce it. The term "template" is used to define the proper tools and techniques required to assess and balance the technical adequacy of a product transitioning from development to production.

The next step is to evaluate and determine the "level" of risk for each area based on the answers to the questions. One method is to use checklists that compare current design information to items that have historically caused problems. Subjective judgment of management and the design team is often used to determine the severity of the risk.

The last step is to identify actions that for reducing areas of high technical risk. This can include design or vendor changes, closer management scrutiny, etc.

Innovative products generally push the limits of current processes and technologies. For example, meeting the especially aggressive size and weight product requirements for a notebook computer might require the use of new or unique technologies and design approaches. The technical risks for this example are:

- Using chip-on-board (COB) technology to reduce size and weight (technical risks are poor quality due to new, unproven manufacturing and testing processes and long term reliability)
- Placing components closer together to reduce size (technical risk is in pushing the capability limits of existing automated component placement equipment and automated test equipment)
- Thermal conditions caused by closer packaging of components (technical risk is higher internal temperatures causing high failure rate and warranty costs)
- Changing to a new vendor for the display (technical risks are the vendor's unknown level of quality and their ability to meet delivery schedule)

2.8   SUMMARY

This chapter has described a product development strategy focused on product teams, concurrent engineering and technical risk management that can be used to foster continuous improvement. This business strategy helps organize projects to promote collaborative product development practices, since integrated product development requires collaborative thinking and integrated technology organization. The ideas presented in this chapter also help disseminate a product development vision of a company to communicate the technical framework that allow administrative and technical people to work together to get the best of their own and each others skill.

2.9   REVIEW QUESTIONS

1. Explain why various technical and non-technical professionals of a company need to understand the importance of product development for the life cycle?
2. Define and discuss the steps of the product development life cycle
3. Discuss the importance of team building and communication for product development and management.

4. Discuss the characteristics of decision making and information models needed for collaborative product development
5. Select a product or a service from the market and identify some of the potential technical risks that a business organization may face to become an economic success.
6. Organize a multidiscipline design team to discuss the following issues: (a) how does a collaborative design team approach apply to a business organization? (b) What is needed to enable collaborative product development? And (c) What is needed to foster collaboration among the different functional areas of a business organization?

2.10   REFERENCES

1. S. Evanczuk, Concurrent Engineering the New Look of Design, High Performance Systems, April, pp. 16-17, 1990.
2. L. Gould, Competitive Advantage Begins with Concurrent Engineering, Managing Automation, November, p. 31, 1991.
3. D. Tapscott, Minds Over Matter, Business 2.0, March 2000.
4. R. Winner, J. Pennell, Bertrand, and Slusarczuk, The Role of Concurrent Engineering, Institute for Defence Analysis, Detroit, Mich. 1988.

Chapter 3

EARLY DESIGN:
REQUIREMENT DEFINITION AND
CONCEPTUAL DESIGN

Focus on Customer Needs and Requirements

Keys to successful product development are to know the customer's needs, and to provide a product or service that meets these needs at a competitive cost. Identifying current and future needs, developing product requirements, determining the best design and technology approach, and developing effective detailed design requirements are important steps to successful product development. Innovation and creativity must be encouraged. Requirement definition and conceptual design are early steps in product development.

Best Practices

- Evolutionary and Collaborative Process
- Customer Needs Analysis
- Product Use and User Profiles
- Technology Capability Forecasting
- Benchmarking and Company Capability Analysis
- Prototyping and Virtual Reality
- House of Quality or Quality Function Deployment
- Emphasize Creativity And Innovation
- Trade-off Analysis of All Design Alternatives
- Design Requirements Are Easy To Understand, Quantified, Measurable, Testable And Updateable Parameters
- Documentation Provides Foundation For Effective Communication

THE DILBERT PRINCIPLE

Scott Adams is the creator of **Dilbert**, which is a popular comic strip seen in many newspapers that makes fun of today's manager's and engineer's personalities, methods and incompetence. For example, *Normal people believe that if it isn't broke, don't fix it, engineers believe that if it isn't broke, it doesn't have enough features.* Dilbert can be contacted at www.unitedmedia.com. As reported in the Wall Street Journal (1995), here are some of his favorite stories on product development, all allegedly true:

-"A vice president insists that the company's new battery-powered product be equipped with a light that comes on when the power is off."

Good luck on battery life!

-An employee suggests setting priorities so they will know how to apply their "limited" resources. The manager's response: "Why can't we concentrate our resources on every project?"

If everything is important or critical, then resources are spread too thin and nothing gets done properly. Focus on a few of the most critical items.

-"A manager wants to find and fix software faults (i.e., bugs) more quickly. He offers an incentive plan: $20 for each bug the quality people find and $20 for each bug the programmers fix. These are the same programmers who created the bugs! As a result, an underground development in "bugs" sprung up instantly. The plan was rethought after one employee earned $1,700 in the first week"! (Wall Street Journal,1995)

The key is to design without mistakes, not to spend time and money correcting the mistakes.

The keys to successful product development are to know the customer's current and future needs and to provide a product that meets these needs at a competitive cost. In this chapter, customer needs will include things that the customer might want and things that the customer may not even know that they would like to have because of a lack of information on new technologies or innovative ideas. Unfortunately customers needs, technology, economy and the competition are always evolving and changing. As a result, these early phases are an evolutionary, iterative process where the best practices must be continuously updated throughout product development. Identifying these needs, developing effective product and design requirements, and selecting the best design approach that meet the requirements are the early steps to successful product development.

Management's task is to provide a creative environment of adequate resources and leadership to make it happen. For long-term survival of a company this must be a continuous process. This chapter reviews the early overlapping phases of product development; requirement definition and conceptual design.

3.1 IMPORTANT DEFINITIONS

Requirement definition is an evolutionary process of identifying, defining, and documenting specific customer needs to develop product requirements for a new product, system, or process. It focuses on "what needs to be done" and is the first phase in product development but sometimes requires updating as changes occur in the market and technology. This activity focuses on "what is needed" to be successful not on "how to" design the product. It directs attention to critical customer, design, technology, manufacturing, vendor and support needs both within and outside the company.

Conceptual design is a systematic analytical process used to 1.) identify several design approaches (i.e., alternatives) that could meet the defined product requirements, 2.) perform trade-off analyses to select the best design approach to be used, and 3.) transforms the product requirements into detailed lower level design requirements based on the selected approach. It focuses on "how to get it done" and begins when a need for a new product is defined and continues until a detailed design approach has been selected that can successfully meet all requirements. In addition, it determines detailed design goals and requirements, allocates them to the lowest levels needed and then finalizes them during this phase. The product development team will make important trade-off analyses and decisions in the following areas:

- Level of innovation and technical risk
- Product's functional, manufacturing, service and aesthetic requirements
- Design, technology, manufacturing, and logistic approaches
- Global considerations
- Technology acquisition strategy and resources
- Partnerships and key suppliers
- Project management, communication and documentation practices

Product requirements focus on "what the product should do". Design requirements focus on "how the product will meet the product requirements". Both of these phases overlap each other during product development. Other phases may also be occurring including detailed design, test and evaluation, manufacturing of prototypes, and logistics planning.

3.2 BEST PRACTICES FOR EARLY DESIGN

A systematic evolutionary requirement definition process identifies what the product should do based on identifying, evaluating, quantifying, prioritizing, and documenting customer needs and encouraging innovation using best practices such as:

- **Customer needs analysis** uses several different methods to insure correct user needs or predict opportunities.
- **Product use and user profiles** are realistic operational requirements that include scenarios, task analysis, functional timelines and all types of environments, including the maximum performance limits.
- **Technological capability forecasts** are made for the short and long term future to ensure that the design will be up to date (i.e., not obsolete) when the product is ready for manufacture.
- **Benchmarking and company capability** studies are performed to evaluate the company's capabilities and the competition, identify best practices, and to finalize design and manufacturing requirements.
- **Prototyping, virtual reality and House of Quality** are all used to systematically identify consumer needs and develop product requirements.

Conceptual design process is an evolutionary process that identifies how the design will meet the product requirements using best practices such as:

- **Collaborative multidisciplinary process** emphasizes creativity and innovation.
- **Identify all possible design alternatives** so the best approach is identified.
- **Extensive trade-off studies** such as benchmarking, design analyses, modeling, simulation, prototyping, and house of quality are used for conceptual design decisions.
- **Design requirements** are developed, allocated, and stated in quantified, measurable, testable, easily understood, design oriented criteria and easily updated format.
- **Documentation** provides the foundation for effective communication between the collaborative team.

3.3   SYSTEMATIC REQUIREMENT DEFINITION PROCESS

Product development is a requirements driven process. Requirement definition is the process of identifying, evaluating, quantifying, prioritizing and documenting specific needs for the development of product requirements for a new product, process or service. It is the start of the product development cycle. The opportunity for a new product can be caused by an innovative breakthrough, new use of an existing technology or service, new technology or process, new customer or product line, or improvement of an existing product.

The major objectives for this step is to systematically and thoroughly identify, consolidate, and document all the features that the system could have into an innovative, feasible, and realistic complete specification of product requirements. Product level requirements or specifications are the final output of this early phase of product development although modifications and updates will probably be required as the design evolves or new technologies are introduced.

This evolutionary process requires several analyses that include customer needs analysis, product use profiles, technological capability forecasting, benchmarking, prototyping, and house of quality. These steps are performed in an iterative manner as the design evolves in more detail and as changes occur in the marketplace. Very few projects are static enough for these steps to only be performed once.

Since the producer's resources and technological capabilities are almost always limited, a product can never completely meet all of the customer's desires. You cannot design a car that gets great gas mileage, is extremely fast, very luxurious, and costs less than $10,000 (U.S.). In addition, different customers may have very different needs based on culture, ethnicity, income, climate, or other aspects.

This requires the design team to make difficult trade-off decisions that can affect the success of a product. To obtain the criteria necessary to make this effective trade-off, the development team must first investigate, identify, document, quantify, and then evaluate what customers really need, want, and expect within the company's objectives and limitations. Without an adequate understanding of the customer's needs and company's capabilities, a design will be sub optimal at best.

3.3.1   Pitfalls in Requirement Definition

Successfully translating customer needs into product level requirements is extremely difficult. There are several common pitfalls in this process.

The first is that a specific solution (e.g., technology, resolution, bandwidth, or part types) is determined too early before conceptual design and trade-off studies have been performed. This is especially true for new "state of the art" technologies. There are few successful products that were the first to incorporate "state-of-the-art" technologies. There is a famous saying in product development that there is a small difference between successfully implementing "cutting edge technology" into a product versus being "cut" by the sharp edge of the cutting edge of the technology.

The second is that product requirement must be extremely innovative. There are many successful products that only have incremental improvements in performance. Too often we set very high goals for ourselves and our company that cannot be met. This can be terrible for a company since unrealistically high goals can make the design team take too many risks and ensure failure.

The third is that requirements can be stated in general terms. If a requirement cannot be measured and tested then it does not allow the product team to measure progress. Engineers depend on mathematical models that require quantitative parameters. For example, what does the product team design to if the design requirements specify that a notebook computer should be lightweight? What is lightweight? Is it 1.0 lbs., 5.0 lbs., or 20lbs? If you cannot describe something in numbers it will not be part of the engineering process.

The fourth is the common temptation to accept customer, marketing, or a consultant's suggestions as the only and final input. The customer or marketing department is certainly closer to the problem, but may have, with good intentions, overstated or misstated the product's requirements. The customer may have preconceived notions about the best solution for the problem, such as using certain software or certain brands of microprocessors in the product, regardless of whether they are appropriate or not. Safeguards against this problem rest with the team's detailed evaluation of user's needs.

The fifth pitfall may appear when the problem statement is continuously changing. This is called a "moving target." Requirements that significantly change over time result in constant design changes as management periodically redirects the technical effort. On the other hand, this may be legitimate, such as the continuous market changes that are common in high technology and consumer industries such as toys and clothing.

The sixth pitfall is when the product's requirements become too complex and detailed. As a product's requirements become more complex, the number of things that can go wrong increases exponentially. Problems such as incompleteness, conflicts, generalizations, and specialization start to occur. Simplicity should be encouraged.

The seventh and final pitfall is trying to develop only one set of requirements for all customers. Customers want products that meet their personal needs not what someone has determined for them. The popularity of mass customization (e.g. Dell Computers) shows that customers want to specify the details of the product. When the customers are different companies, they also may have different requirements. What Honda wants in a product may be very different than Opel or Rolls Royce. This is also true for international products, as common requirements are usually a problem for different countries and cultures. Some of the concerns in identifying one set of product requirements for international markets (adapted from Morris, 1983) are:

- Cultural differences result in variations in the relative importance of service, dependability, performance and costs.
- Customers lack the technical skills required to install, operate or perform maintenance.
- Need to provide timely delivery of spare and replacement parts to distant markets.

- Lack of universal or common standards for use such as power quality (i.e. voltage variation, brownouts) and availability (i.e., 220 volt or 110 volt) or the need to convert to the metric system.
- Products that work well under the physical conditions (e.g., climate, transportation, terrain and utility systems) of one region may break down quickly in another region.

3.3.2 Customer Needs Analysis

Defining the customer's needs can be an extremely complex process resulting in many different and conflicting types of information. It is important to recognize that the customers, themselves, may not be able to tell you completely what they want. The customer may not understand the implications of new technology breakthroughs or innovative ideas. There are several approaches for knowledge acquisition of customer needs. The design team should use several of these methods to insure that the final requirements are representative of the customer. Methods for capturing and documenting customer needs includes:

- **Interviews of customers** including techniques such as surveys and focus groups.
- **Design partnerships or alliances** where the company customers participate in the design process.
- **Computer databases and data mining** of actual customer sales and preferences including internet activity
- **Consultants or experts** who specialize in identifying what the customers want.
- **Brainstorming sessions** solicit input that encourages the generation of innovative ideas that are out of the ordinary and change the norm for this type of product.
- **Personal and company experience** including previous successes and failures.
- **Published information** such as magazines, Wall Street Journal, patents, etc. including the use of the Internet to locate information.
- **Technology capability forecasting** of the future based on historical analysis of product markets and technologies.
- **Market and competitor benchmark analysis** identifies best in class and innovative ideas.
- **Prototyping and virtual reality** are used to study customer responses and design team recommendations
- **House of quality or Quality Function Deployment** is used to identify needs and determine product requirements.

Need, want, desire or innovation are fundamental to all acquisitions or purchases. Need is typically a specific deficiency or lack of something in a current product. Deficiency is an opportunity for improving performance, cost, reliability, producibility, human factors, or a combination of these. Desire is something someone wishes or longs for. Innovation is something new that the customer never thought about before the product became available. A competitor's development of a similar, but more advanced product, may also lead to the definition of a new requirement and a subsequent new product development program.

Unfortunately, there are always more needs than there is money to satisfy them. When we are deciding which product to buy, we evaluate the product's cost and other aspects of the product (e.g., innovation, aesthetics, performance, reliability, etc.) with the importance of the need (e.g., how much do we really need or want the product). Because of this, a customer's needs in requirement definition are often prioritized to determine their relative importance.

Average Performance Meets Consumer Needs and Opens New Markets

In February 1900, George Eastman introduced the Brownie camera as a companion to the already famous but higher priced Kodak lines. Eastman believed that existing cameras were too complicated and expensive for novices and youngsters. A simple, cost-effective camera was needed for these groups. Retailing at $1, the first Brownie was made of heavy cardboard covered with black imitation leather. A 15-cent roll of film yielded six negatives. The performance requirements for the camera were designed to be "average," that is, producing reasonable pictures in average light and range. At first, photographic dealers and professionals regarded the one-dollar Brownie as a toy. Less than a month after the first 5000 had been shipped, however, orders were received for an additional 31,000! This average performance product resulted in the sale of over 50 million Brownie cameras.

Invention Beyond What the Customer Needs

When Aaron S. Lapin died he was best known for developing the can of Redi-Wip. Lapin introduced whipping cream packaged in an aerosol can in 1946 and became rich. As noted by Petroski (1999), "necessity isn't the mother of invention. Successful inventors don't usually search for what the consumer needs – instead they are constantly on the lookout for what we don't need. Once the invention is available, we can't live without it."

3.3.3   Product Use and User Profiles

Product use and user profiles document on a time scale, all the functions that a product and the user must perform, including the various environments that the system will encounter. These profiles are often called scenarios, use cases, task analysis, network diagrams and environmental profiles. They list and characterize each step in product use, its important parameters and how they relate to the environment. Profiles provide the operational, maintenance, and environmental baseline for the definition of design requirements. They need to be accurate and complete since they are used as the basis for the entire design process, including requirement definition, detailed design, stress analysis, testing, and maintenance planning. The degree to which the profile corresponds to actual use directly determines design success. Profile methods include:

1. Scenarios and use cases
2. Task analysis and user profile
3. Network diagrams
4. Product use, mission or environmental profiles

Scenarios and Use Cases

Scenarios and use cases are step-by-step descriptions of how the product will be used for a particular application or task. They are usually formatted as a list or a flow diagram. A product will have several scenarios depending on the number of product uses, features and different users. They provide a focus for communication between users, experts, developers and vendors. Scenario's task descriptions provide a direct link throughout the design process. One or more scenarios are developed to illustrate the desired functionality and problem solving focus for each application. The initial scenario(s) helps the development team learn system requirements and provide the basis for early prototyping. As each scenario becomes more detailed, they are used for developing design and test specifications. Each scenario can later be decomposed to define task responsibilities and constraints.

Scenarios are elicited from users and experts in the domain, and validated by independent experts. Scenario analysis identifies typical and atypical process flows within the system. Results are used to define task responsibilities and often involve time dependent sequencing. Documentation techniques include task lists, timeless topologies or "maps," event traces, scenario matrices and models, and flowcharts. Process diagrams and activity state diagrams can also represent scenarios. A storyboard for each scenario(s) is generated using

1. Group techniques
2. Interactive observation
3. Structured interviews
4. Demonstrations
5. Focus groups

It is important to develop both "as is" and "to be" views of the scenario. The descriptions should include responsibilities, capabilities, and views. The scenario perspective focuses on each user's viewpoint on what activity is taking place, rather than how or what technology is used to support it.

Scenarios define important aspects of the domain i.e. how the product will be used. They are initially made at the highest abstraction level, and are then broken into smaller sub-scenarios. In systems development with multiple users, small scenarios would be developed for each user.

Task Analysis and User Profile

Task analysis (sometimes called task-equipment analysis) is a design technique that evaluates specific task requirements for an operator with respect to an operator's capabilities. The analysis lists each human activity or task required and then compares the demand of the tasks with human capabilities and the resources available. Tasks are derived from the previously defined scenarios and continuously updated as the design progresses. The level of detail for the task analysis is based on the design information required at each phase of the design and the importance of each task. User profiles are descriptions of the users and support personnel's capabilities. This includes descriptions of their physical, educational, training and motivational levels.

During early design, task analysis and user profiles are used to ensure that the product's requirements are compatible with the user's capabilities. The design team must look at the product critically to see that the task requirements do not exceed human capabilities. In addition to the operator's tasks, task analysis should also be performed for maintenance and support personnel. It starts at a very simple, limited level of detail and progressively becomes more detailed during the design phase. Constraints are used to identify potential problems and enumerate reference requirements. Studying task analysis results can identify potential product and domain problems.

Task analysis is a progressive process that breaks down major tasks into detailed descriptions. The level of details and requirements for each task will progressively become more detailed as the design process progresses. The level of details varies from project to project but can include:

• Expertise And Training	• Expected Outcomes
• Organization	• Feedback
• Tools And Equipment	• Resources
• Goals	• Decisions
• Features	• Environment
• Preconditions	• Transaction
• Hazards	• Errors

The task frames may be maintained in a relational database system. For software development, the fields in the electronic records can be parsed to create formal model structures such as object or agent models, entity-relationship diagrams, and process diagrams. The task descriptions within these frames are discrete fields with values. The values of the fields include semantic terms in the domain. When the fields are filled with values for a task, that task information and the values can be converted to a formal domain model.

Network Diagrams

Network diagrams are used to graphically show the interrelationships and sequential flow of how the product will be used and supported. Traditional structured analyses may also be used to identify process flows, events and conditions, and entities in legacy system documentation. Structure diagrams such as data flow, state transition, and entity relation diagrams may be used. Early conceptual analyses elicit "as-is" user information for the definition of standard or common values, metrics, roles and responsibilities, and standard high-level abstract components with their capabilities and constraints.

Product Use, Mission, or Environmental Profiles

These profiles should include both environmental and functional conditions. A common problem occurs when only the operational use of the products is shown in the product use profiles. Many products such as washing machines and notebook computers can experience harsher physical environments in moving and shipping than they experience in actual use. An environmental profile shows on a time scale the significant environmental parameters, including their levels and duration that are expected to occur during the life of the product. It defines the total envelope of environments in which the product must perform, including conditions of storage, handling, transportation, and operational use. Similarly, a functional profile shows, on a time scale, the functions or tasks that will be utilized by the system to accomplish its intended use, including maintenance, storage, and transportation.

The functional mission profile emphasizes how a system must perform in every potential situation in the total envelope of environments. Functional analysis helps the design team describe the system completely at each level of detail. Examples of functional analysis tools include flow block diagrams and time line analysis. Time line analyses show sequences, overlay, and concurrence of functions, as well as time-critical functions that affect reaction time and availability.

3.3.4 Technology Capability Forecasting

Technology capability forecasting is a requirement definition tool that predicts whether a new technology will be ready in time to be used in

the design, its expected performance levels and the manufacturing
capabilities that are expected to be available when the product is ready for
manufacturing. For many products, identifying whether to use a new
technology and determining the future performance level of a key technology is a
major design decision. Product innovation and creativity are often designed
around using a new technology. If the technology is not ready and on schedule
the entire product development process is at risk. For many products, the design
team must also anticipate the technology's performance level that will be
available in the future when the product is manufactured and not the
performance level when the product is being designed. This requires
management and the designer to predict the key parameters of future parts and
software. Designers can often use vendor's advanced information and pre-
specification sheets. These specifications are sent out before the product hits the
market. Internal funding may be required to insure that the technology is ready.

For producibility, predicting future process capabilities involves two
technologies 1.) product technologies and 2.) manufacturing process
technologies. It is essential that production personnel participate in its
development to assure production compatibility. This effort ensures that as the
technology evolves, the requisite process capability can be predicted or
determined. While the technology is embryonic, experiments are conducted with
respect to selected manufacturing criteria. This provides insight for predicting
required processes and related capability requirements

The steps are to identify critical technologies, study their historical
trends, contact experts in the field, and then project or predict the levels that are
expected in the time frame for the design. Potential new capabilities can be
found through industry working groups, trade publications, trade shows,
university research, and supplier strategic partnerships. Methods for projection
include:

- Trend extrapolation – extrapolate trends shown in the historical
  data
- Delphi techniques – surveying a group of experts to determine
  future
- Scenarios – develop stories on how new technologies will develop
  in the future. One method is to identify a future situation and then
  the analyst work backward to identify breakthroughs that are
  required to accomplish the identified future situation.

Technology capability forecasting is often used in semiconductor
design. In an example written by P.E. Ross for Forbes Magazine (1995), Gordon
Moore, from Intel, predicted that the number of transistors that engineers could
squeeze onto a silicon chip would double with machine like regularity every 18
months. This prediction has held for 30 years! The transistor continues to shrink

exponentially, and the count of transistors on a chip continues to grow exponentially. Moore's Law is depicted in Figure 3.1. Ross (1995) notes many predictions of its ultimate demise. Many early engineers thought that the optical steppers that align silicon wafers would be too clumsy to carry the shrinking process much further. However, engineering overcame this obstacle. Then many believed that the light waves that passed through the chip-making mask were too wide to make electronic lines smaller than a certain size. Scientists pushed back this limit by switching to ultraviolet light and then again by using electron beam etching. "The current belief is that the tremendous cost of building the new fabrication factories (i.e., $1 Billion or more per factory) will be the parameter that finally slows down this growth rate" (Ross, 1992).

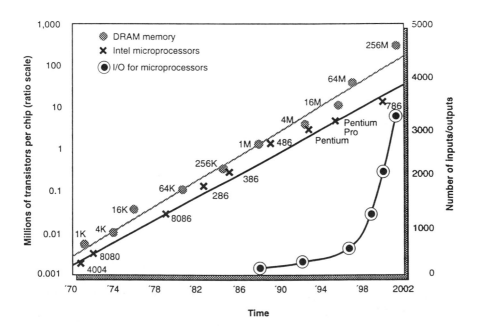

FIGURE 3.1 Technology capability forecasting using Moore's First Law (adapted from Ross, 1995 and other sources).

3.3.5   Benchmarking and Company Capability Analysis

Benchmarking and company capability analysis are processes for studying a company's capabilities and measuring the company's products, services, and practices against the competition or those companies recognized as leaders. The purpose is to determine product requirements and identifying innovative ideas. It is the systematic process of identifying the "best" practices in industry, setting product and manufacturing goals based on results of what the competition has achieved or will achieve in the future and identifying the best vendors. The practice of benchmarking in product development is important for several reasons:

- Better awareness of customer needs, preferences and values
- Better knowledge of each company's strengths and weaknesses
- Better awareness of competitor's product, manufacturing processes, vendors used, etc.
- Identify successful product requirements (e.g., performance, quality, availability, price, service, reliability, support)
- Identify the best vendors and suppliers to use in the design

Benchmarking can be used to identify and evaluate "best in class" parameters and in turn help to develop requirements for a successful product.
The five key steps to benchmarking are:

1. **Analyze all aspects of the competition and other successful companies.**
2. **Determine where your company stands** relative to competitors and leaders on key features, parameters, and processes both on today's products and on predictions of future products.
3. **Establish "Best in Class"** product features and parameters.
4. **Set product, manufacturing, and supportability requirements** based on benchmarking information.
5. **Implement a practice of "innovative imitation"** (i.e., improve on the best ideas and methods that were identified).

A simple benchmarking study to determine some manufacturing design requirements for a computer circuit board assembly is shown in Table 3.1. The company purchased several competitors' products. Each product was disassembled and studied. Note that the "best in class" competitor assembly time, for the printed circuit board (PCB), is the lowest due to the smaller board size, fewest number of screws (4), and no ground plane requirement. As a result, design should attempt to eliminate screws, reduce board size, and eliminate the ground plane.

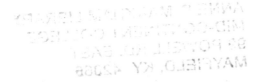

TABLE 3.1 Printed Circuit Board Assembly Benchmark

Model	Board Size (mm)	Number of Screws	Ground Plane	Assembly Time (sec.)
Competitor 1	290mm * 210mm	6	yes	120
Competitor 2	355mm * 195mm	9	yes	125
Competitor 3	255mm * 200mm	4	no	49
Best in class	Competitor 3 (smallest board size)	Competitor 3 (least number of screws)	Competitor 3 (no ground plane)	Competitor 3 (lowest assembly cost)

3.3.6   Prototyping and Virtual Reality

Prototyping and virtual reality in the early design phases use hardware and software conceptual prototypes to study customer responses to new products and to identify areas for improvements for both requirement definition and conceptual design. The basic philosophy behind the use of prototyping is to provide a communications platform for studying consumer needs and refining product use scenarios and task analysis. Evolutionary prototyping can be used for incrementing and refining the design. This can be especially helpful when the product is very different (i.e. innovative) when compared to competing products. Prototyping helps the design team by encouraging the consideration of as many product development issues as possible during the early phases and identifying potential problems. The steps are:

1. Produce prototypes that provide the reviewer a realistic view or feel of the proposed design
2. Develop as many prototypes as economically possible
3. Continuously produce prototypes throughout the product development process
4. Show the prototypes to everyone

The future of virtual prototyping is to merge virtual reality software and CAD/CAE tools into a new kind of software that would allow consumers and the design team to "evaluate a product in a virtual environment". The designer could shape or manipulate a new concept by hand and give it to the customer and other engineers for analysis. The customer and product designer could also "experience" a design concept, walk into the simulation space and examine details or properties not even visible in the real world.

Focus groups are one method to use prototypes and virtual reality for determining product requirements. This is where potential consumers are shown product prototypes to gather their reactions to the product. Consumer reactions are then used to identify design improvements and marketing related information. "In a darkened conference room, 10 consumers stare at a TV monitor displaying a car's center console. It looks like a Rolodex, one viewer says, as the others laugh. If one drawer jams, you will have to tear the whole thing apart. That remark helped kill a new feature for a product. Focus groups can be expensive, ranging from $50,000 for a quick-and-dirty program to $150,000 or so for a full-scale effort. Many companies, however, are willing to pay that price to avoid a costly design blunder" (Crain News Service, 1996).

3.3.7   House Of Quality

House of Quality or Quality Function Deployment (QFD) is both a requirement definition and conceptual design tool that systematically documents customer needs, benchmarks competitors, and other aspects and then transforms this information into design requirements. QFD is a complex process that requires considerable effort. The reader is referred to Hauser and Clausing(1988) for more detailed information. The steps are:

1. Determine and rank customer attributes, i.e. what does the customer want? Which attributes are most important?
2. Document customer perceptions of how well different products meet these attributes. Which products do the customers like for each of the attributes?
3. Determine "measurable" design characteristics/parameters and rate their relationship to the customer attributes.
4. Determine objective requirements and measures (goals) for the design characteristics.

The process starts at very high-level requirements and evolves into more detailed levels as the design progresses. The problem with this method is the large amount of effort that is required to perform the process correctly. Robert Hales (1995) stated in IIE Solutions "the key benefit of QFD is the understanding about the direction they are headed. This common view is gained through the identification and resolution of conflicts arising from the cross-functional team's different perspective. Equally important is the team's customer-supplier focus. The team is developing a product or service that will thrill customers to the extent that they will part with their money. Usually, they are just charged with developing a product that meets the specifications. Instead of being reactive, they are now proactive". A simple House of Quality for a notebook computer carrying case is shown in Figure 3.2 (Stubbings and Yousef, 1996).

An important result is the identification of critical factors for customer quality. Design and manufacturing can then focus on these critical factors.

Relationships

- **! Strong positive**
- ! Medium positive
- X Medium negative
- **X Strong negative**

Engineering characteristics

	Customer requirements	Importance ranking	Physical Characteristics: Length dimension	Physical Characteristics: Width dimension	Physical Characteristics: Depth dimension	Physical Characteristics: Weight	Protective Elements: Exterior material strength	Protective Elements: Shock absorbance	Protective Elements: Water resistance	Customer Perception (poor) 1	2	3	4	5 (good)
Easy to Carry	Light weight	10	**X**	**X**	**X**	**!**	X				Z		E	S
	Compact size	5	**!**	**!**	**!**						Z		S	E
Provides Protection	Durable exterior	20					!		!		E	S		Z
	Padded interior	20						**!**				E	S	Z
	Water/Weather resistant	10							**!**		E	S		Z
Quality	Long warranty	10					!		!					ESZ
Misc.	Inexpensive	15					X	X	X	E		Z		S
	Compatible with other laptops	5	X	X	X							Z		SE
	Extra storage room	5	X	X	X							EZ		S
Objective Measures	Measurements units		in	in	in	lb	psi	G's	psi	1	2	3	4	5
	Our case		16.5	11.25	3.33	5	50,000	11	60					
	Company S		14	12	6.5	5.7	12,000	8	40					
	Company E		18	13	6	5.4	3,000	5	20					
Technical difficulty			1	1	1	2	5	10	5					
Imputed importance (%)			5	5	5	15	25	30	15					
Estimated cost (%)			3	3	3	6	40	35	10					
Design requirements			16	11	4	5	50,000	15	60					

Z ___ Zero (Our Case)
S ___ Company S
E ___ Company E

FIGURE 3.2 House of quality for a notebook computer case.

3.4   PRODUCT REQUIREMENTS AND SPECIFICATIONS

Requirements provide the foundation for the entire design process. The knowledge output from requirement definition is the product's requirement documentation. Requirement definition should produce product level requirements that should document what needs to happen for success. Do not just focus on what you think is achievable or what has been done in the past. Concentrate on creativity, customer value and developing "best in class" requirements that beat the competition without too much technical risk.

To the optimist, the glass is half full. To the pessimist, the glass is half empty. To the design team, the glass is twice as big as it should be.

3.4.1   Special Issues in Software Product Requirements

The emphasis on customer expectations in software design is not limited to the product itself. Other components including documentation, marketing, materials, training, and support are part of the customer's overall expectations. There are two types of expectations in software systems; functional expectations which focus on establishing functional specifications, and attribute expectations which focus on how the system appears to the people who use and maintain it.

Domain models help the software designers understand complex applications and their environment (i.e., the entities, attributes and interrelationships between the entities). Models include concept diagrams, scenarios, use cases, task descriptions, diagrams, and event-traces. Models highlight important aspects of how the domain operates and what is important to the user.

Requirements begin as informal statements of what the users want. There may be many possible ways to implement these requirements in software, but any implementation must satisfy the requirements. Requirements describe the expected behavior of the program and constraints (e.g., security and reliability), irrespective of how that behavior is actualized. Prototyping is especially important for defining attribute expectations. For small projects, a few sentences are sufficient to completely describe the user's desires (e.g., a teacher wants a small utility program to calculate student averages). For large projects, the requirements are many and are heavily interrelated. It is usually quite difficult to detect requirements that are inconsistent with one another by simple inspection. Formal methods can be applied to map the requirements to a formal notation.

Requirements are entered into a requirements traceability system that directly traces requirements to the statement of work. The software group can then prepare software development management plans and facility specifications.

3.4.2 Notebook Computer Product Requirements

Notebook computer design is a good example where product requirements are a moving (i.e. changing) target. The continuous improvements in integrated circuits, memory, displays, etc. require computer companies to constantly introduce new products. Company capability analysis is used to identify strength and weaknesses. This can be used to identify what partners and key suppliers should be part of the development team. Benchmarking would be used to evaluate the competition and identify the "best practices". This information helps to decide what the company will design, build, or support. The house of quality or some other method is used to systematically document customer needs. Since the computer company does not manufacture semiconductor devices, technology capability forecasting would have been used to establish the integrated circuit and memory requirements, (see figure 3.1, shown earlier). Innovative technologies would be considered such as voice recognition and large active matrix displays. Voice recognition systems have improved and cost have lowered where they might be practical for high all product models. Large active matrix screen prices are dropping to allow a larger and higher quality color display.

Prototyping and focus groups could be used to evaluate customer responses to styling issues such as case size and color, location of features, weight, etc. For a 1999 notebook computer case, some "best in class" product requirements are shown in Table 3.2.

3.5 CONCEPTUAL DESIGN PROCESS

The conceptual design process (1) identifies all design approaches (i.e., alternatives) that could meet the defined requirements, (2) performs trade-off analyses to select the best design approach to be used and (3) transforms the product requirements into lower level design requirements based on the selected approach. It begins when a new product is defined in the requirement definition process and continues until the final design approach has been identified. Requirements are allocated down to the lowest levels needed and documented during this process. This is the phase where the size of the design team will grow. As more and more specialists are added, effective communication and teamwork is essential.

3.5.1 Identify All Design Approach Alternatives

The first step is to start identifying potential design solutions to be used in trade-off analyses. Many people are involved in this collaborative effort to insure all possible options are considered. Creativity and innovation must be encouraged not only for design but also manufacturing, logistics and other areas. Identifying alternatives are often performed in "brainstorming" sessions.

TABLE 3.2 Some Notebook Computer Product Requirements

Ergonomics	
User	Business user including use on airplanes
Functional Performance	
Integrated Circuit	800 MHz or faster processor
Display	10.4", active matrix color SVGA
Weight	2.0 lb. With Lithium-Ion battery
Size	8" × 11" × 2"
Battery Life	6 hours without charging
Software	
Office suite	Provided when shipped
Key Partnerships and Suppliers	
Disk drives	Company name
All repair outside of US	Company names
Logistics	
Warranty	1 year
Repair Service	Free phone response for 1 year
	Internet self diagnostics
Reliability	1200 hours MTBF
Order Methods	Internet and dealer networks
Shipping Density/Cost	$5.00
Recyclabilty	95% by weight of all parts and packaging
Retail Price	$1500
Schedule	
First Shipment	5 months from starting design date

Design is about anticipation. The team anticipates new technologies and styling trends to envision how they might be translated into a good-looking, useful, easy-to-use and desired products. The team also anticipates styling and social changes and identities new customer need and desires as a result of those changes.

The widest possible range of potential solutions should be examined early in the program. In this way, the opportunity is optimized to take advantage of recent advances in technology and new styling trends to avoid being locked into out of date or preconceived solutions. Typically, the technologies should range from the mature, that are usually lower in technical risk but also have lower payoff potentials, to the leading edge (i.e., newly emerging) technologies, that have higher technical risks and costs but also have the potential for a more significant impact on product success. Styling and features are also important. The challenge is to eliminate all of the bad ideas. A common decision a team must address is whether particular parts (or software programs) should be specially developed for the system, or whether they should be purchased off the shelf. This is called a "make or buy" decision.

Possible Design Alternatives for a Notebook Computer

For the notebook computer example, several new or innovative technologies should be considered to meet the product's requirements. In the design of a notebook computer (i.e., based on 1999), there are several technologies which could be used in the design to increase performance but would increase the level of technical risk and cost. New technologies to evaluate for increasing performance include:

- **Digital Video Disk (DVD)** DVD's can give a seven fold increase of mass data storage on a 5 inch disk. Full-length movies can be stored on a single disk.
- **RISC Processors** Reduced Instruction Set Computer (RISC) processors can replace the current Complex Instruction Set Computer (CISC) processors. The RISC chip is faster while using less power, but none of the x86 operating systems will run on a RISC CPU in its native mode.
- **Multiple Processors** Redesign of the traditional motherboard (i.e., the main printed circuit board) will allow more computing power and ability to run multiple 32 bit application by using multiple processors. Each processor is in charge of a specialized task with a separate processor designated as the control processor.

Aesthetics and styling could also be part of the design alternatives. Having a titanium metal case rather than a traditional plastic case could be studied. Can manufacturing produce titanium cases at a reasonable cost? Will customers perceive the titanium case as being preferred to a traditional plastic case? If the case is plastic, what color should it be?

For manufacturing, the design trade-off could focus on using new technologies/methods versus existing technologies. For example, the design team could evaluate new manufacturing technologies such as chip on board (COB) or chip in board (CIB). The evaluation might focus on possible reliability issues and manufacturing start up costs. For example, start up costs may be so high that the circuit boards may be outsourced to another company. Service trade-off for customer sales and service could consider Internet use, phone banks or both for consumer information. Self-diagnostics and self-maintenance using the Internet could be considered.

The company must trade-off the benefits of these new technologies versus the increased technical risk that comes with their implementation. Technical risks would include cost increases, schedule delays, poor quality due to the potential problems of designing for the new technologies, developing new manufacturing processes, or using new vendors.

3.5.2   Extensive Trade-Off Analyses

The next step of conceptual design consists of evaluating each of the identified design approaches. **Trade-off studies examine alternative design approaches and different parameters with the purpose of optimizing the overall performance of the system and reducing technical risk.** This includes both innovative and traditional approaches. A trade study is a formal decision-making method that can be used to solve many complex problems. Trade studies (also called tradeoff studies or analyses) are used to rank user needs in order of importance, develop cost models, and identify realistic configurations that meet mission needs. That information then helps highlight producible, testable and maintainable configurations with quality, cost, and reliability at the required levels.

Trade-off studies are directed at finding a proper balance between the many demands on a design. The trade-off studies should include all-important parameters such as cost, schedule, technical risk, reliability, producibility, quality, and supportability. Utility metrics can be also used to quantify the different alternatives such as by Sanchez and Priest (1997) and Pugh (1990). As recommended by the BMP Producibility Task Force (1999), the steps are to:

1. Form a cross-functional team. The team may be a completely independent group, with augmentation by functional experts.
2. Encourage customer involvement and innovation.
3. Define the objectives of the trade study alternatives.
4. Determine the approach and resources required.
5. Evaluate and select the preferred alternative.
6. Validate the study results through testing and/or simulation.
7. Iterate more detailed trade studies throughout the design process.
8. Document the study and results.

Benchmarking, trade-off studies, mathematical models, and simulations verify that the optimum design approach has been selected support the best design approach.

Functional Allocation

One important trade-off to be made is what tasks will be performed by the product, which tasks by the user and which tasks by both. **Functional task allocation is the process of apportioning (i.e. dividing) system performance functions among humans, product, or some combination of the two.** It is usually performed early in the design process during requirement definition or conceptual design. Failures occur, when a functional task is assigned to a human, but the necessary information is not provided to the user. This methodology will be further described in the Chapter on Human Engineering.

The major activities involved in functional task allocation are as follows:

- Determine the product's performance objectives.
- Determine functional requirements necessary to meet product objectives.
- Allocate these functional tasks to persons, product, software, or some combination based on analysis.
- Identify alternative design configurations.
- Verify, for each design alternative, that the human and the product can perform the assigned functions and satisfy all design requirements.

A critical but common mistake occurs when the design team attempts to automate every function that can be automated. Only functions that cannot be automated through hardware or software are then left for the human operator. This design approach results in excessive complexity, schedule delays, and assignment of tasks in a manner that the human cannot perform properly. Appropriate functions for humans and products are occasionally provided in handbooks, but they are only guidelines for evaluating specific functions in real-world design.

Software Trade-off Analyses

For software, prototypes provide valuable information for trade-off analysis. Software designers perform feasibility studies on possible alternative solutions and evaluate their projected cost and schedules. Trade-off analyses can disclose significant life cycle cost savings through a clear allocation of hardware and software requirements. The results of the conceptual design are the selection of an overall software approach that is then used to develop detailed design guidelines and requirements. Methods include data flow analysis, structured analysis, and object-oriented analysis.

Notebook Computer Trade-off Analyses

Some trade-off analyses for the notebook example are shown below.

- Plastic case vs. titanium case
- Chip on board technology vs. conventional surface mount technology

- Vendor X hard drive vs. Vendor Y hard drive
- Purchased power supply from a vendor vs. design and manufacture a new power supply inside the company
- Display size (readability) vs. increased cost, size, weight of larger displays
- Full size keyboard size vs. reduced size keyboard based on cost and user preference
- Where should the product be manufactured or options added?
- Which important suppliers should be included in the design process?
- Method for service help (web based, phone service, retail, etc.)

International trade-off could include:

- Is conversion to different voltages necessary? If yes, what design approach is best?
- Is conversion from AC to DC power necessary? For which countries?
- Can production for export be done within the firm's existing physical capacity or must capacity or new processes be added?
- Should some components be selected from or made in the buyer's country or a third country?

For the power conversion problem in the global market, there are three design solutions that could be considered in the conceptual design process. Three options are:

1. Different battery chargers are manufactured for different voltages.
2. Voltage two way, an "internal" switch is added to the product or battery charger for either 220-volt systems (50 HZ and 60 HZ); 110-volt (60 HZ) systems would still require a different product.
3. Voltage 3 way, user accessible switch for all 3 voltage conditions.

Producibility Trade-off Analyses[*]

Producibility's goal in conceptual design is to develop the most effective design and manufacturing approach prior to detailed design. This section is an adaptation of the recommendations from the 1999 BMP Producibility Task Force (BMP,1999).

---

[*] From BMP, Producibility Systems Guidelines For Successful Companies, www.bmpcoe.org, 1999

Key tasks are:

1. **Identify key characteristics and manufacturing resources –** focus on design features with greatest impact.
2. **Identify Design to Cost and other manufacturing requirements** - set appropriate and realistic goals
3. **Perform trade studies on alternative product and process designs including rapid prototyping** - rank requirements and identify alternative strategies and technologies
4. **Develop a preliminary manufacturing plan and strategy** - identify detailed plans for manufacture
5. **Perform a complexity analysis and simplify where possible** - determine how complex the product and required processes are to schedule and cost impact. Simplify where possible.

Task 1   Identify Key Characteristics and Resources

The goal of an efficient key characteristics (KCs) effort is to identify and control the design features that have the greatest impact on manufacturing time, cost, and overall product performance. These key or often called critical characteristics are those that can most affect the product's performance or manufacturing. This allows the design team to focus on what is most important. Identifying key characteristics begins with product requirements and flows down to lower level requirements.

Task 2   Identify Design-To-Cost And Other Manufacturing Related Goals

Design to cost (DTC) is a methodology that focuses on minimizing unit production costs. This method reduces product cost through a rigorous approach of identifying and implementing cost-reducing design and manufacturing improvements. A design-to-cost (DTC) goal is established for the product in requirement definition to provide a measure of the level of producibility for any proposed design. Setting appropriate and realistic cost targets is critical to successfully controlling product cost. DTC is further discussed in the next chapter.

Task 3   Perform Trade-off on Alternative Product and Process Designs

Producibility, reliability, and other disciplines can be either an independent or dependent variable depending on the requirements, but should always be a documented variable. Producibility measurements can be related to cost, schedule, quality, complexity, and risk. The trade study's quality depends on the quality of the input data. The results will be unreliable if the input data comes only from peoples' memories, estimates, or "best guesses."

Task 4   Develop A Preliminary Manufacturing Plan and Strategy

A manufacturing plan identifies the processes and vendors used to create a product. During the plan's development, the design team agrees on the critical features, resources, technologies and the processes required. The manufacturing plan can help identify and highlight risk areas throughout product development. It is created early during conceptual design and updated frequently.

Task 5   Perform A Complexity Analysis and Simplify Where Possible

A complexity analysis of product and process alternatives should be performed to reduce manufacturing risk, schedule and cost impact. New or complex design attributes or features may require the acquisition of new technologies, machinery, processes, or personnel capabilities. The steps for complexity analysis are to:

1. Assemble the independent review team.
2. Determine complexity metrics to be analyzed (design attributes or features, part count, process required, schedule, cost, tooling, etc.)
3. Analyze design against complexity metrics.
4. Incorporate suggestions into a modified design and manufacturing plan.

This will be discussed in greater detail in a later Chapter on simplification.

3.5.3   Design Requirements are Developed and Allocated

After identifying the "best" design approach, the next step in conceptual design is to translate the high-level product requirements into lower level design requirements. Since these design requirements will provide the performance baseline for each design team member they should be:

1. Easy to understand
2. Realistic
3. Detailed and measurable for the selected design approach.

Design requirements are an important method of communication and provide the foundation for the design effort. It is used to develop program organization, funding, partnerships, and guidelines (including part selection, producibility, and reliability).

Design goals and requirements should be sufficient in detail to:

- Communicate essential requirements to all the members of the design team including vendors.
- Permit complete technical control of the design process in all aspects of the program.
- Minimize loss of continuity resulting from personnel changes.
- Provide a quantified baseline for design trade-offs, design reviews and measurement of technical progress.
- Provide quantified testable requirements for test and evaluation.

Even when the design requirements are well defined and documented, the design team must ensure that the stated requirements are reasonable and appropriate for the end user and within the limits of existing technology or that the technology could be developed.

One problem is where there are too many design requirements. For example, using extensive lists of standard design requirements from previous projects can often be counterproductive to the overall program.

Another problem is when design requirements are written in contractual or "lawyer like" terms. These are often not easily understood, difficult to use in everyday design decisions, and structured so that their achievement is laboratory oriented. Management can assist the design team by providing measurable and easy to understand design guidelines.

After the higher-level design requirements are defined, these requirements are then further defined for and allocated to lower level subsystems. Effectively establishing lower level design requirements from the higher-level product requirements is a difficult task. This is often called requirements allocation or partitioning. Requirement allocation identifies how to decompose and allocate the system-level requirements to subsystems and then to components, e.g., hardware, software, personnel, technical manuals or facilities. Subsystems and components receive technical requirement budgets that together add up to the total system requirement. A total of all subsystem requirements should equal or be less than the system requirement. This allows each designer to have assigned goals and responsibilities for a particular item of the design. This allocation should identify and compensate for areas of unusual technical problems or risk.

All design requirements should be measurable and "testable". Figure 3.3 shows the top down (requirements) - bottom up (testing) process. That means that each requirement at every level should be defined to allow the evaluation as to whether the design is satisfying the requirement. If a requirement cannot be measured and tested, it is not a requirement. Table 3.3 illustrates an example of the requirement allocation process and its subsequent test requirement definition for an automobile's fuel economy.

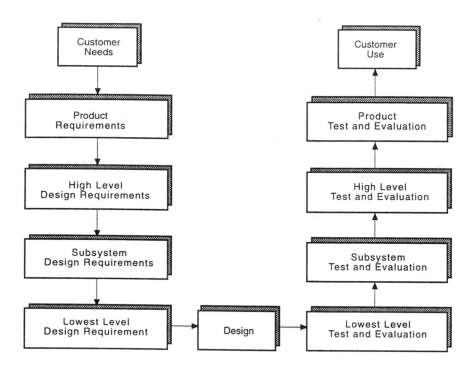

FIGURE 3.3 U - Shape development: flow down of design requirements
and flow up of testing.

TABLE 3.3 Requirements Allocation and Verification for Fuel Economy

Top down allocation of customer/product/design requirements

Step 1	Customer Preference/Requirement: Increased fuel economy
Step 2	Product Concept: Good mileage
Step 3	Product Requirement (automobile): 35miles/gallon using govt Testing procedures, 2000 lb vehicle
Step 4	System Design Requirement (engine): .03 gal/min at 4000 rpm producing 200 hp
Step 5	Subsystem Design (carburetor): .03 gal/min $\pm$ .001 at 4000 rpm at 30^0 C

Bottom up test and evaluation verification

Step 6	Measure fuel consumption in laboratory – test of carburetor
Step 7	Measure fuel consumption in laboratory and field – test of engine
Step 8	Measure fuel consumption in laboratory and field – test of automobile
Step 9	Measure customer perception using customer focus groups
Step 10	Measure customer satisfaction through interviews, product success

In summary, well-defined requirements should be clear, unambiguous, understandable, concise, stable, testable and measurable.

3.6 REQUIREMENT ALLOCATION FOR A NOTEBOOK COMPUTER

One of the notebook's product requirements was to weight 2.0 lb. with the battery installed. This weight requirement is allocated down into weight requirements for each major design area or module. An example of allocating design weight requirements between the various major parts in a notebook computer is shown in Table 3.4.

3.7 SOFTWARE DESIGN REQUIREMENT ISSUES

Design requirements for software usually focus on the human-computer interface, hardware requirements, data flow, and interfaces with other software products. Requirement definition and the conceptual design process for software products are similar to hardware. Important differences, however, do exist.

TABLE 3.4 Notebook Design Requirements for Weight

Subassembly	Weight Design Requirement (oz.)
Display	4
Printed Circuit Board	3
Plastic Case	3
Battery	10
Hard Drive	4
CD ROM Drive	3
Fax/Modem	3
Management Reserve*	2
	32 oz.

*Management reserve (i.e. extra budget) is determined arbitrarily to allow management flexibility to compensate for weight problems that may be found later in the design process.

These differences include unique product life cycle characteristics, increased use of prototypes in requirement analysis, and software making redesigns and modifications late in the development process. A difficult task is to ensure that all software requirements are testable.

Some software requirements for a notebook requirement would include:

- Meet all requirements for being IBM compatible
- Compatible with the latest and projected versions of Microsoft's Windows
- Software interfaces between different hardware components and software modules
- Built in test, self-diagnostics, and self-maintenance

3.8 SERVICE AND ELECTRONIC COMMERCE DESIGN REQUIREMENT ISSUES

As mentioned earlier, service industries design and build systems that provide service to customers. This process is similar to products and in fact, many companies sell both products and services. Service requirements depend on the service provided and often focus on customer response such as average and maximum response times, number of responses necessary to resolve/complete service action, cost per service action, shipping time and costs, etc.

For example, electronic commerce focuses on the transfer of information. Requirements focus on bandwidth, traffic volume, types of traffic, and computing resources. A key design requirement is "scalability" or reproducibility for meeting changes in demand. Does the design allow for the system's capability to be enlarged or updated quickly and efficiently without interrupting service?

Service and repair methods are changing. Future products will include much higher levels of built in test, self-diagnostics, and self-maintenance. Design requirements for the level of capability must be determined. This capability will affect whether additional processors will be required and the amount of extra circuitry required. Modem capability will be required for products that can directly contact factory service. Self-maintenance might require even more computing capability and actuators for activating hardware repair actions.

3.9   RELIABILITY DESIGN REQUIREMENT ISSUES

Reliability is a major design parameter and depends on actions taken during, not after, design efforts, manufacturing, and testing. Reliability is usually defined in terms of the probability that a product operates successfully over a specified period of time in a defined environment. Typical reliability measures such as mean time between failures (MTBF) are explained in detail in the chapter on reliability.

Developing and tailoring requirements and guidelines that achieve the user's reliability needs and at the same time provide adequate guidelines for designers is a difficult task. A major element of this effort is to base reliability requirements on realistic and accurate profiles of the system's use in the operational, maintenance, and storage environment.

It is also very difficult for designers and managers to translate reliability requirements into everyday design decisions. Reliability goals, requirements, and guidelines must be translated into design-related parameters. The parameters should be easily understood, provide a basis for design decisions, and have the ability to be continuously measured by designers and management at all stages of the design process. Tests to measure and verify reliability should be specified in terms of the items to be measured, methods of measurement, failure definitions, testing environment, and test procedures.

Design guidelines have shown to improve reliability. Examples of reliability oriented design guides include part selection, derating criteria, and design practice. Sources used for these guidelines included allowable parts lists, military specifications, standards, and other design guides. In all cases, the guidelines were provided only as a starting point for design objectives.

After the total failure rate for a product has been established, failure rates are then apportioned or distributed among all identifiable lower level elements of the system. Apportioning the failure rate allows the development team to have a quantitative design requirement that can be measured for technical progress, similar to other design parameters. Although the requirements could be divided equally among the subsystems, most programs divide the requirements based on the complexity of each subsystem.

The predicted failure rates of different design modules, piece parts, or software modules can then be compared to the allocation to ensure that the reliability requirements are continually met as the design progresses. Conversely, if any of the subsystem failure rate allocations are exceeded, the designer is alerted that corrections must be made. A number of alternatives are available to the designer to correct the problem:

- Reduce part count or software complexity
- Use more reliable parts, materials, and software
- Reduce stress levels
- Design for variability and increase robustness of the design
- Add redundancy, self corrective, or self maintaining

Derating Criteria, Safety Factors, and Design Margins

The product development team must consider the uncertainty of many of the assumptions and parameters used in their design analyses and in manufacturing the product. To compensate for the uncertainty of these assumptions, design margins, derating criteria and safety margins are used. This requires the design to meet stress requirements that are greater than expected (i.e. over-designed) to compensate for uncertainty. **Design and safety margins compensate for uncertainty by forcing the design to meet requirements higher than expected.** They are usually expressed as a multiplier such as 200% or 2X. This would mean that the design must be more than twice the required stress. **Derating is used in electronics where the maximum operating envelope stresses on the part can be no greater than some percentage of the maximum rating as determined by applicable inflection points.** The purpose of setting derating criteria, safety factors, and design margins is to provide a design with the extra capability to compensate for any unforeseen problems or stresses. This "safety net" or "over design" established by the customer, management, or the designer, increases the strength and precision of the parts. This criterion must be carefully developed since over-ambitious derating and design criteria can significantly increase design costs, schedule, and risk. Because of this problem, criteria should be developed only after thorough analyses of the project's goals, product environment, and other applicable guidelines.

3.10   DESIGN REQUIREMENT ISSUES FOR MANUFACTURING, PRODUCIBILITY, AND PROCESS DESIGN

Design requirements should be set to allow the economical manufacture of a high quality product at a specified production volume or rate that is capable of achieving all performance and reliability inherent in the design. Important manufacturing parameters are initial fixed cost, recurring or variable cost, quality, lead-time, cycle time, and technical risk. A key factor is to set manufacturing requirements based on the premise of optimizing design requirements that are critical to the product's functional performance and minimizing requirements that are trivial or not important. Critical parameters or key characteristics are those that directly affect customer satisfaction.

Often the company's existing manufacturing processes are not capable of meeting the new design's requirements. When this occurs, manufacturing must develop new manufacturing processes or significantly improve an existing process. New process development occurs concurrently with the design process.

Some design requirements for the notebook computer that could affect manufacturing includes:

- Total manufacturing unit cost $1000 including overhead
- Assembly time   5 minutes
- Total number of parts in
  final assembly   25
- Number of fastener types   1
- Number of new vendors   3
- Percentage of parts from
  ISO 9000 certified vendors   90%
- Purchased part lead times   2 weeks maximum
- Predicted defects per million
  opportunities (DPU)   3.2
- Time to test   2 minutes
- Percentage of computer
  tested by built in test   95%
- Time to repair including retest   10 minutes
- Time to disassemble   57 minutes

Comparing specific design requirements to the capability of a particular manufacturing process or part will provide a defect distribution with a plus or minus sigma capability. Manufacturing typically uses the metrics of $\pm 3\sigma$ or $\pm 6\sigma$ as acceptable measures for sigma. Producibility design requirements often use Cp and Cpk. **Cp** is a ratio of the width of the distribution (i.e., process width) to the width of the acceptable values (i.e., design tolerances). **Cpk** refines Cp by the amount of process drift (i.e., the distance from the target mean to the process distribution mean). These are important and popular statistic based metrics for balancing design and manufacturing requirements. They measure the ability of a manufacturing process to meet specified design requirements. For the notebook computer Cp > 2.0 and Cpk > 1.5. These measures are further discussed in later chapters.

3.11   REQUIREMENT ISSUES FOR GLOBAL TRADE

When a product is sold and used in other countries, design requirements must include the customer's unique characteristics, such as language, culture, and government requirement of the countries involved. The specific requirements would vary depending on the countries and cultures. Design requirements that could be affected include labeling, colors, options, technical manuals, packaging and maintenance plans.

Design for international sales will increasingly be an added value that makes a competitive difference. As companies continue to enter international markets, designers must design to the broad specifications of international requirements for technical standards, safety and environmental regulations. More and more will be demanded from designers to appeal to the narrower definitions of consumer preference based on demographic characteristics and cultural differences (Blaich, 1993). Some design requirements for global trade could include:

- Except for the company's name and model number, only symbols will be used to convey information on the product.
- Instructions will be printed in English, Spanish, French, Japanese, and Chinese. Other languages will be translated at a later date.
- Packaging must be sufficient and efficient for international shipping.

3.12   COMMUNICATION AND DESIGN DOCUMENTATION

Documentation is the most important form of communication in product development. It is the foundation of the product development process. Although it may be impossible to document every conceptual thought, reason, calculation, or decision during the design stage, it is possible for the design team to provide better documentation during the design process. Businesses have found that few people adequately document their work. For some designers, deciding what and how their design should be documented seems more difficult than the design work itself. They argue that user tasks and support techniques are, for the most part, straightforward and known. Documentation principles are then left up to someone else to understand.

The purpose of design documentation is to provide a path of communication within the design team and with those who must direct, review, manufacture, and support the design. Personnel must be able to correctly interpret the designer's intentions.

- Production must be able to accurately purchase materials, components, and tooling for efficient manufacture of the proposed design.
- Assemblers must use the drawings to correctly assemble the product.
- Technicians must be able to use the drawings to install, test and troubleshoot completed assemblies and software.
- Technical writers use engineering drawings to produce the product's operation and maintenance manuals.

The design drawings are the start of a long line of communication in a product's life cycle. Drawings that are well thought out, concise, and articulate will help all members of the project team.

Some of the designers may be in another country or have a different cultural background. One effective method to overcome potential language and cultural differences is to use applicable standards, more illustrations, and common symbols in the design documentation.

One reason good documentation is difficult to produce is that each discipline seems to have their own language and terminology (i.e. ontology). The areas of responsibility for generating the documentation also tend to overlap between the different design disciplines. For instance, a designer may believe that it is the quality or production personnel's responsibility for planning the inspection of a product. Without the designer input of information, the production personnel may not know enough about the product to adequately inspect it.

For example, consider an airfoil-shaped fin. Aerodynamic behavior of the fin is controlled by varying the contour and thickness of the fin's airfoil. The design engineer specifies the appropriate airfoil that controls the outside shape of the fin and calculates the allowable thickness tolerance at any point. Since the airfoil section is basically a spline, it is nearly impossible to inspect the thickness of the entire fin with a typical gauge. Therefore, the fin thickness must be manually inspected at every point. Yet, how many points should be checked? Even though the fin's thickness requirements are clear and concise, the inspection requirements are not. Here, the design engineer can help by choosing a finite number of specific locations to inspect. The locations chosen represent the most critical or meaningful points for performance, thus guaranteeing well-manufactured fins.

The benefits of good design documentation are immediate and long term. In addition to the documentation helping other team members, as discussed earlier, accurate and readable design notes benefit the individual in successive iterations along the path to a finished design. Once completed, a properly documented design can survive long after its concept and serve as a database of ideas to improve future designs and ensure continued profitability for the company.

For design documentation to be effective, it must have the following characteristics:

1. Accuracy
2. Clarity
3. Conformity to policy, convention, and standards
4. Completeness
5. Integrity
6. Brevity
7. High quality
8. Retrievability
9. Proper format

Methods of complying with these requirements vary from one project to another since each project has its own organization, requirements, and resources. Because of their importance, these characteristics are discussed here.

Accuracy

Accuracy in documentation is important since several people from different disciplines depend upon the documentation to do their jobs. Therefore, the design team should use care when generating documentation. If certain facts or areas are still undefined, this should be so stated, or the fact should not be stated at all. For example, dimension tolerances on subassembly engineering drawings should be stated after much thought is given to their multiplying effect on the total product dimension limits. If previous data becomes outdated, engineering change notices should be written as soon as possible.

Clarity

Documentation that is clearly stated will be more useful (and less ambiguous) than documentation that is not clearly written. Verbal documentation that is clear requires fewer words to express thoughts and concepts. Drawings should state values clearly.

Conformance to Policy, Convention, and Standards

Established policies, conventions, and standards have evolved for a reason and should be used whenever possible to keep things running smoothly. An example is using the proper numbering system to identify drawings. An established numbering system makes drawings easy to find and reference. However, since conventions are an evolving process, suggestions for improvement should not be discarded but, rather, discussed with superiors. Some suggestions may prove valid and become new policy after going through the proper formal change channels.

Completeness

Good design documentation is complete documentation. Because so many other disciplines rely on the documentation, it is important that the design team carefully think through what data may be needed down the line to help get the product out efficiently and cost effectively. As previously mentioned, specifying a finite number of specific locations to inspect on an airfoil-shaped fin and documenting them on a supplementary information drawing ensures that the inspector can properly inspect the product.

Integrity

Integrity in documentation is important since several people from different disciplines depend upon the documentation to do their jobs. The individual who consistently uses care in generating documentation is the person whose source material has integrity. Drawings, for example, maintain integrity if change notices are written as soon as the changes occur.

Brevity

Brevity is sweet. People using the design documentation want to get the needed information quickly. Therefore, documents should be generated using language and documentation conventions that are concise and to the point.

High Quality

Documentation should be produced using the latest and best equipment the company can offer. This assures not only that the documentation is clear and cost effective but also that it is more or less uniform and standard with other company drawings of the same type.

Retrievability

Documentation that is not easily found is often overlooked. Therefore, documentation should conform to numbering systems and be catalogued according to company policy, convention, and standards so that it is easily retrievable when needed. When applicable, each drawing should reference other related drawings so that other personnel can locate all of a design's documentation.

3.13   SUMMARY

Requirement definition and conceptual design organization are "first steps" in product and process development. Properly determining a product's requirements and identifying the best design approach are critical for a company's long-term success. Ensuring success is accomplished through a systematic design process that utilizes the expertise of many disciplines. Design goals and requirements are the foundation for successful product development.

3.14   REVIEW QUESTIONS

1. Briefly explain why the requirement definition stage is important for a new product design project.
2. Identify and describe the key steps for the requirement definition process.
3. Identify and describe at least three design methodologies for identifying and evaluating user needs.
4. Describe several international aspects for the requirement definition stage.
5. List and describe three design requirements for producibility.
6. Since nothing is accomplished without some degree of planning, what could be the key issues to be considered for a program organization or a system design process?
7. For the following items prepare a list of product requirements that you think are necessary to have before any manufacturing commitments are made.
    a. stapler
    b. bicycle
    c. flash light
    d. tape recorder
8. Using one of the products listed in question 7, prepare a list of technical risks that should be evaluated in product development.

3.15   SUGGESTED READINGS

1. J. R. Hauser and D. Clausing, The House of Quality, Harvard Business Review, p. 63-68, May-June, 1988.
2. C.C., Wilson, M.E. Kennedy, and C.J., Trammell, Superior Product Development: Managing the Process for Innovative Products, Blackwell Publishers, Cambridge, Mass., 1996.

3.16   REFERENCES

1. R Blaich and J. Blaich, Product Design and Corporate Strategy, McGraw Hill Inc., 1993.
2. BMP, Producibility Systems Guidelines For Successful Companies, NAVSO P-3687, Department of the Navy, www.bmpcoe.org, December 1999.
3. Crain News Service, Focus Groups Playing a Larger Role with Makes, Dallas Morning News, p. D-1, April 6, 1996.
4. R. Hales, Adapting Quality Function Deployment to the U.S.A., IIE Solutions, p. 15-18, October 1995.
5. Hauser and D. Clausing, The House of Quality, Harvard Business Review, p. 63-68, May-June, 1988.

6. H. Morris, Industrial and Organizational Marketing, Merrill Publishing Company, 1983.
7. H. J. Petroski, Invention Is The Adopted Child Of Necessity, Wall Street Journal, July 26,1999.
8. S. Pugh, Total Design, Addison Wesley, New York, 1990.
9. P.E. Ross, Moore's Second Law, Forbes Magazine, March 25, 1995, p. 116.
10. J. Sanchez, Priest and Soto, Intelligent Reasoning Assistant for Incorporating Manufacturability Issues in the Design Process. Journal of Expert Systems with Applications Vol. 12(1), p. 81-89, 1997.
11. H. Stubbings and M. Yousuf, Unpublished Student Report, The University of Texas at Arlington, 1996.
12. Wall Street Journal, Dilbert, p. b-1, 1995.

Chapter 4

TRADE-OFF ANALYSES: OPTIMIZATION USING COST AND UTILITY METRICS

Analysis Provides A Foundation For Design Decisions

> *Trade-off analysis is an important method for developing information to help the design team in making design decisions. Every member of the design team uses it in every stage of the process. Cost is the most useful and popular measure for these trade-off and optimization studies due to its universal nature and its flexibility. Almost any design or service parameter can be converted to a currency-based measure. This allows the development team to perform analyses of different parameters based on a single performance metric. Although simple in concept, cost analysis is frequently difficult to perform because it demands current and future knowledge of a wide spectrum of product development disciplines.*

Best Practices

- Systematic Trade-off Process
- Design Improvement Focus
- Extensive And Accurate Models
- Incorporates A Realistic Assessment Of User Needs, Market Requirements, Product Performance, Manufacturing Capabilities, Prototypes, Logistics, And Other Factors
- Design To Cost
- Design To Life Cycle Cost

COST RELATED TRENDS IN THE AUTOMOBILE INDUSTRY

Customers evaluate many parameters including a product's price, operating, warranty, service, and other related cost parameters when choosing which product to purchase. Managers and designers most often use cost as their metric in design requirement and trade-off analyses. Cost is the most popular performance measure used by customers, management and designers.

Key design cost trade-offs induces technological advancements that might improve performance, or reduce a product's price or the consumer's operating costs. Foe example, one best practice in the automotive industry for reducing price is part standardization and common chassis/engine platforms between different auto models.

Ideas to reduce customer's operating costs include increasing fuel economy, special engine parts to reduce tune-ups, paints with rust inhibitors, galvanized body panels, and clear gel coats that protects the paint. Some technological advances, as reported by Car and Driver magazine (1994), are even advertised to be money savers. One is the Goodyear Eagle GS-C Run Flat tire, which is on the 1996 Corvette. The tire design incorporates a matrix layer internally within the tire to render small punctures ineffective in the depletion of air pressure. Another design example reported is Hyundai's "self-tuning" engine. This employs a board computer that incorporates a fuzzy logic memory integrated circuit (IC). The computer interprets a driver's particular driving habits and adapts accordingly to provide the corresponding transmission and engine operation that will provide the optimum fuel economy and mechanical longevity. A third design development is in the Cadillac Northstar engine. It uses a special spark plug that allows the car to travel for 100,000 miles before requiring a tune-up. The spark plug houses a dual platinum spark tip that retards deterioration of the tip. The dual spark tip configuration increases fuel economy to offset the plug's higher part cost. These are now found on many cars. More recent advances include night vision systems, GPS positioning and mapping, cameras to view lane markers, and laser range finders to maintain safe distances between cars.

The product's design should insure the highest degree possible of performance, quality, and reliability at the lowest possible total cost. Total cost includes all of the costs for designing, manufacturing, testing, purchasing, operating, maintaining, environmental issues, and disposal of a product. Concurrent engineering uses design trade-off studies to decide what combination of performance, cost, availability, producibility, quality, reliability and environmental factors will make the most successful product.

This chapter reviews methodologies for using cost metrics to assist the development team in performing design trade-off and optimization analyses. Two commonly used methodologies are design to cost (DTC), which minimizes unit production costs and life cycle cost (LCC), which minimizes the cost of the product over its entire life.

4.1   IMPORTANT DEFINITIONS

Design trade-off studies examine alternative design approaches and different parameters with the purpose of optimizing the overall performance of the system and reducing technical risk. Trade-off studies are directed at finding a proper balance between the many demands on a design. Common metrics for trade-off analysis is a currency or a utility metric. Cost is measured in a currency denomination of a particular country and is the most often used metric. Utility measures are non-dimensional scaled values (e.g. 1 to 10 or 0 to 1.0) that rate how well a parameter compares to a norm.

Cost is the most useful and popular measure for trade-off studies due to its universal nature and its flexibility as a measure. Cost is a flexible measure since almost any design parameter can be converted to a cost measure. This allows the development team to perform analyses of different parameters based on a single metric. Although simple in concept, cost analysis is frequently difficult to perform because it demands current and future knowledge of a wide spectrum of disciplines. Cost analysis also calls on the ability to envision future changes in technologies and to forecast their effects on the product's cost. A major decision when using cost metrics is the types of costs and the length of time to include the costs in the study, and whether to modify the cost metric for the cost of money or inflation.

Design to Cost (DTC) is a cost analysis technique aimed at reducing or minimizing a product's price or cost, which results in increased sales volume. This analysis focuses on the product's purchase price. As a product's price decreases, the sales for most products increase. Examples of this can be seen with color televisions, hand-held calculators, personal computers, and integrated circuits. This can result from more customers being able to afford the product and/or from taking market share (i.e., sales) from the competition. Reduction of product cost is accomplished through a rigorous approach of identifying and implementing cost effective design decisions.

Life cycle cost (LCC) is a cost analysis discipline that develops a model of the total cost for development, operation, maintenance and disposal of a product over its full life to be used in design trade-off studies. The model is used to optimize product costs and predicting future costs of maintenance, logistics, and warranties.

4.2   BEST PRACTICES FOR TRADE-OFF ANALYSIS

The best practices for performing effective design trade-off studies are as follows:

- **Systematic decision making process** that addresses all possible impacts of various design decisions.
  - **Design improvements** are identified and implemented through an action-oriented approach.

- **Models** are accurate and based on a realistic assessment of user needs, market requirements, product performance, manufacturing capabilities, prototypes, logistics, and other factors.
- **Parameters** used in the model are up-to-date, accurate.
- **Design to Cost** aggressively lowers product costs in order to increase sales and profit.
- **Life Cycle Cost models** are used for in-depth trade-off studies of design, manufacturing, operation, maintenance, logistics, environmental, and warranty parameters to improve the design.

4.3   SYSTEMATIC TRADE-OFF ANALYSIS PROCESS

Analysis is a technique for gathering additional information in order to make better design decisions for improving the design. All analyses need to address the possible impacts of their results on other areas/disciplines in product development. This includes all aspects of a product at the appropriate level of detail. Effective trade-off analysis requires a systematic process. The steps of a successful trade-off analysis procedure at one company are to develop: (BMP, 1999)

- Clear problem statement
- Identification of requirements that must be achieved
- Ground rules and assumptions
- Decision criteria
- Schedule
- Potential solutions and screening matrix
- Comprehensive array of feasible alternatives
- Comparisons of alternatives using decision criteria
- Technical recommendation of trade study leader

In collaborative and concurrent engineering, support personnel provide a critical function in the design analysis and trade-off study process because of their expertise in areas not familiar to the designers. Staying current on technological advances in their specialized area leaves little time to remain current in the other engineering, business, accounting, and other disciplines that also are advancing at an extraordinary rate. Since knowledge and expertise about our ever-changing technology base are critical for design success, only a team approach can adequately evaluate all design parameters. Depending on the situation, the team may consist of consultants, company employees, and vendors. The team may be very large or small. Even in a team approach, however, final responsibility for a design must always lie with the designer. If everyone is responsible for the design, then no one is!

4.4   TRADE-OFF ANALYSIS MODELS AND PARAMETERS

Models provide information to the design team. The quality of the model and its parameters determines the quality of the information provided. Models and their parameters need to be accurate, up to date and based on a realistic assessment of user needs, market requirements, product performance, manufacturing and support capabilities, prototypes, logistics, and other factors. In the future, the Internet will provide much of this information using technologies such as agents. Good models provide quality information that reduces technical risk and are cost effective, accurate, and timely. The best model depends on the application and the resources and time available. Thus, the level of sophistication can vary from simple formulas to discrete event simulation to chaos theory.

One example of a cost model is used for estimating the manufacturing cost of a bearing by the bearing's configuration, material, length, diameter, and tolerance on diameter (Δ_D) as modeled by (Tandon and Seireg, 1989).

$$C_m \quad = \quad \gamma(k_0 + k_1/\Delta_D{}^a)(k_2 + k_3 LD^2)$$

Where

γ	=	Machining cost factor based on the material hardness
k_o through k_3	=	cost coefficients (see article)
a	=	1/3
Δ_D	=	Tolerance diameter
L	=	Length
D	=	Diameter

This model shows that tolerance, diameter and material hardness are cost drivers for predicting the cost of a bearing.

4.5   DESIGN TO COST

Design to Cost is a technique aimed at reducing or minimizing a product's price or cost, which results in increasing sales volume. This analysis focuses on the need to reduce a product's purchase price. As a product's price decreases, the sales for most products increase. This cycle of lowering cost to improve sales is shown in Figure 4.1. Reduction of product cost is accomplished through a rigorous approach of identifying and implementing cost reducing design and manufacturing improvements.

There are three steps in developing an effective DTC program:

1.   Determine critical product price goals or targets using market elasticity research or contractual requirement information for various levels of sales.

2. Establish realistic product cost goals based on projected sales
    volumes and learning curve improvements in design and
    manufacturing that accomplish the established product price goals.
3. Reduce costs to meet these cost goals through an action-oriented
    approach using trade-off studies.

4.5.1.   DTC Step 1: Determine Product Price Goals

The first step is to identify the pricing goals necessary to meet the
company's business goals (i.e. sales, market share, contract requirements etc.)

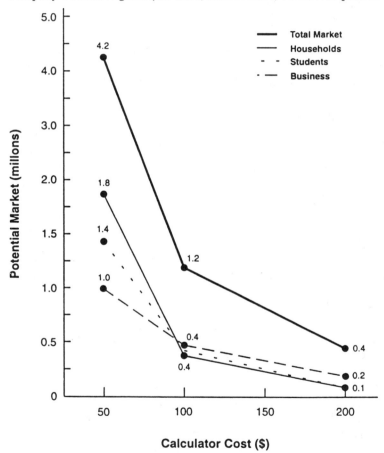

Note: Price break at $100 shows household
       and student use growing significantly

FIGURE 4.1 Potential customers

for the product throughout its forecasted production run. DTC is based on the relationship that when the price comes down, sales will increase. The curve that relates increased volume to reduced price is called the price elasticity curve. Realistic product price goals are then determined from the price elasticity curve. The elasticity curve also serves as a basis for determining the level of production volume and the price at which the company can obtain a significant share of the market.

4.5.2   DTC Step 2: Establish Product Cost Goals

Using the product price goals, unit production cost goals can then be established. The step is to determine the required unit production cost that the design team must meet for different periods of time. Working with product prices and volumes from the price elasticity curve, the forecast of manufacturing costs is based on the predicted various levels of sales. Using standard data and estimates of manufacturing costs, the future cost goals of the design are established and then adjusted based on the learning curve technique.

The basics of learning curve theory have been known for years, but it was not until the 1930s that they were documented by the aircraft industry. Aircraft manufacturers had learned that building a second plane required only 80% of the direct labor needed to build the first plane. In turn, the fourth plane needed only 80% of the direct labor required for the second plane. The eighth plane could be built 20% faster than the fourth, and so on. This improvement factor, based on volume, is called a learning curve.

Many companies use a log-log scale when plotting a learning curve because it results in a straight line, reflecting a constant rate of reduction. This makes cost reductions easier to forecast. On a typical 80% learning curve, such as that shown in Figure 4.2, the direct labor is measured on the vertical scale and the cumulative number of units produced is on the horizontal scale. When determining DTC goals, most learning curves are drawn through two important checkpoints: the manufacturing cost of the first unit and the required cost of the product at maturity. The necessary improvement in the learning curve is then transformed into design cost goals. This curve becomes a cost timetable. In addition to direct labor, manufacturing costs, unit costs, or raw materials can also be measured.

4.5.3   DTC Step 3: Reducing Costs to Meet Goals

The next step is to reduce the actual manufacturing costs of the current or proposed design through design improvements and manufacturing improvements.

Costs decline as designs and manufacturing processes are improved. Learning curves are constantly influenced by ongoing redesigns or manufacturing improvements. For example, training production operators typically results in a 10% improvement to the learning curve slope. Another 5-10

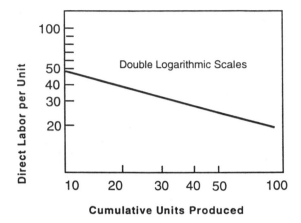

NOTE: This log-log scale graphed with
       a straight line reflects a constant rate of reduction

FIGURE 4.2 Learning curve (80%)

percent contribution can be gained through automation or more efficient use of machines. Product design improvements for producibility result in the largest savings! Producibility has reduced costs on some products by more than 60%. These cost reductions come from the actions of designers, manufacturing personnel, and management. Product simplifications, innovation, and manufacturing improvements together are necessary to substantially reduce cost and affect the downward slope.

Design to cost improvements do not end with a product's entry into the marketplace. Continual efforts are often made by management to further reduce costs, since lower prices will continue to increase sales. The sales then result in higher manufacturing volumes, leading to still lower costs and prices. Successively, design to cost cycles usually require increasing efforts and result in diminishing savings. This regeneration fuels the growth of the company, as the resulting profits are plowed back into the development of more new products. Although manufacturing improvements such as the training of production workers and automation are critical, the largest reductions are usually realized through product redesign (i.e., producibility), simplification of the process, and quality improvement.

4.5.4   Calculators

An example of DTC is seen in the first electronic calculators that were introduced at a price of over $1000. A calculator retailing between $50 and $100 would open a large consumer market. However, significant manufacturing improvements and technological advancements were necessary to reduce the price of a calculator from the pre-1970 level of $1000, to a price goal of $100. Integrated circuitry was first used in business calculators in 1968. This resulted in a 10:1 reduction in parts, causing a significant decrease in parts handling and assembly costs. Using more powerful circuits that further reduced the number of parts (20:1) resulted in a sales price decrease to the range of $300-$500. This made calculators affordable to almost any business. By 1971, a single large-scale integrated circuit could perform all of the logic and memory functions, further reducing the price of four- and five-function calculators below $200. This low price started the development of a consumer-oriented calculator market. Over the next 6-7 years, the price decline continued to be very dramatic: from 89 electronic parts costing $170 to a single part at $20. Design to cost was extensively used to promote the technological advances necessary for the continual decline in price.

4.6   DESIGN TO LIFE CYCLE COST

Life cycle cost (LCC) is a discipline that develops a model of the total cost for acquisition, operation, maintenance and disposal of a product over its full life to use in design trade-off studies. The model is used for analytical trade-off studies, identifying overall cost of a product and predicting future costs of maintenance, logistics, and warranties. A major decision to be made when using cost metrics is the types of costs to include, length of time for the study, and the cost of money i.e. inflation in the study.

An effective effort requires a realistic LCC model, valid input data, extensive design trade-off studies, and the implementation of design improvements identified in the trade-off analyses. The limitations of the model are as important as its strengths. An important limitation to remember in life cycle costing is that the results are estimates, and as such are only as accurate as the inputs.   Because of this, interval estimates are often more practical than single point estimates. For trade-off studies, relative ranking and general trends are often the best decision parameters.

There are three steps in developing effective life cycle cost models for design trade-offs.

1. **Develop cost models that accurately describe the costs associated with a product**.
   a.   Define parameters and collect data
   b.   Develop LCC model (parametric or accounting)
   c.   Perform baseline analysis using the model

2. **Perform verification analyses, trade-off analyses and identify cost drivers.**
   a. Vary LCC model inputs and iteratively evaluate its effects to verify the model and to identify cost drivers
   b. Perform trade-off analyses
   c. Identify design improvements
3. **Reduce costs to meet these goals through an action-oriented approach using design trade-off studies.**
   a. Implement improvements

4.6.1   LCC Step 1: Develop Cost Models

The first step in LCC is to develop the cost model and clearly define its parameters. All phases and disciplines of the product's life cycle should be reviewed, and factors should be identified which could have an impact on the LCC. All affected factors should be included in the LCC model. Many design decisions are made early in a program (such as technologies, life-limited or wearout items, maintenance concepts, and the human-machine interface) which have a great potential impact on the product's total life cycle cost. Cost models should be appropriate to the degree of definition needed and the level of data available

A product's parameters such as purchase cost, speed, quality, reliability, serviceability, safety, usability, marketability, etc. must be included. Metrics can include mean time between failures, failure rate, etc. Examples of metric calculations might include:

- Manufacturing scrap cost= (failure rate) x (% of failures that result in scrap) x (number of units produced)

Another category is costs that are expended regardless of the product's performance. These costs include design labor, production labor, overhead, material and parts cost, inventory, equipment and tool, environmental, etc. Examples of measurement calculations include:

- Inventory cost = (number of different parts) x (administration/accounting cost in dollars per part) + (cost of all parts in inventory) x (interest holding cost %)

Time can also affect cost. Schedule is the length of time to be used in the study. This can include conceptualization, design, manufacture, market, operational use and disposal. Measures include design lead time, manufacturing lead time, cycle time, test time, lost market opportunity due to delays, market share, hours of use, time to dispose, etc. Examples of measurement calculations include:

- Manufacturing lead time = (number of processes) x (days per process) + longest delay of setup factors (parts ordered, tooling, process availability)

After the parameters are identified, the model is developed. Estimates for the parameters are then made based on the data available. For effective design trade-off analyses, the models must be developed in sufficient detail to identify parameters that the design team can directly affect.

The two types of models most often used are parametric and accounting. Parametric models are a sophisticated form of regression analysis in which the cost experience and performance level of past systems become a baseline for estimating the cost of future systems, based on their projected level of performance. As the program progresses and system data becomes better defined, the designer can use accounting or "bottom-up" models. Accounting models with their detailed algorithms can directly relate design decisions and their effects on cost. Most trade-off studies use accounting models.

Parametric Models

Parametric models are often used early in the program before many of the design decisions have been made. Parametric costing is a forecasting approach in which a product's cost is regressed against physical or performance parameters (e.g., weight, speed, and thrust) of past products similar in type to the new system. The method is used to analyze the relationships between historical data and certain design parameters available during preliminary design. Comparisons are then made using multiple variable regression analysis.

Independent variables are chosen on the basis of explanatory power from among the set of a product's characteristics that can be estimated with reasonable accuracy. Predicted values for the relevant parameters of the new system are then used with the equation whose coefficients have been estimated from the historical data. For example, Large and colleagues (1975) developed a series of parametric equations for estimating the overall aspects of aircraft airframe costs based on performance. Their estimate for the total cost of aircraft airframes based on a sample of 25 military aircraft with first flight dates after 1951 was:

$$C_{100} = 4.29W^{0.73} S^{0.74}$$

Where:

C_{100} = total cost for 100 airframes (Thousands of $)

W = airframe unit weight (lb.)

S   =   maximum speed (knots)

With the

R^2 =   0.88 and F = 79.
Both parameter coefficients were significant at the 1% level.

Accounting Models

Accounting models are the other major type of cost models. They use detailed algorithms that relate basic data particular to the subject system. This method has more potential for prediction accuracy and design trade-offs than parametric models. It also provides more detailed visibility of the various data sensitivities. The design team can take design-related parameters, such as part quality, producibility, redundancy, reliability, and maintenance concepts, into account. As the design progresses and becomes more defined cost data can become more exact. The predictions also afford a means of optimizing and tracking design progress. The major emphasis, however, continues to be on the design trade-off application.

A high level life cycle cost model is:

LCC = nonrecurring costs (NRC) + production recurring costs (RC)
       + operating and support costs (O&S)

Nonrecurring costs are often called development and set-up costs and are one-time costs that are not directly affected by the number of products manufactured. It includes one-time costs of research and development, purchase or acquisition, manufacturing set-up, tooling, environmental, test equipment and software, initial training, and support setup. Recurring costs are costs directly affected by every product that is manufactured. It includes costs of manufacturing, operating facilities, labor, logistics support including transportation and repair, warranty, and environmental disposal. Operating and support costs are simply all costs of operating and supporting the equipment.

For effective design analysis, the models must be developed in sufficient detail to identify parameters that the design team can directly affect.

An example of some recurring manufacturing costs is:

Assembly cost per product (unit cost) = Parts Costs + Assembly Labor
               Cost + Facility Usage Cost + Test Cost + Repair
               Cost

Parts Cost   =   Parts/materials purchase cost + prorated handling
                     and shipping cost + unit cost of purchasing and
                     inventory

Models vary of course for different companies and products. An example of assembly cost models for predicting part insertion cost in printed circuit board assembly is shown in Table 4.1. The reader should note that the model shows that many different disciplines affect the insertion cost. For example, designers affect variables N (number of component types) and R (number of components) and manufacturing quality and vendor reliability affects A (number of faults).

4.6.2   LCC Step 2: Perform Verification Analyses, Trade-Off Analyses and Cost Driver Identification

Verification analyses are performed to verify the model's accuracy and completeness. Sensitivity feedback is vital in deriving the maximum benefit from the analysis. Data estimation is another major part of the trade-off process. If the input data or model does not accurately reflect the system, the model is useless as a decision tool. The actual values can vary according to the objectives of the analysis. Figure 4.3 illustrates several types of estimate parameters. The best approach depends on the specific situation.

A key emphasis for the development team is to identify and focus on those design parameters that significantly affect life cycle costs. These are often called cost drivers. Design parameters that are commonly found to be cost drivers include unit cost, mean time between failure (MTBF), and mean time to repair (MTTR). Common cost drivers found for different products are listed in

TABLE 4.1 Assembly Cost Equations

Assembly costs (C_a) are obtained from the following equations:

$$C_a = R_p(C_{ai} + C_{ap}/B_s + C_{rw}A_f) + N_t N_{set}$$

Where:

A_f = average number of faults requiring rework for each auto insertion

B_s = total batch size

C_a = total cost of operations

C_{ai} = cost of auto insertion

C_{ap} = programming cost per component for auto-insertion machine

C_{rw} = cost of rework

N_{set} = estimated number of set-ups per batch

N_t = number of component types

R_p = number of components

Source: Adapted from Boothroyd et al, 1989

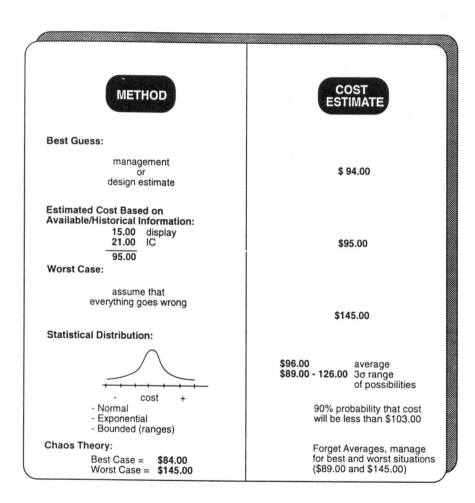

FIGURE 4.3   Methods to determine cost inputs.

Table 4.2. If the major drivers can be optimized through design trade-offs, the product's total life cycle cost can be minimized.

The next step is performing trade-off analyses to identify areas for cost improvements and evaluate design alternatives. The analyst exercises the model, studies the results, varies the inputs, and interprets the results for each input parameter level. The design team evaluates the results, defines any required changes for evaluation, evaluates other criteria, (i.e., performance, operational readiness, and schedule), and makes a decision on the preferred choice. The most recent design should be a candidate in every trade-off analysis, as well as the baseline system, so that each trade-off analysis candidate is essentially a variation from the baseline.

TABLE 4.2 Examples of Major Cost Drivers

Machined Parts
  Material costs
  Process set-up costs
  Machining Costs
  Tolerances requirements
  Part features and geometry
Plastic Parts
  Tooling costs
  Time and cost to develop
  and build tooling
Mechanical Assembly
  Number of parts
  Ease of assembly
  Precision and tolerance
  requirements
Electronics Assembly
  Design density and
  tolerances
  Test requirements
  Number of parts
Software Products
  Complexity
  Time and cost to develop
  software
  Quality problems that
  require revisions in product
  Customer service

Global Sales and Service
  Shipping
  Overseas maintenance and
  service
  Government regulations
  Design for other countries
Logistics and Other Costs
  Service and logistics support
  Projected yield
  Product liability
  Reliability and repair cost
  Production volume
  Environmental
  Product warranty
Electronic Commerce
  Traffic volume
  Computer startup
Overhead Rates
  Facilities usage
  Labor usage
  Inventory level

4.6.3   LCC Step 3: Reducing Costs

Once the development team has identified a preferred trade-off candidate, the team must implement the choice. It should be pointed out that the lowest LCC candidate is not always the preferred choice. For example, styling, performance or technical risk could outweigh cost, resulting in the implementation of an alternative approach. An emphasis on reducing the cost drivers and implementing the design improvements is basic to the entire trade-off process. Most companies develop their own models, which can describe their unique processes.

4.7   PRODUCIBILITY EFFECTS ON LCC

Many parameters can be positively affected by producibility, including cost, quality, reliability, schedule, and technical risk. Although performing producibility analyses have some associated cost; this is usually offset by its benefits. Producibility improvements can be measured using detailed LCC models.

4.7.1   Producibility LCC Analysis

An electronic instrument manufacturer considered using a new power supply. The new power supply, being offered by a small vendor, is considerably smaller than the standard power supply, which has been provided by a large company for several years. The new power supply would weigh less than half as much and take up less instrument frame space. The weight reduction reduces ergonomic risk to the material handlers and assemblers in the production of the instruments, as well as to the field service engineers replacing them at the customer location. The size reduction would improve producibility by allowing more access room for other components.

The new power supply, however, might increase inventory costs by increasing the number of parts in storage and part numbers, which must be monitored by purchasing. It will also introduce technical risk to the design process because the vendor has never produced it and the power supply has not been used in a product. This risk might affect quality as well as the project schedule. The estimated cost avoidance of reducing occupational injuries due to the manual material handling of the current power supply was $55.58 per instrument. This is based on the injury incidence of 3.4 injuries per 200,000 hours of production with the average incident cost of $52,000 including lost time and medical costs. The production hours per instrument are 64 hours. Injury cost avoidance = [($52,000 per injury x 3.4 injuries) / (200,000 hours x 64 hours per instrument)]

The assembly cost reduction, due to improved producibility, is estimated to be $11.88 per instrument based on a reduction of .8 production hours at an hourly rate of $14.85 per hour. (.8 hours per instrument x $14.85 per hour). The risk of delay in the project schedule was eliminated when the vendor delivered a working prototype within three months of request that was prior to completing the trade-off analysis. The quality risk cost was estimated to be $4.11. This was based on a 10% increase in failure rate. The current power supply failure rate was 2 per 100 units with associated cost of $187 per failure to replace and refurbish. (1.1 x 2 failures / 100 units x $187 per failed unit). There was no inventory cost increase because the new power supply replaced the previous parts so that the inventory costs canceled each other out.

The LCC model results for this trade-off analysis looked like this:

Cost savings ($64.35) = injury cost avoidance ($55.58) + reduction in assembly cost ($11.88) - quality risk cost ($4.11)

In this case, it was decided to use the new power supply.

4.7.2 Producibility Cost Trade-offs

Some important cost trade-off models that illustrate the effects of producibility are shown in the following discussion.

Part Reduction and Standardization

Part reduction is one of the most effective methods for reducing costs. A rule of thumb is that a 10% reduction in parts should result in a 10% reduction in manufacturing cost. The goal is to (1) reduce the total number of parts, (2) reduce the total number of different parts, and (3) reduce the total number of new parts. The cost areas affected by these goals are:

PPC = purchased part, material, or software costs
MPC = manufacturing process and assembly costs
O/H = overhead or indirect costs including inventory, purchasing administrative and facility costs
QC = quality costs including test, inspection, and repair
TR = technical risk problems expressed as a percent chance
SRC = cost of schedule delays
PRC = cost of reliability/warranty problems

Reducing Precision, Tolerances and Manufacturing Requirements

Another effective method of simplification is to reduce the level of precision, tolerance, or requirements for manufacturing. A more precise part may require a more expensive machine or process to produce the part and more precise test equipment to evaluate the part. Higher precision parts increase the number of defective parts that are caused by the incapability of the current process. Moreover, the part may require longer setup and processing times. If additional precision is not required in the design requirement, less precise design requirements are desirable to reduce the cost of producing the part. Lower requirements will allow manufacturing to:

- Use cheaper processes
- Use fewer processes
- Provide higher levels of quality (i.e., fewer defects)

Lowering requirements must be traded off with the possible effects of lowered performance!

Quality, Test, and Repair Costs

Quality costs are usually classified into three categories:

1. **Prevention** costs
2. **Appraisal** costs
3. **Failure** costs

Prevention costs are activities that prevent failures from occurring. These include design techniques such as simplification, standardization; mistake proofing, etc. and quality control of purchased materials, training and management. It also includes the costs of reliability activities during design and test. Appraisal costs are activities that could identify failures in manufacturing. This includes all test and inspection, process control and other quality costs. Failure costs are the actual costs of failure. Internal failure costs occur during manufacture and external occurs in the field. Internal failure costs include scrap and rework costs (including costs of repair, space requirements for scrap and rework, lost opportunity costs due to poor quality, and related overheads).

One interesting cost trade-off is determining the optimum level of testing which affects the type of test equipment, operational effects, and production repair methods. Better test equipment will cost more to purchase and built in test will cost more in design. These two steps will catch more defects, which in turn will cost more to repair. Hopefully, these cost increases are more than offset by the reduced number of defects reaching the customer. This will result in lowered warranty costs and product liability cost and improved customer satisfaction.

Environmental Costs

Environmental costs can be the hardest costs to model due to the constantly changing environmental regulations. Although estimating a product's environmental cost with today's requirements is possible; projecting the costs for future environmental laws when the product will be disposed of is much more difficult. Environmental costs include manufacturing wastes and pollution, damage to the environment when used, waste disposal costs such as landfills, and any hazardous materials costs.

Reduce Lead Time

Producibility can reduce manufacturing lead time by reducing the number of processes needed, designing for more available processes, designing to reduce setup time, or purchase parts from vendors with shorter lead times.

$$\text{Reduced lead time} = \text{LDC (FP + AP + SU + SL+RT)}$$

Where

    LDC = cost saved per lead time day
    FP = time saved using fewer processes
    AP = time saved using more available processes
    SU = time saved designing for reduced setup
    SL = time saved using shorter lead time parts
    RT = time saved in test

For example:

| Initial design | 6 processes, each takes 1 day; longest lead time part has a 2-month lead time |
| More producible design: | 4 processes, each take 1 day; longest lead time part has a 30 day lead-time. |

This producibility change results in savings of 32 days!

Technical Risk

Technical risk is increased when using new or unproven technologies, parts, processes, or vendors. Unproven technologies and processes will often result in longer lead times that is due to extensive testing for validation. Using unfamiliar technologies or processes may result in poor quality parts, delays in production scheduling, and high scrap due to high number of defective parts. These possible results will lead to increased cost. Technical risk can be modeled using probabilities of different occurrences. For example, quantifying the technical risk of a company's first time use of a new technology called Ball Grid Array (BGA) might be modeled as:

Number and Level of Problems	Probability of Occurrence	Problem Cost per Product	Effect on Lead time
Minimal	10	$2.00	no impact
Few minor	50	$5.00	no impact
Many minor	20	$5.00	one week
Many major	20	$10.00	one month

Warranty Cost

Warranty costs are driven by the warranty conditions, number of failures in the field (i.e., reliability) and the cost to repair or replace them (i.e., repairability). For design and manufacturing, it is important to minimize the probability of defects by designing the product for high levels of reliability and catching as many defects as possible in test. The key cost drivers are number of failures times the cost per warranty repair. Where the number of warranty failures is based on manufacturing quality (e.g. number of defects), level of test (i.e., number of defects not caught), reliability of the design (i.e., MTBF), and length of the warranty (i.e., time). The average cost per warranty repair is based on part selection (i.e., cost of parts to be replaced) and cost to repair (i.e., repairability).

$$\begin{aligned}&\text{Average cost}\\&\text{per warranty} = PC + TEC + SC + TTR(RLC)\\&\quad\text{repair}\end{aligned}$$

Where

PC	=	replacement part cost
TEC	=	proportion of test equipment cost
SC	=	shipping cost
TTR	=	time to repair
RLC	=	repair labor cost

4.7.3   Reliability Design Trade-offs

Many design analyses can improve reliability, but these come with an associated cost. On programs with limited resources, managers may want to evaluate which analyses are the most beneficial when compared with their subsequent cost. Evaluating this trade-off requires essentially two pieces of information:

1. An estimate of the amount of reliability improvement
2. Cost and time for performing the analysis.

One method is to formulate a ratio of the amount of reliability improvement over the cost incurred to achieve those improvements (Sheldon, 1980). By ranking these ratios, an optimal method of implementing changes can be selected. For example, an analysis for design simplification may reduce the number of parts by 10%. This would result in a 10% reduction in failure rate, but would cost a total of 200 hours of engineering labor. This can then be evaluated and translated into a numerical ratio where the most effective analysis has the lowest ratio.

Design Requirements Versus Life Cycle Cost

There are often trade-offs between LCC and design requirements. Changes in design requirements should be recommended when relaxing of certain specifications could result in a significant cost reduction. Design requirements in this case can include reliability, maintainability, as well as size, weight, power, range, and resolution. Cost variables can include cost per unit weight, cost per unit volume, etc.

Unit Production Cost Versus Reliability

A strong case has been established showing that selecting highly reliable parts can reduce costs. The added cost to the customer for purchasing highly reliable part is, however, in conflict with the customer's price goals. The use of more reliable products can be measured in terms of less repair, labor, and replacement costs resulting from failures. The decision to select parts with higher reliability should be based on analysis. If the part's additional cost is less than the expected increase in failure and support costs, then the part with higher reliability should be selected. This decision should be based on the results of the life cycle cost model for each alternative. Design to life cycle cost accomplishes this objective by employing sound economic principles and reliability fundamentals to generate a balanced product design that performs at an acceptable cost to the customer.

Repair Versus Throwaway

Many items, such as printed wiring boards, are often difficult to repair in a cost-effective manner. A decision must be made between low-cost modules that lend themselves to a throwaway or discard-on-failure philosophy and the higher cost for more reliable modules. The cost of these discarded modules must be traded off against the cost of repairing the module. This type of trade-off analysis is used when determining the optimum level of repair.

Built-in Test Versus Conventional Testing Methods

The incorporation of built-in-test (BIT) increases the design and unit production cost owing to increased design effort and added circuitry. BIT, however, can reduce production, test, and maintenance labor costs due to automatic fault isolation. It can also reduce the cost of production and maintenance test equipment. These costs must be traded off against the lower unit cost.

4.8   DESIGN FOR WARRANTIES

A warranty is a written or implied promise between a company and customer to assure a certain level of performance and support after the sale. In general, the manufacturer gives a written warranty in terms of services (such as free maintenance) under a stated condition and period of time. When determining the total life cycle cost of a particular design, design team must estimate and evaluate the cost of warranty services that will occur in the future. **Design for warranties uses life cycle cost models to identify warranty cost drivers and identifies the best choice of design and support parameters that can minimize future warranty costs.** Some persons incorrectly think that this cost is minimal or that warranty service can be added to the product's price. Due to competitive pressures, adding warranty costs to a product's price hardly ever happens. For most companies, price competitiveness results in all warranty costs directly affecting the company's profits. The amount of warranty cost depends on the type of warranty, the period of time for which the warranty is extended and key product parameters such as reliability, ease of maintenance, parts costs, and supportability.

The type of warranty can be specified in contract requirements or strictly left up to the manufacturer. There are many different types of warranties. The following is a partial list of warranty types:

- Hold harmless from product liability
- Warranty of performance
- Warranty of quality and reliability
- Availability guarantee
- Basic commercial warranty

The data for determining (i.e., estimating) the cost of warranty failures is found in the areas of reliability, logistics, maintainability, and life cycle cost. Key cost drivers for warranties include failure rate (i.e., MTBF), part life (wear out), number of uncorrected manufacturing defects, time to repair (i.e., MTTR), parts needed per repair, and expected product use. General repair cost categories to be included are

- Labor cost
- Facility cost
- Supply cost or the cost of replacement parts
- Test equipment cost
- Shipping cost
- Travel cost
- Administrative cost
- Training cost

The most important warranty design element for the design team is usually the number of product failures, including those caused by reliability problems and those caused by manufacturing defects. Manufacturing quality is especially critical since field repair costs are generally higher than factory repair costs by an order of magnitude. It is very important to catch defects in manufacturing before they reach the customer. Design for inspection (i.e., inspectability), design for test (i.e., testability), and selecting high quality parts and vendors are key design techniques for reducing warranty costs.

Finally, warranties are one of the most important design considerations in the automotive industry. Over the past twenty years, warranties have increased in yearly and mileage coverage from 1 year or 12,000 miles to in some cases 7 years or 70,000 miles on some models. Even in inexpensive models such as the Hyundai Elantra, warranties perform a role in lowering its LCC. As reported in Road and Track (1993), "After 23,242 miles, to the Elantra's credit, we spent little on repairs and maintenance, thanks to its excellent warranty. Our average cost per mile was only 23 cents - that's about ten cents less than average" (Road and Track, 1993). That translates to a 30% savings in the cost per mile for the consumer due solely to its warranty (Knight, 1996)!

4.9   SUMMARY

A life cycle design approach in product development and manufacture is needed. The cost considerations must begin upon initiation of the design process and continue through the development process. Three important cost control analyses are design to cost, design to life cycle cost, and design for warranty. Practical trade-offs must be made between design performance, cost, manufacturing, service plans, logistics, warranties, and schedule. Cost analysis is a major part of design. This chapter has reviewed the methodologies of design to cost, design to life cycle cost, and design for warranty.

4.10   REVIEW QUESTIONS

1.   How does design to cost reduce manufacturing costs?
2.   What is the most effective method to reduce manufacturing costs?
3.   Develop a life cycle cost model for an inexpensive calculator.
4.   For a calculator, how would the design team decide whether to repair or throwaway products returned to the store?
5.   For different cost drivers, identify appropriate products.
6.   Explain how producibility can reduce different life cycle costs.

4.11   REFERENCES

1.   BMP, website, www.bmpcoe.org, 1999.
2.   G. Boothroyd, W. Knight, and P. Dewhurst, Estimating the Costs of Printed Circuit Assemblies, Printed Circuit Design, Vol 6, No. 6, June 1989.
3.   Car and Driver, Hachette Magazines Inc., p. 95-95, February 1994.
4.   M. Knight, Trends in the Automotive Industry to Lower Life Cycle Costs, Unpublished Student Report, the University of Texas at Arlington, 1996.

5. Large, H.G. Carnpbell, and D. Cate, Parametric equations for Estimating Aircraft Airframe Costs, RAND, R- 1 693-PA&E, Santa Monica, California, 1975.
6. Road and Track, Hachette Magazines Inc., p. 125, April 1993.
7. M.K. Tandon and A.A. Seireg, Manufacturing Tolerance Design for Optimum Life Cycle Cost, Proceedings of Manufacturing International, p. 381-392, 1992.

Chapter 5

DETAILED DESIGN:
ANALYSIS AND MODELING

Complex Analyses Consider All Disciplines

Product development pushes the limits of innovation and creativity. Design analysis, modeling, and simulation are design techniques used to help the development team make more informed decisions. They increase the chance of a correct design and reduce the technical risk in product development. These analyses need to address all possible impacts of their results on design decisions. This includes every aspect of a product at the appropriate level of detail. To be most effective, design analyses must be an integral and, timely part of the product development process.

Best Practices

- Collaborative, Multi-Discipline Design Analysis Process
- Design Synthesis and High-Level Design Tools
- Prototypes in Detailed Design
- Modeling and Simulation
- Parameter Variability, Uncertainty, Worst-Case Values, Statistical Analyses and Aging
- Stress Analysis
- Thermal Stress Analysis
- Finite Element Analysis
- Environmental Stress Analysis
- Failure Mode Analysis (FMEA, PFMEA and FTA)

BEAN GUESSING WITH AN ENGINEERING APPROACH

Dr. George A. Hazelrigg in 1985 wrote a fascinating paper on how engineers solve problems. His illustration focused on the simple task of guessing, or estimating, the numbers of beans in a sealed glass jar. He examines the different engineering methods that could be used. For example, if there is no prize for a good estimate, a person will only guess at the number of beans. When a prize is rewarded, however, people will develop methods and models to improve the quality of their guess and thereby increase the probability of winning. "The best guess depends on factors that go beyond traditional engineering analysis to take into account the nature of the prize and the personal preferences of the bean counter" (Hazelrigg, 1985).

He notes that there are many methods to analyze the number of beans in a jar. The methods can be classified into two groups: prototype testing and mathematical modeling. Using the prototype method, someone can buy a similar jar, fill it up with beans, and then count the number of beans. This method cannot provide a perfect answer as the purchased jar or the beans may not have the exact same dimensions or the beans may pack differently. In product development, this is called prototyping and physical testing.

The more common engineering method is to develop mathematical models of the process. Models are developed which estimate the size and shape of the jar, bean and the packing factor of the beans in the jar. The more accurate the model is, the more likely the estimate will be closer to the actual value. Since every bean varies in size and the packing factor may vary, this does not give a perfect answer either. "A good model is one that yields an answer that reduces uncertainty. Any improvement in the state of knowledge will generally have value to the decision-maker. The cost of model improvements, however, escalates rapidly as one attempts to approach perfect information" (Hazelrigg, 1985).

Most engineering problems are like bean counting in that they require information beyond getting answers directly out of a book. "Good" guesses require the development team to perform detailed design analysis, modeling, simulation, prototyping and testing. A major question is how much information is needed and how good the information must be before a decision is made? Gathering information takes additional money time to develop more sophisticated and accurate models or to perform more testing. The dilemma is determining the optimal amount of information that is needed for reaching a decision within the time and resources available.

This chapter identifies and reviews key design techniques for the detailed design tasks that reduces technical risk.

5.1 IMPORTANT DEFINITIONS

Detailed design is a group of tasks used to finalize a product design that meets the requirements and design approach defined earlier. This requires decisions, even though some technical information may not be available. The design team must use "best estimates," otherwise known as assumptions, to develop the design. Unless the design is thoroughly analyzed, this situation increases the probability that the design is inadequate or incorrect. Good analyses and models can remove much of this uncertainty. Design analysis, modeling, and simulation are design techniques used to assist the development team in substantiating those assumptions, which will increase the chance of a correct design and reduce the technical risk in product development.

Design analysis is the use of scientific methods, usually mathematical, to examine design parameters and their interaction with the environment. The purpose of analysis is to gather enough information to improve our knowledge of a situation so to make better decisions. Its goal is to reduce technical risk. Since the team uses so many assumptions, design is often thought of as an iterative or continuous process of design, analysis, and test that utilizes the knowledge available at a given time. Examples of knowledge include rules of thumb, published standards, textbooks, databases, and results from analysis, modeling, simulation, and testing. The processes of design analysis, modeling, and testing are used to ensure that a design is appropriate.

Modeling and simulation are tools for evaluating and optimizing designs, services and products. Their purpose is to assist the design team in the development of a product. They constitute a process in which models simulate one or more elements of either the product or the environment. The metrics for modeling depends on the analysis being performed.

5.2 BEST PRACTICES FOR DETAILED DESIGN

Practices discussed in this chapter are as follows:

- **Design analyses and trade-off studies** are systematically conducted in a collaborative manner to ensure that a design and its support systems can meet or exceed all design requirements
- **All disciplines** including manufacturing, reliability, testability, human engineering, product safety, logistics, etc. are included
- **Design synthesis and high-level design tools** are used to increase design quality and efficiency.
- **Modeling and simulation** are extensively used for design analysis, trade-off studies, and performance verification
- **Analyses contain sufficient detail** to accurately model the "real world" including:
  - **Variability and uncertainty**

- **Worst-case, parameter variation, and statistical analyses of parameters**
- **Aging**
- **Stress reduction including mechanical, thermal, and environmental** improves reliability and quality.
- **Failure modes analysis such as failure modes and effects analysis (FMEA) production failure modes analysis (PFMEA) and fault tree analysis (FTA)** are used to identify and then correct or minimize potential problems.

5.3   DESIGN ANALYSES

Design analysis disciplines may include digital circuit, analog circuit, printed circuit board, software, mechanical structure, plastics, etc. Support disciplines include manufacturing (producibility), testing (testability), logistics, reliability, etc. Effectively coordinating these disciplines is a difficult process. Computer-aided design, knowledge bases and networks are areas of technology that is being used by many companies to assist in the transfer of knowledge and information between the design team. These systems have access to databases and the Internet that can contain:

- CAD drawings and parts, software and materials data
- Vendor history and information
- Design rules and lessons learned (both corporate and product specific)
- Design and support specifications and guidelines (scenarios, product use profiles, performance, producibility, reliability, supportability, and design to cost)
- Detailed producibility criteria (capabilities of special and standard processes, testability, and estimated production quantity)
- Detailed reliability criteria (reliability models, failure history, physics of failure, failure mode information)
- Results from prototype testing

Advanced CAD and design automation systems allow users to create concept models easily and quickly using digital sketching or mathematical models. Networks allow the design team to evaluate many concepts in a short period of time. In the future automated design advisors and agent based analysis technologies will allow product generation and evaluation to be completed almost instantaneously. Paperless designs automatically determine how the parts could be manufactured and assembled.

Data from projects that have implemented computer-aided engineering tools indicate that design cycle time can be reduced as much as 60%, while producing equal or superior product quality (Swerling, 1992). Daimler-Chrysler saved months and $800 million off the development cost of the 1999 Cherokee using advanced design tools.

Design trade-off studies examine alternative design approaches and different parameters for the purpose of optimizing the overall performance of the system and reducing technical risk. Trade-off studies are directed at finding a proper balance between the many demands on a design. Some of the many design elements and analyses involved are shown in Figure 5.1.

To be most effective, design analyses and trade-off must be an integral, timely part of the detailed design process. The goal is to prevent problems rather than fixing them later. Otherwise, the analyses merely record information about the design after the fact. Changes made later in a program are more costly and less likely to be incorporated.

Detailed Design Process For Semiconductors[*]

The semiconductor design process is used as an example. First to market (i.e. lead-time) is probably the most critical element in semiconductor product development. The semiconductor detailed design process is a unique and very iterative design endeavor using a building block approach as shown in Figure 5.2. All semiconductor designs are extensively modeled through the use of detailed modeling and simulation. This is due to the complexity of the design, and the high cost and long lead-time of building prototype designs in actual silicon. Most designs are based on previous designs and modules using the iterative application of the historical lessons learned and best design practices. For example, the design of dynamic random access memory (DRAM) devices has progressed from 16 K to 64 K to 256 K to 1 M to 4 M to 64 M to 256 M to 1 G generations over the last 20 years. Although each new product used some new technologies and innovations, they all used the previous design as a starting point i.e. baseline.

The Simulation Program for Integrated Circuit Evaluation (SPICE) is the most common detailed design analysis tool for evaluating and simulating a semiconductor's design for performance parameters. The basis for SPICE is a set of parameterized equations that describe the behavior of the identified type of semiconductor technology. Individual parameters affecting performance characteristics are variable in SPICE modeling so that they can be set to behave like a particular device under specified conditions. The equations and its parameters are developed by universities or companies and are based on empirical data. As new semiconductor designs push the technological limits of semiconductor technology, the information quality in existing SPICE models is declining.

---

[*] This section was adapted from Rogers, Priest, and Haddock, Journal of Intelligent Manufacturing, 1995.

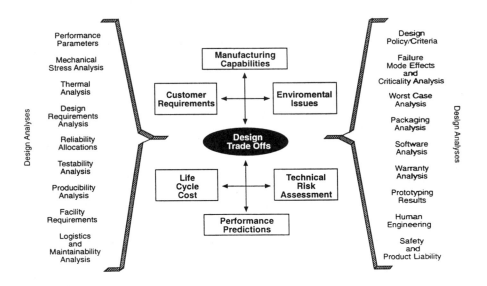

FIGURE 5.1   Relationship of design elements in trade-off studies.

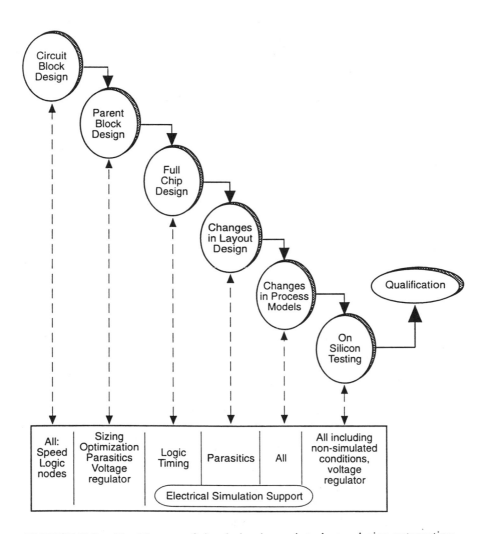

FIGURE 5.2   Significance of simulation in semiconductor design automation (Rogers, Priest, and Haddock, 1995).

Designing for all of the performance variations that are caused by process and layout variations is an extremely complex process. This results in many design problems that require redesigns. Modeling accuracy is additionally affected by environmental factors (e.g., temperature), usage factors (e.g., power-up, multiple cycles, operational) and voltage/current factors. Circuit path lengths in turn affect performance parameters in the form of performance degradation. This is modeled in SPICE as a parasitic parameter. If the layout changes, the circuit's performance is affected. Unknown and changing layouts have caused designers to typically 'overdesign' for worst-case conditions. This practice results in costly, less efficient designs. What are needed are more intelligent, statistical methods for modeling manufacturing variability for product design.

Another problem is the immense amount of data that is produced for each computer evaluation. A typical run can produce hundreds of pages of results. The data can be so overwhelming that problems are hidden, and only experienced designers are able to properly evaluate the results. Although semiconductor devices have the highest levels of product sophistication and technology, their iterative detailed design process is similar to most products.

5.4   DETAILED DESIGN FOR SOFTWARE

A collection of best software design practices recommended by many authors for product development include:

- Customer and user focused design
- Technical risk management and measurement
- Simplification
- Design for change and revisions
- Scalability and reproducible design
- Module reuse and standardization
- Evolutionary prototyping
- Prototyping and human factors to define user requirements, user interfaces and software interfaces
- Traceability to requirements
- Emphasis on testability
- Independent design reviews and testing
- Good documentation
- Design reliability emphasizes robustness, fault avoidance, tolerance detection, and recovery

During the early or preliminary design the inputs and outputs and a general idea of the architecture required are identified. This includes the identification of technologies, data management, utilities, hardware, and communication protocols. Software architecture alludes to two important characteristics of a computer program: hierarchical structure of procedural components (modules) and structure of data (Pressman, 1992).

There are two design strategies now being used in software design as described by Somerville (1996). They may be summarized as follows:

- **Functional design.** The system is designed from a functional viewpoint, starting with a high-level view and progressively refining this into a more detailed design. The system state is centralized and shared between the functions operating on that state.
- **Object-oriented and agent design.** Collection of objects rather than functions based on the idea of information hiding. The system state is decentralized and each object manages its own state information. Objects have a set of attributes defining their state and operations that act on these attributes.

Detailed design focuses on refining the conceptual design and develops the program modules. The objective is to group activities together that relate to each other. Module coupling is the degree of interdependence between two modules. Module cohesion is the degree to which activities within a single module are related to one another.

Design documentation includes event trees, task analysis, entity relationship diagrams and object diagrams. As in hardware design, software testability is a key design practice and is discussed in the chapter on testability.

5.5 DESIGN SYNTHESIS AND HIGH-LEVEL DESIGN TOOLS

As computer aided design tools become more powerful, the development team will be able to spend more time on higher-level design tasks and less time on repetitive, simple tasks. One extension of simulation and analysis in detailed design is called "design synthesis" (Waddell, 1996). Design synthesis uses high abstract level description to develop physical representations. This allows the designer to evaluate more design approaches quickly. Synthesis is currently used in integrated circuit (IC) and application specific integrated circuit (ASIC) design, but the idea is relatively new to other areas. This method requires powerful intuitive, interactive, hierarchical design tools that allow high levels of design flexibility and control. In IC design the higher-level description uses a hardware description language (HDL), such as Verilog or VHDL (Waddell, 1996). The HDLs were developed to address increasing logic complexity by reducing the detail of analysis from the transistor level to a higher switch or gate level. The switch-level simulators are optimized for digital logic functions.

In circuit board design, the design team inputs high-level descriptions such as a schematic, list of parts, component models, and design rule constraints.

For a notebook computer, the board descriptions would include component parts, board size, schematics, and design goals. The design synthesis software then uses a database of component models and design rules to develop the board design.

Many advances are being made in the field of artificial intelligence especially in automated program understanding and knowledge-based software. New interface technologies will reduce the amount of information a user must learn (and retain) to effectively perform design analysis, and instead put this burden on the machine itself. A major concern is how well an expert's knowledge is incorporated into a system. Thus, a producibility knowledge base would include company resources and availability, cost data, schedules, process capability data, vendor data, quality information, and lessons learned. The computer tool would then synthesis one or more design approaches. After some review and modification by the design team, the system would generate the many detailed drawings and documentation needed by manufacturing, vendors, testing, warranty, and logistics.

5.6 PROTOTYPES IN DETAILED DESIGN

Prototypes play a large role in all phases of development especially in detailed design. Physical models and software models (virtual reality) are used to gather information to reduce uncertainty optimize parameters and test the design. Prototyping provides information that is especially important for:

1. Information that is not available
2. Software and software interfaces
3. Global and cultural design aspects
4. Innovative or creative products that are very different from the norm
5. Data for unknown uses or environments

5.7 MODELING AND SIMULATION*

Modeling and simulation are analysis tools for evaluating and optimizing designs and products. Their purpose is to assist the design team. They constitute a process in which models simulate one or more elements of either the product or the environment. Simulation and modeling can be low cost and effective methods to gather and verify information when compared to full-scale prototypes. Modeling allows a designer to experiment with requirements, optimize design decisions, and verify product performance. Reasons for simulation include:

---

* Jim Hinderer, TI Fellow, Systems Engineering, Texas Instruments, Dallas, Texas wrote this section.

1. Increase the level of knowledge of how the product interacts with its environment
2. Assess the benefits, costs, and attributes of each requirement
3. Perform design trade-off studies to optimize various design elements, such as performance, producibility, and reliability
4. Verify that the design can meet all requirements

Almost any design or process can be simulated through the use of computers or scale models. Most design simulations involve mathematical models generated on computers. These models evaluate system requirements, including performance, reliability, cost, environment, and design trade-offs. It is, however, possible to model in other ways, such as prototypes, breadboards, or scale models. Simulation with scale models is a method often used in aeronautics. For example, a scale model in a wind tunnel can simulate a full-sized airplane in flight. The following discussion reviews the design concepts of modeling and simulation.

Simulation is often used for experimenting with different design requirements to ensure that the requirements are feasible and lead to the desired end product. In this use, performance standards are established and approaches proposed to meet producibility and reliability goals. Sometimes design decisions made to enhance producibility or reliability result in losses in performance, weight, or power. Simulation allows the design team to perform trade-off studies before the product is built so that the design can be optimized. Simulation can also be used to increase the understanding of how the product interfaces with the environment. This understanding gives the designer a better appreciation of the benefits, costs, and attributes of each design requirement.

Development of the simulation has many steps that are analogous to actually building the system. For example, in the case of an electro-optical sensor that automatically detects objects, there are many parameters to be modeled even if the sensor is purchased rather than developed. Purchased sensors are advantageous in that the risks of design are removed and reliability, cost, weight, power, and size are precisely known. Modeling may still be required, however, to resolve such design questions as sensor look angles, optics diameter, field of view, software drivers and compatibility with other elements of the system. The exercise of analyzing these trade-offs often reveals new ideas about the product that had not been apparent before.

Simulation is also used as a tool for design analysis, such as optimizing the choice of parts. Simulation can resolve questions about whether a part or a particular technology will improve performance, reliability or producibility and about its potential effects on other parameters. Simulation allows the product to be exercised more extensively than is possible in reality. Indeed, simulation of environments to provide external inputs and effects is one of the most important uses of simulation. In order for any simulation to be useful, it must be an effective design tool that all users (people who are going to use the simulation) can trust. Several properties characterize an effective simulation. Some of these properties are:

- Realistic and correct
- Useful and usable
- Well-planned, well-managed, and well-coordinated
- User acceptance
- Favorable benefits-cost ratio
- Modular, flexible, and expandable
- Transportable

Realistic and Correct

An effective simulation must be realistic and correct. An incorrect simulation is worse than no simulation at all. It leads to erroneous design decisions and causes users to lose confidence. Verifying the correctness of the simulation requires the same rigor used in the verification of software. Verification includes calculations, tests of limiting cases, such as the response of the simulation to step or sinusoidal inputs, and tests using observed responses to known inputs. Correctness of the simulation should be formally demonstrated.

Useful and Usable

Simulation must be designed to solve problems that the users need solved. A simulation is usable if the designers understand how to use the simulation and, in fact, use it as a design tool. Historically, many very powerful simulations have fallen into disuse because they were too complicated to use or their intended use was not clear. Coordination with all users is the best method to make the simulation both useful and usable. Each user has experience and prejudices about what the simulation needs to do and how the simulation should work. Understanding the potential users is critical. In effect, the simulation then becomes "theirs" (the users'), not an academic exercise by the simulation team. User considerations to be evaluated when designing the simulation system are given in Table 5.1.

Well-Planned, Well-Managed, and Well-Coordinated

Simulation requires extensive planning. Planning starts with initial identification of a need for simulation and continues through the last instant the simulation is needed. Lack of planning and management are principal reasons given for simulation failures.

The planning starts with the system evolution plan. The first step is a written simulation plan that guides the effort. The second step is generation of a written specification precisely defining the requirements for the simulation.

TABLE 5.1 Design Considerations

Who are the users?
        Simulation staff
        Designers
        Management
        Customer
        Support engineers
What will the simulation be used for?
        Experimentation
        Development of requirements
        Design trade-offs
        Simulating environments
        Test
        Design verification
        Documentation
What outputs are required from the simulation?
        Statistics and plots
        Design optimization
        Hardware signals
        Software signals
        Human interactions
How does the user want to use the simulation?
        Interactive or batch
        Operator in loop
        Hardware in loop

User Acceptance

An effective simulation is characterized by user acceptance. Defining what impresses a user is not a science. However, three principal areas can be addressed. First, is the user interface acceptable? Is this point of interaction with the user not only adequate, but also first class within the budget constraints? Does it provide significant niceties and performance parameters that are beyond what the user demanded? Second, is the technical performance of the simulation sufficient for the project? Third, is the simulation a cost-effective tool? Each of these three areas contributes to the perceived excellence of a simulation.

Favorable Benefits-Cost Ratio

An effective simulation produces benefits in fair proportion to the cost. "This simulation is taking 90% of my software budget" or "this simulation is a Cadillac where a Chevrolet is needed" are statements that reflect a problem often encountered in simulations: the simulation can often cost more than it can save in design trade-offs. A capability versus cost analysis is very important in defining what is worthwhile in increasing the scope of the simulation and what can be eliminated to cut costs. A simulation that is too expensive to be run will eventually not be run at all, and most of the work put into its development will be lost.

Two techniques are used for keeping the cost of a simulation reasonable. First, make the simulation modular and maintain a written evaluation of the usefulness of each module relative to its cost. Second, make the simulation fit the problem. In implementation of the simulation, there are many cost-saving options. Several areas in which trade-offs can be made are listed in Table 5.2.

Often, the problem being solved requires only a limited amount of realism. Models with 6 degree of freedom are very expensive and require a lot of time and skill to make it work accurately. Although an actual aircraft experiences movement with 6 degrees of freedom, many aircraft components can be effectively simulated with fewer degrees of freedom. If the performance of the system being simulated is not sensitive to the vehicle motion, a simple motion model is sufficient. In addition, some systems are modeled for less cost by many small, special-purpose.

In striving for a cost-effective simulation, do not sacrifice the flexibility of the simulation with respect to growth and change. Design the simulation and its inputs and outputs to be able to easily add or drop modules.

Modular, Flexible, and Expandable

An effective simulation is modular so that it can be assembled and used in parts. It is flexible to accommodate changes in purpose and adaptable to different phases of system evolution or to new programs, and expandable to accommodate new applications. A top-down approach to the simulation design is helpful in building a control structure that accommodates changes. As new applications appear, modules are added to the core.

TABLE 5.2 Simulation Options That Affect Complexity and Cost

Degree of realism
A few complex simulations versus many simpler simulations
Digital computer versus analog computer versus hybrid computer
Choice of computer manufacturer and software packages
Discrete time versus continuous time
Periodic steps versus event-driven steps
Integration method and sample interval
Deterministic versus random inputs
Use of Monte Carlo techniques
Data availability and collection methods
Training requirements

Planning for growth is essential. Two typical modes of growth are expansion across phases of system evolution and expansion to new programs. Frequently, a simulation is developed to support a proposal to acquire new business. When the new business contract is won, the simulation is adapted to gain system insight and to define requirements. When the simulation proves to be an effective system engineering tool, it is pushed into the design phase and used for trade-offs and integration. Eventually, the simulation is asked to generate data to support tests. Ideally, the simulation team comprehends that a simulation can support all phases of a system evolution and plans accordingly.

Transportable

Transportability is an important factor because it allows the tool to be used for new applications and design efforts without major changes. Two factors that influence the transportability of a simulation are the computer and the choice of programming languages. Among the options available for a simulation is the choice of digital, analog, or hybrid computers.

5.8 METHODOLOGY FOR GENERATING AND USING A SIMULATION

The basic steps for creating an effective simulation are discussed here.

Step 1. Define the System and Environment

The definition should include system and environment appearances at each phase of system evolution so that the simulation can be used to support each phase.

Step 2. List the System Unknowns and How They Are to Be Resolved

List the system unknowns and how to be resolved. This step is the point at which quality, reliability, and producibility considerations are introduced.

Step 3. List Effects of the Environment

Identify all points at which the system must interact with the remainder of the world during normal operation. These points suggest elements that may need to be simulated.

Step 4. Define How the Environment Affects the System

The environment may affect the system in only simple ways. This may introduce step changes, or it may cause a more long-term change, such as drift in a parameter, as in the case of temperature. Sometimes the environmental effects may be complex and may require extensive modeling.

Step 5. Define Needed Simulations and a Simulation Plan

Identify what simulations are needed and define what phases of system evolution each simulation is to support. In addition, this step defines which elements are to be modeled. It defines which system elements are to be incorporated into the simulation as elements of hardware in the loop.

Step 6. Create the Simulation

Start with a core of basic functions, and add to them in a top-down fashion. Starting a simulation before the whole simulation is ready is a learning tool and provides valuable insights into what the simulation needs to do and how it is to do it.

Step 7. Verify the Simulation

Verification involves checking all the simulations by whatever means are available in such a way that all users are confident that the simulation works.

Step 8. Implement and Maintain the Simulation

Using the simulation should provide the answers anticipated in the simulation plan. This includes interaction with the customer, and other audiences and implementation items such as training and maintenance.

5.8.1   Aircraft Landing System Example

The following discussion describes the simulation of an aircraft landing system and its environment. This example illustrates how simulation can be a tool in the areas of hardware, software, and signal processing. The landing system's purpose is to detect and locate airports in all types of weather. A sensor on the aircraft scans the scene and makes decisions about whether a landing strip is present. When it detects an airport, it identifies the type of airport and indicates where the landing area is located.

The input to the system is a video signal that comes from a sensor and is similar to the video signal from a home computer to a monitor. The video signal contains frames of imagery. A frame is a snapshot of the scene currently in the sensor field of view. A new frame is input several times per second. The proposed sensor is laser radar, which is an electro-optical sensor. The laser illuminates objects, and the radar receives the reflected energy. The laser beam is very narrow, so in order to illuminate an object the laser must scan at high speed across the field of view.

The output from the system is a decision about whether a landing strip is present in the frame and where in the scene it is located. The environment is everything that influences the operation of the system, including the objects visualized, aircraft used, and environmental conditions. Several key design issues must be resolved:

1. How many frames of imagery are required?
2. How many picture elements (pixels) are needed per frame?
3. What algorithms can most effectively extract information from the video?
4. How does the type of airport, landing strip, aircraft, or environment affect the choice of processing?
5. What effect does the sensor signal-to-noise ratio have on the detection?
6. How do other objects in the scene influence the detection system?

Building a prototype of this complex system and associated sensors would be extremely expensive and time consuming. Too many parameters must be investigated prior to finalizing the design concept. The solution is to first model the system and then use the simulation to resolve these design issues.

Several considerations must be included in the planning for the simulation. Although the initial use of the simulation might be to support a proposal effort, the simulation should also be designed to cover the later phases of product development. The simulation should be designed for a general user community that may be interested in different types of sensors such as laser or radars. The simulation is planned to be user friendly and generally useful to a wide range of potential users. In this example, the simulation was programmed in FORTRAN for transportability to other computers.

To save money, sensitivity analyses were run to determine what is and is not significant to object detection. Atmospheric considerations were studied and found to not be a major factor in detection. In addition, since the sensor head is stabilized with gyros, aircraft dynamics have little effect. Thus, a decision was made to save money by simulating neither the atmosphere nor the aircraft's 6 degree of freedom dynamics.

The simulation was composed of three models and various design options, as shown in Figure 5.3.

The models are the (1) imagery model, (2) sensor model, and (3) detection system model. The imagery model produces imagery that represents the appearance of airport landing strips and other objects. The orientation of each object and the location within the field of view can be varied as part of the simulation conditions. The image is three-dimensional in that it contains width, height, and depth. In addition, it contains information for the sensor model to use in producing a video output. This information includes object reflectivity and temperature distribution.

The three-dimensional imagery from the imagery model is input to the sensor model. The sensor model converts the imagery input into a video output that goes to the detection model. The original sensor model was designed to mimic the operation of laser radar and to be used for other aspects of sensor development as well. As a result, a model of a millimeter-wave radar sensor, an infrared sensor, or a television sensor can replace the laser radar model.

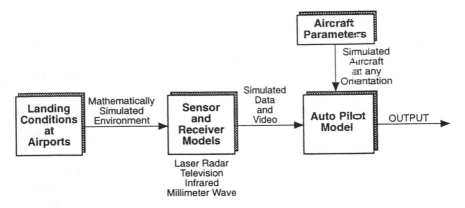

FIGURE 5.3   Simulation model for a landing system

The detection system model accepts the sensor model video signal and generates the location of any detected landing strip. In order to accommodate different phases of system evolution, the simulation was required to accept inputs from several sources. The detection model can accept inputs from the sensor model, from a tape recorder replaying real imagery, and from an actual laser radar sensor itself. This simulation, which uses actual hardware, is called a "hardware in the loop" simulation. This type of flexibility must be planned early in the design. The simulator is then put together in modules and assembled in a top-down manner. As a result the cost impact from the added versatility is negligible.

Several reliability and producibility issues are to be resolved. Reliability has an impact on this design, since detection is a statistical process. Some landing strips are always detected; some may never be detected; some are detected some of the time and not detected at other times. These statistical considerations contribute to the probability of detection, or how reliably objects are detected. The simulation can be used for designing optimizing algorithms to improve the chances that the system properly detects landing strips as reliably as the customer desires. Producibility can also be studied. The producibility of the sensor is dependent upon the complexity of the design, which is related to its optical resolution and scans rate. The complexity affects the tolerance level and the amount and type of processing it must do. The simulation allows the design team to experiment with each of these variables.

5.8.2   Simulation And Modeling For Producibility

Software tools can provide manufacturing process analysis simulation and modeling capability for use in product design and producibility analyses. Networking and solid modeling of the product provides a common path for information sharing and simulation of producibility information between design and manufacturing. For example, tool and fixture concepts are developed and functionally simulated during design to highlight configuration or concept problems. Vendor quality, cost, and technical risk data can be reviewed to identify the best vendors. Detailed part information such as tolerances and features are evaluated for economic machining, assembly, and part selection. Solid modeling of products, parts, and processes allows evaluations of interference (e.g., assembly fits) and ease of assembly. Assembly efforts are simulated using discrete event simulation software to assess efficiency and highlight improvement opportunities. Variation simulation analysis is also used. Part geometry, assembly tolerances, and process capabilities are used to define the assembly sequence as a tree structure. A Monte Carlo simulation is used to randomly vary design parameters to identify the effects on manufacturing.

Producibility analysis and their knowledge databases can be integrated into the CAD system. Virtual reality tools enable the team to evaluate the producibility and affordability of new product and/or process concepts with respect to risk, impacts on manufacturing capabilities, production capacity, cost, and schedule. Knowledge databases are developed to provide accurate and realistic predictions of schedule, cost and quality; address affordability as an iterative solution; and improve collaboration between design and manufacturing in an interactive fashion. Producibility guidelines can be inserted into the design database as rules or constraints. Developing an effective and comprehensive producibility knowledge database, however, is usually difficult and costly since much of the data may be unknown.

5.9   VARIABILITY AND UNCERTAINTY

The only thing constant is change itself. Change causes variability in the design/process and uncertainty in our decisions. The design team must take these variables into account. Design parameters can vary due to:

- **Design variability** (e.g. precision, interaction between parts)
- **Manufacturing variability** (e.g. tolerance, vendor)
- **Effects of stress and time** (e.g. stress failures, aging)
- **Uncertainty** (e.g. unknown data, incorrect models, unknown product uses, part interactions, environment)
- **User and use variability** (e.g. number of customers, skill level)

The uniqueness or newness of the design increases uncertainty in product development and it's required processes. Uncertainty can be caused by many things such as new, unproven, or unique: designs, parts, materials, software modules, features, technologies, options, vendors, logistic support methods, users and operators, manufacturing or software programming processes, design tools and testing methods used in development. The effects of this potential variation and its uncertainty must be considered.

Worst-Case, Parameter Variation, And Statistical Analyses

Three methods are often used in design analyses to compensate for variability. These methods are worst-case analysis, parameter variation analysis, and statistical analysis methods, which includes root sum square, moment, and Monte Carlo techniques. As the name infers, these methods determine whether the various variability distributions or long-term degradations can combine in such a way as to cause the product's performances to be out of specification. Very sophisticated models may be used, but these are usually based on the methods described below. A major decision in the product development process is to select which models will be used for different design parameters to ensure an optimal design without "over designing".

A worst-case analysis is a rigorous evaluation of the ability of a design to meet requirements under the worst possible combination of circumstances. This is accomplished by determining the worst-case values of critical design parameters, high and low, slow and fast, small and large, and long term degradations that could affect performance, reliability, producibility, and so on. Design parameters are then evaluated for both the highest and lowest conditions. If the overall performance of each part or software module under these conditions remains within specified limits, then the design is reliable over the worst possible conditions. To perform a worst-case analysis, the designer must identify variability limits, including problems such as manufacturing and vendor capabilities and degradation due to stress or time, and external problems due to environmental extremes. Worst-case analysis is a very conservative approach that generally increases the manufacturing and software cost of a product.

The parameter variation analysis method is a less rigorous methodology that determines allowable parameter variation before a design fails to function. Parameters, either one at a time or two at a time, are varied in steps from their maximum to their minimum limits, while other input parameters are held at their nominal value. Data is thus generated to develop safe operating envelopes for the various parameters. These parameter envelopes can be plotted and are called Schmoo plots. If each parameter or plot is kept within the safe operating limits, the design will perform satisfactorily. Parameter variation results in a design that can meet a high percentage of conditions before a failure could occur. It generally produces a design with lower manufacturing or software costs than that produced by a worst-case analysis.

Statistical analysis models use statistical methods for determining allowable parameter variations. It is used to prevent "over designing" the product and to lower costs. In statistical analysis each individual parameter may be described as an independent random variable. The allowed distribution of the dimensions depends on the process and vendor used. A commonly used distribution is the normal distribution. With this assumption, the tolerance stack up can be expressed as follows:

$$\delta_R = \left(\delta_1^2 + \delta_2^2 + \ldots + \delta_n^2\right)^{1/2}$$

Where

δ_R = design tolerance
$\delta_{1..n}$ = tolerance of "n" parameter

This model is often called the root sum square (RSS) model, because the tolerances add as the root sum squared. When compared with the worst-case approach, the statistical approach results in larger tolerances resulting in lower manufacturing costs.

Tolerance Example

One popular tolerance software system was developed by Texas Instruments and is available with PRO/ENGINEER. Tolerance modeling starts with the contact points between each part in the assembly model. One-dimensional tolerance models are set up using straightforward dimensional-loop methods. Complex problems involving 2D and 3D tolerance models use proven kinematics vector-loop methodology that comprehends not only dimensional and geometric feature variations resulting from natural manufacturing process variations but also kinematic variations resulting from assembly processes and procedures.

As described in their literature, the statistical tolerance analysis engine is based on the Direct Linearization Method (DLM). The fully automated analyzer module generates multiple analysis summaries:

- Worst Case – 100% fit at assembly under all tolerance conditions.
- Root Sum Square (RSS) Statistical – Based on normal probability distributions.
- Assembly Shift and Component Drift – Based on 6-Sigma analysis approach.

Tolerances for a full range of manufacturing processes are provided to improve the tolerance model definition, producing more realistic "as-built" results. You can modify or add new process definitions as needed.

One optimization method is to set the target assembly sigma to the desired value and let the automatic allocate function synthesize a set of tolerance values that will satisfy the quality target. The assign function adjusts the tolerance values to achieve the targeted Cp quality level for each tolerance element. The system shows the optimized tolerance allocations generated. Processes are selected from the Process Library for each element, and setting the Cp and Cpk values for each element.

5.10   AGING

Another concern of variability is the part or product's performance changes with respect to time and environmental conditions (e.g., stress). **Aging is the change in performance over long periods of time.** An example of the effects of aging on the resistance values for a resistor is shown in Figure 5.4. This figure shows the average and standard deviation of the change in resistance over time from its initial value. Sufficient margin in the design should exist so that no combination of stress and aging distributions causes failure or significant performance degradation over time.

There are two ways to design for these effects. One method is to control the product's use to preclude any parameter stress outside specified limits. Another method is to design sufficient margins or flexibility into the design. The first method usually requires controls by the manufacturer and continued maintenance to replace parts as needed. The second method is to design the part into the system so that there is sufficient margin or flexibility in the design to tolerate the expected effects of aging. The designer needs knowledge of typical part variation, both initially and with time, and the effects of various levels of stress on the variation. If they are not taken into account, failures occur due to the parameter drift. There are circuit analysis techniques that are suitable for aging or degradation analysis. These techniques use mathematical models describing circuit output variables in terms of several interrelated input parameters.

5.11   STRESS ANALYSIS

As a basic design philosophy, a design's reliability or failure rate can be improved in anyone of the following ways:

- **Increase average strength** of the product by increasing the design's capability for resisting stress.
- **Decrease the level of stress** placed on the design through system modifications, such as packaging, fans, or heat sinks.
- **Decrease variations of stress and strength** by limiting conditions of use or improving manufacturing methods.

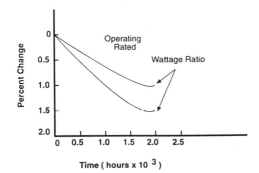

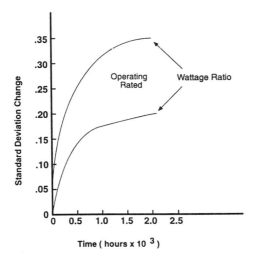

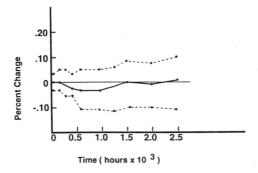

FIGURE 5.4   Aging - resistor parameter change with time and stress.

Since stress causes most of the hardware failures during a product or system's useful life, stress analysis is a critical detailed design analysis method for high reliability. A major goal for all design analyses must be to reduce the effects of stress. Stress in this context is used in the broadest sense. This includes stresses introduced by manufacturing, test, shipping, environment, users, disposal, and other products. Stress can have quantifiable parameters such as temperature, current, corrosion, weight, number of customers, and bandwidth, or general product parameters, such as maintenance procedures.

Stress analysis is the systematic method of examining a design to determine the magnitude of operating stresses on the individual parts and their effects on performance and reliability. Its purpose is to identify potential problems early in the design cycle. The major design steps for reducing the effects of stress are as follows:

- Determine and understand all stress potentials that may be caused by product use and environment.
- Eliminate or reduce the level of stress when possible.
- Design a system that can adequately withstand all levels of stress.
- Use environmental and developmental testing to ensure that the design can adequately meet all areas of stress.

For hardware, the three major types of stress analysis are mechanical, thermal, and electrical. Loads induced from the interaction of two physical interfaces often cause mechanical stresses. Forces studied in mechanical stress analysis include axial, bending, shear, and torsion. Common types of mechanical failures are deformation, buckling, creep, fatigue, wear, and corrosion. Thermal stresses are caused when the environmental temperature of a product (or its components) changes. Problems include creep, deformation, material instability, temperature cycling stresses, and burnout. Electrical stresses are induced in much the same way mechanical stresses are. The level of stress is usually caused by the level of power, current, voltage, and power cycling.

Significant benefits in reliability and cost are realized when the stress on a design is reduced. It has been shown that high temperatures cause the early failure of electronic components. Conducting tests with actual equipment, the Office of Naval Acquisition Support (DOD, 1985) determined that an annual operating savings of $10 million could be realized for just 200 aircraft when the operating temperature is reduced by only 5°C!

For software and electronic commerce, stress analysis focuses on bandwidth, user traffic volume, types of user traffic, and computing resources.

5.12   THERMAL STRESS ANALYSIS

Since thermal stress is a major cause of electronic failures, this section will review the reliability and producibility issues of thermal analysis. **Thermal analysis is an engineering discipline that minimizes the effects of temperature stress through thermal analysis, modeling, testing, and component qualifications.**

Failures induced by temperature stress include faulty performance caused by device "burn-up", solder cracks, stress fracture, blistering, and delaminating. These result in common failure modes, such as surge currents, intermittent shorts, changes in resistance or capacitance, and mechanical failure. Thermal analysis attempts to improve reliability by controlling two major types of temperature related failures: high temperatures on electronic components and thermal expansion between different materials resulting in stress fractures. Low temperatures and extreme changes in temperature, often-called thermal shock, are also evaluated in most thermal analyses. Producibility is a major concern since thermal design recommendations and solutions can reduce manufacturing-induced failures.

Thermal design techniques identify potential problem areas and minimize their effects through component selection, material selection, circuit design, board layout, heat sinks, air movement, or other techniques.

The steps for thermal analysis are:

1.   Determine thermal goals and requirements.
2.   Perform thermal modeling and analysis.
3.   Perform thermal test and evaluation.
4.   Design to improve thermal performance.

The importance of thermal analysis is apparent when the failure rate for a common type of electronic part can increase 70% as it's junction temperature rises from 100 to 125°C. A rule of thumb is that each temperature rise of 10°C reduces the electronics life by 50% (Ricke, 1996). Reliability is also influenced by fluctuations in device temperature. Temperature cycling in excess of 15°C about a specified average temperature has been found to reduce reliability almost independently of the temperature level. Thermal stresses induced by differences in thermal expansion throughout the device are the probable cause for this effect. Thermal expansion results in out-of tolerance conditions and mechanical stress points causing stress fracture. Stress problems also result when two dissimilar materials are joined because of a difference in their thermal expansion coefficient. This type of thermal expansion can break soldered connections between the electronic component or chip carrier pins and the printed circuit board. This stress is caused by different thermal expansion rates for the different materials, which results in deformation and, possibly, separation of the solder joint.

5.12.1   Thermal Requirements

The first step is to develop realistic thermal design goals and requirements for the product. These must be tailored for each product, since temperature profiles are unique for each design. These criteria apply to case and hot spot temperatures. The thermal stress guidelines have been instrumental in reducing the failure rate of electronic equipment by a factor of up to 10 over traditional handbook design criteria. In one program involving 200 aircraft, each 5°C reduction in cooling air temperature was estimated to save $10 million in electronic system maintenance costs by reducing failure rates (DOD, 1985).

Although standard thermal guidelines and criteria are available, the unique considerations of each product's design, application, and environment must be evaluated and incorporated in the thermal requirements. This tailoring results in thermal design goals and requirements that accurately model the system's environment and required performance.

5.12.2   Thermal Modeling And Analysis

The critical element in the success of thermal analysis is the quality of the model. Thermal analysis requires an empirical model that can predict the thermal characteristics of various system configurations and design alternatives. In a typical air-cooled electronic system, such as the notebook computer, the temperature can be thought of as decreasing from local hot spots to other, cooler locations. Power is dissipated as the heat is transferred through the component case, from the case to a mounting board to a card guide, from the card guide to an average equipment wall location, from the wall to a local, somewhat heated air mass, and finally out to the local ambient air. The energy added to a system to be carried out as heat can be transferred from the system through a combination of conduction, convection, or radiation heat transfer.

The thermal design of a product will focus on the different modes of transfer as shown in Figure 5.5. Forced convection coefficients are significantly larger than free convection for most products. Radiation often turns out to be a minor contributor for many products and is often neglected when forced convection is used. When forced convection is not used, the analyst frequently considers only the parameters of free convection. However, the contribution from radiation is usually on the same magnitude as that for free convection and should not be neglected. Designing without considering the effect of radiation can result in significant and costly over designs. On the other hand, introducing design changes that greatly decrease the radiated heat transfer, such as using low-emissive surfaces, can reduce the design margin and may actually result in increased failures.

An electronic product that capitalizes on conduction heat transfer frequently relies on enhanced conduction paths along printed wiring boards or component frames to a temperature controlled surface, a so-called cold plate or

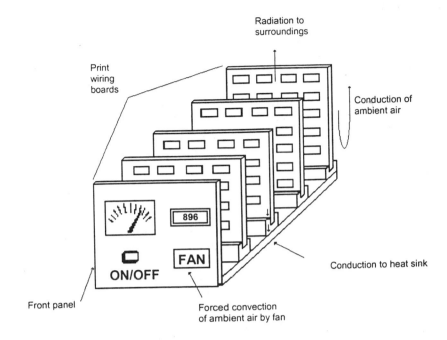

FIGURE 5.5 Typical modes of heat transfer

cold wall. This surface may be controlled by gas or liquid convection or by conduction.

It is difficult to use rules of thumb for recommended operation for various cooling modes. The reason is that the operation of a design is determined by the interplay between numerous parameters that are not specified by the use of a few general classifications.

The practice of predicting thermal behavior by using computerized numerical models has proved to be the most practical tool for all phases of design. The goal is to create a numerical model that can be used to describe spatial and temporal temperature variations of the system. The two major modeling approaches are finite-difference and finite-element techniques. In the finite-difference technique a detailed description of system geometry is fed to a finite-element analyzer that formulates appropriate equations to define heat transfer processes internally. The input to a finite-element analyzer, on the other hand, is a description of the heat paths between nodes.

Frequently the preliminary reliability check focuses on the maximum device junction temperature. Knowledge of device types, chip sizes, and mounting details allows the thermal analyst to characterize device and circuit board thermal behavior. Device thermal resistance is frequently expressed in the form of a junction-to-case thermal resistance in degrees Celsius per watt. A model of a circuit board with component power inputs is then constructed to predict temperatures. The initial design inputs must be complete enough to allow specification of all major thermal paths and boundary conditions. A typical three-dimensional plot is shown in Figure 5.6.

5.12.3   Thermal Test And Evaluation

Many uncertainties exist in modeling even simple systems. An accuracy of plus or minus 10-20% is usually thought to represent good accuracy. In practice, uncertainties in the thermal analysis are compounded by uncertainties in total system power dissipation, and even more so by uncertainties in the distribution of the power dissipation. The result of this situation is that although detailed thermal analysis is an invaluable tool early in the design process, testing is necessary to verify the design.

The objectives of thermal test and evaluation are to ensure that the thermal model is adequate, that the operating temperatures are within specified tolerances, and reliability growth is improved using test, analyze, and fix methodologies (see the Chapter on Test and Evaluation). The tested equipment is usually a full-scale engineering or production model. These tests are often part of a test analysis and fix, reliability development, or qualification test program.

These tests subject the equipment to the environmental parameters that might be expected. Thermocouples or thermographic imaging can be used. Several areas to be considered are:

- Surface temperatures of all critical components
- External case surface temperatures
- Air temperature in the local vicinity of the unit
- Air temperature inside the unit
- Air pressure in the local vicinity of the unit
- Fan inlet and outlet temperature
- Fan outlet temperature and flow rate

Although the primary goal is to ensure that all temperatures are within acceptable ranges (i.e. requirements), a more aggressive goal is to reduce temperature affects as much as feasible through design improvements. This is critical for designing highly reliable products.

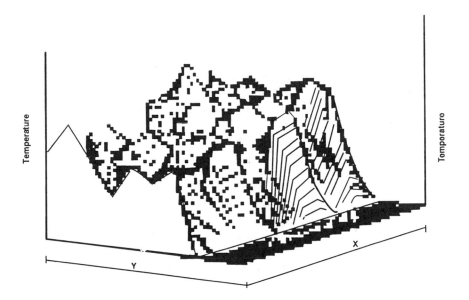

FIGURE 5.6 Thermal plot of a circuit board

The primary design methods to improve thermal performance are:

1. Part derating
2. Part selection and location
3. Improve thermal conduction by reducing thermal resistance
4. Lower surrounding temperatures using convection

5.12.4   Electronic Part Derating

One method for high reliability is to set requirements high enough to compensate for uncertainty, variability, and aging. Each company and design project should derive a set of design margins and derating criteria that meet the needs of their systems. Both analysis and measurement determine these rules. Stress limits are set at or below a point at which a slight change in stress causes a large increase in the corresponding part failure rate. It is, however, possible to be too conservative in setting stress limits, resulting in significantly higher costs.

Derating electronic parts has been proven to be an extremely effective method for ensuring high levels of reliability. Although parts can function at their maximum ratings, historical experience has demonstrated a relationship between operating stress levels and part reliability. At some point in the operating range, the failure rate increases at a much higher rate for a given level of stress than below that point. Whenever possible, choose an operating point well below this inflection point. Derating criteria are established to ensure sufficient margins for the effects of stress.

Derating criteria are based on those stresses that affect reliability. For example with a transistor, the two dominant parameters that affect transistor failure rates are junction temperature and voltage. Power and current are secondary effects that influence junction and hotspot temperatures and may or may not have derating criteria. Examples of some high level derating criteria that could be used for the notebook computer as listed in the U.S. Department of Defense Directive 4245.7-M (1985) are:

Electrical parts (except semiconductors and integrated circuits)
    ≤ 3 W: 40°C rise from the part ambient with a maximum absolute
       temperature of +110°C
    > 3 W: 55°C rise from the part ambient with a maximum absolute
       temperature of +110°C
Transformers: 30°C rise from the part ambient with a maximum
    absolute temperature of +100°C for Mil-T-27 Class S insulation
Capacitors: 10°C rise from the part ambient with a maximum
    absolute temperature of +85°C
Semiconductors and integrated circuits: junction temperatures
    should not exceed 110°C regardless of power rating

5.12.5   Part Selection and Location Considerations

The best design approach is to select parts that are 1.) not heat sensitive, 2.) do not produce abnormally large amounts of heat and 3.) can help improve thermal conduction. by minimizing a component's case to thermal sink resistance. The thermal sink of interest may be the component's case, printed wiring board, a metal plate, cooling air, or other component. It is worth noting that the junction-to-case temperature often represents a significant portion of the junction-to-sink rise. When practical, this value can be influenced by changing package types (i.e., going from a plastic to a ceramic case).

Another major design consideration is the location or layout of the different parts in the design. Heat-sensitive parts should be located near cooler air. Parts that produce large amounts of heat should be located far from the heat-sensitive parts. When a fan is used for forced convection, parts that produce large amounts of heat should be placed near the exhaust outlet rather than the fan (Riche, 1996).

5.12.6   Improve Thermal Conduction

The next design method is to get rid of the heat. Thermal conduction is one effective for heat reduction. Thermal resistance needs to be reduced to improve thermal conduction. For electronics, resistances include junction to case, case to board, board middle to edge, card guide resistance, wall-spreading resistance, and case wall to ambient. A study of principal thermal resistance is used to indicate the particular areas causing the largest temperature drops and, therefore, the prime candidates for improvement. The thermal contact resistance from a component case to a mounting surface is frequently a significant design problem. This resistance is a function of:

- Mounting pressure
- Surface characteristics
- Surface materials
- Surrounding environment

Values for contact thermal resistance are reduced as the pressure is increased. Values can be located in reference books for most applications. For example, the variation of aluminum changing from 0 to 250 psi contact pressure approaches a factor of 10. The variation in going from aluminum to steel at the same pressure approaches a factor of 3.

Specifying close (i.e., tight) manufacturing tolerances of the gap between the component case and the board can reduce thermal resistance. This, however, requires a very tight tolerance fit of the parts, which can cause producibility problems. The gap resistance is a small fraction of the device-to-sink thermal resistance so that large changes in gap resistance do not appreciably change the total resistance. A considerable reduction in thermal resistance and in variability can frequently be achieved through the use of joint filler materials.

Convection cooling is sometimes used to remove heat directly from the components. The thermal resistance from devices to cooling air can be decreased by increasing the convection coefficient, probably by adding or increasing forced convection, or by increasing the surface area for the transfer. Several vendors manufacture small finned heat sinks for attachment to individual components. These heat sinks are available for a range of devices.

5.12.7 Lower Surrounding Temperatures Using Convection

The third design method is to reduce the surrounding temperature level. Increasing the airflow around the area of concern using convection and designing the enclosure for better heat dissipation can accomplish this. Common methods include increasing the size of openings to allow more air and to utilize the principles of forced convection by adding fans. Other methods that can be used are heat pipes, forced liquid, and liquid evaporation.

Enclosures are also important for heat dissipation. Painted steel and fiberglass enclosures dissipate heat better than unfinished aluminum, even though, the aluminum's thermal conductivity is higher (Riche, 1996). For outdoor applications, the preferred color is white.

5.12.8 Circuit Board Level Considerations

Heat can be removed from circuit boards in a variety of ways. The simplest approach for conduction cooling is to use only the board as a thermal path. The thermal conductivity of most polymer board materials is very low relative to that for aluminum or copper. Even a rough calculation to define the allowable board power must include the effects of copper-etch signal layers and especially of ground layers, which have a high percentage of copper coverage. The carrying capacity of unaided boards is the lowest of any conduction-cooled approach to be considered here.

Another approach is to add an additional conduction path over that provided by the basic board. The added path, often of metal, is frequently referred to as a thermal plane or cold plate. The form of the thermal plane depends on the type of components on the board. The backside of the board is clear so the board can be bonded to a solid metal plane.

The joint between the board and mounting wall is frequently a source of thermal resistance. A range of card guides is available to improve thermal performance. A wedge clamp type of card guide, which gives a thermal resistance six times less than the worst performer, is also the most expensive.

There are several cooling options available if thermal problems are greater than allowed with a standard metal thermal plane. One approach is to add a fan or blower to increase the, amount of airflow over the thermal plane. A fan can improve the heat transfer rate by a factor of 10 (Ricke, 1996). Other methods include heat exchangers, chillers, and coolants. An easier approach may be to use extra thick thermal planes. The weight penalty resulting from this approach precludes its use for many applications. Higher thermal conductivity materials can also be substituted for the normal thermal plane materials. Thermal expansion problems require that the thermal plane material have a specific thermal coefficient of expansion (TCE), greatly limiting the design choices.

Heat pipes may be considered for some types of high-power, conduction-cooling problems. Heat pipes are sealed enclosures filled with a working liquid in which heat is transferred along the heat pipe by first evaporating liquid at the area of heat input, flowing vapor to a cooler pipe section, and finally condensing vapor at the cooler section. A wick in the heat pipe returns the working fluid to the heat-input zone by surface tension or capillary effect. Heat pipes give a very high effective thermal conductivity for their weight.

All board conduction cooling approaches must overcome a thermal resistance associated with transferring heat from the board area to a heat sink. The heat sink is frequently an air-cooled extended area (e.g., a finned metal surface). One approach for handling high-power heat dissipation involves placing a finned heat sink on the backside of a board. The conduction resistance along the board is eliminated in this integral heat-exchanger approach. Higher thermal capacity from the integral heat-exchanger approach can be obtained by providing more heat-exchanger area on the back board surface by increasing fin density or height or by increasing cooling airflow rate, or both

The ranking of preferred cooling approaches is inappropriate because many different assumptions must be made before a particular approach can be selected. The constraints provided by a particular design environment probably tend to boost a particular approach.

5.13   FINITE ELEMENT ANALYSIS

Finite element analysis (FEA) is a design technique, which uses mathematical techniques for predicting the stress and its effect on the physical behavior of a system. A set of simultaneous equations is developed that mathematically describes the physical properties of a product. Solutions to the equations are then obtained which approximate the behavior being studied.

The concept underlying the FEA is to divide a complex problem into small solvable problems called elements. Equations are developed that define the elements and define the entire product's response. These equations are then numerically evaluated iteratively going from element to element. Stress analysis, heat transfer, fluid flow, vibration, and elasticity problems often use FEA techniques. It is an important analysis when extensive testing is too costly or time consuming.

5.14   ENVIRONMENTAL STRESS ANALYSIS

How well a product performs is to a great extent dependent on the designer anticipating the expected environmental stresses and their effects on both individual parts and the product. This is especially important as products are now being sold all over the world with many different climates.

Temperature is the most common environmental stress that affects electronic parts. As noted earlier, reliability of electronic components is improved with a reduction in operating temperatures. Temperature changes also have a profound effect because of the differing thermal expansion properties of various materials. Extended or multiple thermal cycles cause materials to fatigue, resulting in open solder joints, cracking of seams, loss of hermeticity, and broken bond wires. The addition of vibration to thermal cycling increases these effects. Derating reduces these strains by ensuring that all parts are rated in excess of the expected environmental conditions and all materials are selected with thermal expansion coefficients matched as closely as possible.

Another concern for designers is high-humidity environments, which are extremely corrosive. Problems caused by high-humidity environments include corrosion of metallic materials, galvanic dissolution when dissimilar metals are in contact, surface films that increase leakage currents and degrade insulation, moisture absorption that causes deterioration of dielectric materials and changes in volume conductivity and the dissipation factor in insulators. One method of designing for a high-humidity environment is to seal the component in a hermetic package. A hermetic package has a cavity where the component element resides, sealed off from the outside environment. Other design considerations include reducing the number of dissimilar metals (especially those that are far apart on the galvanic table), using moisture-resistant materials, using protective coating or potting the assembly with moisture-resistant films, sealing the assembly in a hermetic package, or controlling the humidity environment around the system.

The environmental factors pertinent to each design must be identified along with their intensity. Some CAD systems have the ability to simulate the effects of environmental extremes on a product. Extensive lists of the main and secondary effects of typical environments are also available in books. The synergistic effect of various environmental combinations must also be considered.

5.15   FAILURE MODE ANALYSIS

Murphy's Law is that "If something can go wrong, it will". Product failures and manufacturing problems will occur so we must minimize their number, minimize their effect, and be ready for them when they occur. Product failures affect reliability, safety, manufacturing, product liability, logistics, and most important customer satisfaction.

A failure mode and effects analysis (FMEA) is a technique for evaluating and reducing the effects caused by potential failure modes. It can be used for analyzing the design, user interface, or its manufacturing processes. The design FMEA is a design analysis technique that documents the failure modes of each part, signal, or software module and determines the effect of the failure mode on the product. The manufacturing process FMEA (PFMEA) focuses on the processes and vendor's failure modes. For design, critical failure modes are eliminated through design improvements that can include component/vendor selection, redundant circuit or software paths, alternative modes of signal processing, and design for safety. For manufacturing, design improvements can include new processes, preventative maintenance, mistake proofing, operator training, etc

Failure mode analysis starts early in the design process to minimize the number and cost of design or manufacturing changes. Owing to the complexity of many products and the large number of failure modes, this analysis technique can require considerable effort and cost. Early in the development cycle, cost and schedule trade-offs must be made to determine the level of detail at which the analysis should be performed.

The analysis is a "bottom-up" approach. Knowledge of the failure modes of each item or process is then used to determine the effect of each failure mode on system performance. The key benefits to be derived from a FMEA are:

- Identification of single-point failures
- Early identification of problems and their severity
- Information for design trade-off studies

The analysis usually assumes that conditions are within specification, all inputs are correct, and no multiple failures have occurred. For example, if a computer is being analyzed, it is assumed that the environmental conditions, such as humidity and temperature, are within specified limits. The design under analysis must be clearly defined, including block diagrams. The function of each block and the interface between blocks should be defined. The second important aspect of FMEA is to establish the baseline or ground rules for the analysis.

A seven-step method for FMEA circuit analysis was developed by Wallace (1985). An adaptation for all types of hardware and software is as follows:

1. Identify critical areas (i.e., areas that must function) of the design.
2. Identify the major failure rate contributors. This is done by researching the history of failure rates of items such as parts, software modules, users, signals, etc.

3.   Examine output responses to input fluctuations, and identify the critical input parameters (i.e., those inputs that would produce critical problems such as out-of-tolerance output or none at all).
4.   Examine the effect of anticipated user and environmental extremes on performance.
5.   Look at tolerance build-up and parameter variations.
6.   Identify areas for improvement and change the design.

A manufacturing process PFMEA is similar except that the analysis focuses on process failures. To assure a systematic and consistent approach, information is usually logged on a common form as shown in Figure 5.7 The failure mode lists column represents each possible failure mode of the item under consideration. This data should be available. Examples of some major failure modes and their percentage of occurrence are shown in Table 5.3.

Additionally, this data can be combined with the measure of the severity of the effect to quantify criticality. A criticality analysis is then performed. The criticality analysis provides a quantitative measurement of the criticality of each failure mode based upon the qualitative results of the FMEA. Criticality is the probability that a failure mode will create an adverse effect, whereas a reliability prediction states the probability that a failure will occur when it occurs. Each failure mode is uniquely coded and its code number entered. Considering predetermined analysis guidelines, failure modes falling into specified classifications are declared critical and must be corrected.

For very important failures, a fault tree analysis may be needed. **A fault tree analysis develops a model that graphically and logically represents various combinations of possible events (i.e., failures or faults) that could cause or lead to a particular undesired situation.** A fault tree is composed primarily of a top event with the various situations or sub-events that must occur before the top event would occur. Several different types of symbols are used to describe the events. A circle depicts a basic event and a rectangle represents a failure or fault event. Logic gates are used to link various combinations of events. There are three basic logic gates used in fault tree construction. The AND gate describes a situation in which the output event occurs if all the input events happen at the same time. The OR gate is a situation in which the output event occurs if any of the input events occur. The INHIBIT gate exists when both the input event (bottom) and conditional event (side) both need to occur if the output is to be produced (top).

The fault tree is constructed to determine the possible causes of failures, single point failures, or to determine the adequacy of a design. A fault or failure can be evaluated quantitatively or qualitatively or both, depending upon the extent of the analysis. A key purpose of fault tree analysis is to determine whether the design has an acceptable level of reliability, availability, and safety in the proposed design. Should the design be inadequate, trade-off studies are then performed to determine what design changes must be made to minimize the

A. Design FMEA

SYSTEM: Notebook Computer

SUBSYSTEM: Printed Circuit Board

Item	Code	Function	Failure Mode	Failure Effect	Loss Probability	Failure Mode Frequency Radio	Failure Rate (Per Million Hours)	Total	Corrective Action
Resistor	R1	Voltage Drop	Open	No Output	1.00	.80	1.1	.880	100% Functional Test
Resistor	R1	Voltage Drop	Change of Value	Wrong Output	.10	.15	1.1	.022	6 sigma design
Resistor	R1	Short	Short	Wrong Output	1.00	.5	1.1	.055	6 sigma design

B. Process or Manufacturing PFMEA

Product: Notebook Computer Printed Circuit Board

Process: Circuit Board Assembly

Item	Function	Failure Mode	Failure Effect	Loss Probability	Failure Mode Frequency Radio	Defect Rate (Per Million Tasks)	Total	Corrective Action
SMT Part Insertion	Insert SMT Components	Jammed	Machine Down	1.00	.1	.2	.020	ISO 9000 Vendors
		Wrong Part	Quality	1.00	.1	.2	.020	Mistake proofing
		Mis-alignment	Quality	1.00	.7	.2	.140	Computer Vision Alignment
		Damaged	Quality	1.00	.1	.2	.020	New packaging

FIGURE 5.7 FMEA and PFMEA Worksheets

TABLE 5.3 Sample Failure Mode Distributions

Part type	Major failure modes	Occurrence (%)
Bearings	Lubrication loss	45
	Contamination	30
	Misalignment	5
	Brinelling	5
	Corrosion	5
Capacitor	Open circuit	35
Electrolytic	Short circuit	35
	Leakage	10
	Decrease in capacitance	5
Connectors	Shorts	30
	Solder joint (mechanical)	25
	Insulation resistance	20
	Contact resistance	10
	Miscellaneous mechanical	15

critical events. When the design has been updated, then another fault-tree analysis should be conducted to again determine whether the proposed system is at the desired level of safety or reliability. A simple example of a fault tree is shown in Figure 5.8.

Fault tree analysis is also used for program logic to determine the set of possible causes of a hazard, or show that they cannot be caused by the logic of the software (Glass, 1996). The results of the analysis are used to guide further design, pinpoint critical functions, detect software logic errors, guide the placement and content of run-time checks, and determine the conditions under which fail-safe procedures should be initiated (Littlewood, 1987).

5.16   DETAILED DESIGN OF GLOBAL SOFTWARE

A dilemma is the development of software for people all over the world to use. Software must take into account differences in cultural conventions and language. Language translation needs to include a country's slang, number format, date representation, paper dimension standards, controls on the hardware equipment, and monetary format. Internationalization translation includes "the process of building in the potential for worldwide use of software as a result of the efforts of programmers or software designers during the development or modification process" (Madell et al, 1994).

The following discussion is a student's description of his work as software translator in Brazil (Sabino, 1996). There are at least three important aspects of globalizing a product: 1.) Language translation, 2.) Cultural differences, and 3.) Government regulations.

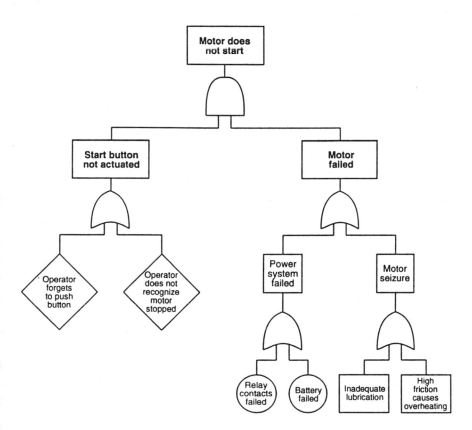

FIGURE 5.8 Fault tree analysis

The most significant words of the software are identified and listed. This glossary is approved before the translation starts. To keep consistent, all translators follow the glossary. Regular meetings are held to discuss the terminology and to keep the glossary up to date.

The translated text is then submitted to a reviewer to analyze the translation for grammar, linguistics, style and property. After the initial translation is done, the software is put back together, installed, and tested. Software translation requires the involvement of specialists from different areas in the process. The file format (e.g., rtf, txt, rc, bitmaps, dlg, shg, msg), editor type and platform type (e.g., Windows, UNIX, Novell) must all be considered.

Some software design will probably be needed for compiling, capturing images, and translating screens with the necessary adjustments for the fields. Most cases require adaptations to the fields. This is more difficult when programmers (developers) forget that their work may be adapted to another language. One example is how they define the restricted fields for words in the original language. When the company decides to translate and localize the program for another country, the translator may have serious problems if the new word has a different number of alphanumeric from the restricted field. Often the new word does not fit in the field, which forces the translator to find ways to adapt the new word.

The problem of adapting fields is sometimes easy in one language but very frustrating in others. English to German can be very hard because of the length of the words in the German language. An example is the word CLOSE (5 bytes, English); FECHAR (6 bytes Portuguese); SCHLIESSEN (10 bytes, German). This problem can be solved if the programmers design the fields to account the average length of words in different languages.

Another software design aspect is the adaptation of reports and screens to the laws and customs of the new country. Some examples of problems that software developers face are the use of colors, female voices, (which is not accepted in some countries), and the pointing-finger cursor which in some cultures they associate with thieves. Another good example is accounting software that needs to be adapted to the particular tax laws and forms of that country.

In many countries it is required by law that any software to be sold in that country must be translated before it goes to the store shelves (e.g., Brazil, Spain, France). They believe that their citizens do not have an obligation to know other languages and they want to preserve their language for future generations. Even if it is not required by law, if you offer software in English and your competitor offers a lower quality package in the native language, you will probably lose sales.

It is important that the translation be done in the country where it is going to be sold. Countries with similar languages like Spain and Mexico have problems because the software owner may think that the different Spanish dialects are the "same". Most companies use a professional agency from the country that they want to sell the software package.

5.17   DETAILED DESIGN FOR A NOTEBOOK COMPUTER

Almost all of the design analyses discussed in this chapter would be used for our notebook computer example. The major concerns for this type of product and the design analyses that would be used to reduce their technical risks are shown in Table 5.4.

TABLE 5.4 Design Analyses to Reduce Technical Risk

Major Technical Risks for a Notebook Computer	Detailed Design Analysis
Thermal concerns because of the compactness of the electronic components	Thermal stress analysis of internal components and may use worst case values. Design options include cooling fans for case, using a cooling fan or heatsink for the microprocessor locating the hotter running components, and choosing different components
Shock concerns of dropping notebook computer	Mechanical stress analysis of case and internal components, using FEA techniques
Limited time to develop product	Design modeling, design synthesis, and extensive prototyping in detailed design. Use vendors with proven delivery.
Use of new technologies such as Digital Video Disks (DVD) and new screen technologies	Simulation and "hardware in the loop" simulation of the new technologies. Use best vendors.
Battery life and power consumption	Circuit and power analysis

5.18   SUMMARY

Detailed design uses design analysis, modeling, and trade off analysis to optimize design decisions. As the design process progresses, analytical techniques guide the continuing effort to arrive at a mature design. Design analysis evaluates the ability of the design to meet performance specifications at the lowest possible risk. There are many analyses for reducing design risk. By extensively and effectively using design analysis, the development team can develop a producible and reliable product. This chapter has provided the background and reviewed design analysis practices for implementing these techniques.

5.19   REVIEW QUESTIONS

1. What are the differences between design analysis and design trade-offs?
2. Describe the future role of computer design tools in product development.
3. Describe the properties of an effective simulation.
4. List advantages and disadvantages between worst case, parameter variation, and statistical analysis.
5. Describe the major steps in stress analysis and reduction.
6. List several unique environmental stresses found in some countries.

5.20   SUGGESTED READINGS

1. G.A. Hazelrigg. Bean Guessing and Related Problems in Engineering. Engineering Education. p. 218-223. January 1985.

5.21   REFERENCES

1. Department of Defense (DOD), Transition from Development to Production. Directive 4241.7M, Washington, D.C., September 1985.
2. G. Glass, Software Fault Analysis, Student report, The University of Texas at Arlington, 1996.
3. G.A. Hazelrigg, Bean Guessing and Related Problems in Engineering, Engineering Education, January, p. 218-223, 1985.
4. B. Littlewood, Software Reliability, Achievement and Assessment, McGraw-Hill, Boston, 1987.T.
5. Madell, C. Parsons, and J. Abegg, Developing and Localizing International Software, Englewood Cliffs, New Jersey, Prentice Hall, Inc., 1994.
6. S. McConnell, Rapid Development, Microsoft Press, Redmond, Washington, 1996.
7. R. S. Pressman, Software Engineering, McGraw-Hill, New York, 1992.
8. J. Ricke, Sr., Managing Heat in Electronics Enclosures, Electronic Packaging and Production, p. 87-89, February 1996.
9. K.J. Rogers, J.W. Priest, and G. Haddock, The Use of Semantic Networks to Support Concurrent Engineering in Semiconductor Product Development, Journal of Intelligent Manufacturing (6), p 311-319, 1995.
10. M. Sabino, Personal Experiences, Astratec Traducoes Tecnicas Ltda., Sao Paulo - SP - Brazil, unpublished Student Report, The University of Texas at Arlington, 1995.

11. I. Sommerville, Software Engineering, Addison-Wesley Publishing, New York, 1996.
12. Swerling, Computer-Aided Engineering, IEEE Spectrum, November: 37 1992.
13. Waddell, Design Synthesis, Printed Circuit Design, p. 14, 1996.
14. Wallace, A Step By Step Guide to FMECA, Reliability Review, Vol. 5, June 19, 1985

Chapter 6

TEST AND EVALUATION: DESIGN REVIEWS, PROTOTYPING, SIMULATION, AND TESTING

Validate and Verify!

When an innovative product or service is being developed, early ideas and designs will probably not meet all customer expectations and design requirements or be ready for production. Test and evaluation is the "design team's tool" for improving the design, identifying and correcting problems and reducing technical risk. Starting with the earliest design, it is continuously used throughout the design process. The goal is to both "validate" that the design will satisfy the customer and "verify" that the design meets all specified requirements. A mature design is defined as one that has been tested and evaluated to ensure validation and verification prior to production

Best Practices

- Effective and Efficient Test and Evaluation Strategy
- Design Reviews
- Prototyping and Rapid Prototyping
- Design Verification Using Modeling and Simulation
- Design for Test
- Reliability Growth Using Test, Analyze and Fix
- Software Test and Evaluation
- Environmental and Design Limit Testing
- Life, Accelerated Life and HALT Testing
- Vendor and Part Qualification Testing
- Qualification Testing
- Production Testing
- User and Field Testing

RUSHING TEST AND EVALUATION COST MILLIONS OF DOLLARS

This was the case in the Wall Street Journal article by O'Boyle in 1983. A major manufacturer of refrigerators had started a completely new design that incorporated rotary compressors. However, a rotary compressor had never been used in a refrigerator and would require powder metal parts that had never been manufactured before by this company. The new refrigerator needed to be developed as soon as possible. Unfortunately, "it is difficult to strike a magical balance between getting it right and getting it fast." Too fast a development cycle can cause problems.

Although a five-year warranty was going to be offered, they could not wait for five years of testing. Accelerated "life testing" was used to simulate harsh conditions to find design problems in a short period of time. It is a complex testing method for predicting a new product's future reliability. In the past, these tests were supplemented with extensive field testing. The original plan was to put models in the field for two years of testing but that was reduced to nine months, as managers tried to meet schedules.

As told by O'Boyle the compressors were run continuously for about two months under temperatures and pressures supposedly simulating five years' operating life. In the fall of 1984, senior executives met to review the test data. The evaluation engineers had "life tested" about 600 compressors, and there was not a single documented failure. "It looked too good to be true". Executives, therefore, agreed to start production.

Although they had not failed, they did not look right. A test technician disassembled the compressors and inspected the parts. About 15% of them had discolored copper windings on the motor, a sign of excessive heat. Bearing surfaces appeared worn, and some small parts indicated that high heat was breaking down the sealed lubricating oil. The technicians' direct supervisors, a total of four in three years, discounted the findings and apparently did not relay them up the chain of command. The supervisors wanted to believe in the air conditioner design. "Some dispute this story, but nearly all agree that the pressure to produce influenced the test results." (O'Boyle, 1983)

The first refrigerator in an unventilated closet in Philadelphia failed after little more than one year of use. At first, executives thought that the failure was due to being operated in a closet. Over the next few months, as the number of failed refrigerators greatly increased, engineers worked furiously to diagnose the problem. The solution showed that the rotary compressor would not work as designed. This required an extensive redesign effort.

The main goal of every test and evaluation program should be to identify problems and areas for design improvement. The purpose of this chapter is to review important techniques for a successful test and evaluation process.

6.1   IMPORTANT DEFINITIONS

Developmental test and evaluation is an integrated series of evaluations leading to the common goal of design improvement and qualification. All reviews and tests are organized to improve the product. All identified problems and detected failures result in analysis and corrective action. A planned program requires that all available test data be reported in a consistent format and analyzed to determine reliability growth and the level of technical risk.

Validation is the process of insuring that the design meets the customer's expectations. Will the product make the customer happy? The level of verification is directly related to how well the requirement definition phase was performed. Using prototypes, this testing starts in the earliest phases of requirements definition, conceptual design and detailed design.

Verification is the process of insuring that the design and manufacturing can meet all design requirements. Will the design and support system meet the design requirements? The level of verification is related to the quality of the design process, test and evaluation phase and manufacturing test.

Design reviews are used to identify problems and technical risks in a design's performance, reliability, testing, manufacturing processes, producibility, and use. A successful design review will identify improvements for the product. A major problem occurs when design reviews are conducted as project reviews, where only a simple overview of the design is given.

A mature design is defined as one that has been tested, evaluated, and verified prior to production to meet "all" requirements including producibility. Unless the design's maturity is adequately verified through design reviews, design verifications, and testing, problems will occur because of unforeseen design deficiencies, manufacturing defects, and environmental conditions.

Reliability growth derives from the premise that as long as a successful reliability effort continues to improve the design, the design's reliability will improve. This is reflected by an increase in a reliability index such as the MTBF of the system, approximately proportional to the square root of the cumulative test time. Reliability growth requires an iterative process of design improvements. A product is tested to identify failure sources, and further design effort is spent to correct the identified problems. The rate at which reliability grows during this process is dependent on how rapidly the sources of failure are detected, and how well the redesign effort solves the identified problems without introducing new problems.

6.2   BEST PRACTICES FOR TEST AND EVALUATION

The goal of every test and evaluation method is to identify areas for design improvement. The key practices are:

- **Test and evaluation strategy** effectively coordinates all tests to verify a design's maturity in a cost-effective manner.
- **Design reviews** use a multidisciplinary approach for evaluating and improving all parameters of a design including producibility, reliability, and other support areas.
- **Prototyping, design modeling and simulation** are used to both validate and verify the design, identify problems and solicit ideas for improvement.
- **Design for test** is used to design the product for easy and effective test.
- **A test, analyze, and fix methodology** is used to identify areas for design improvement to maximize the reliability growth process.
- **Software test and evaluation** uses proven methodologies to ensure effective verification and identify areas for improvement.
- **Environmental, accelerated life, and HALT testing** of critical components is initiated early in the program.
- **Qualifying new parts, technologies, and vendors** are started early and used for improving the product and reducing technical risk; not as a means of identifying poor design.
- **Production testing** considers all quality control tests including incoming testing and environmental stress testing.

6.3   TEST AND EVALUATION STRATEGY

An effective test and evaluation program can reduce the technical risk associated with product development, and actually improve the performance, producibility, reliability, and other aspects of a product. Its purpose is to make the design "visible" so that problems can be identified and resolved. In the early stages of product development, test and evaluation techniques are used to validate customer requirements, evaluate design approaches and to select alternative design solutions for further development. As the design matures, the tests become more complex and verify that the design can meet the requirements. This provides confidence that the system will perform satisfactorily in the actual operational environment.

Developmental test and evaluation is an integrated series of evaluations leading to a common goal of design improvement and qualification. It must be emphasized that this improvement comes only as a result of a dedicated, in-place, effective test and evaluation program. This type of program is accomplished by coordinating a series of evaluations and tests through the direction of an integrated test plan. This test plan is initiated early in a design to ensure proper levels of testing while minimizing costs. The plan becomes progressively more comprehensive and detailed as the product matures. Some of the important considerations when developing a test plan are shown in Table 6.1.

Another purpose of developmental test and evaluation is the evaluation and maturing of newly emerging technologies, manufacturing processes, and vendors. Ever-increasing breakthroughs in science and technology continue to pressure the design team to incorporate new technologies, vendors, and ideas in product design as soon as possible. Failure to use a new technology or process can result in lost sales, as competitors are the first to introduce innovative products. On the other hand, using a new technology, process, or vendor before it is ready, may be a critical mistake that results in cost overruns, poor quality, and high warranty costs. Since using new and often "unknown" technologies is becoming an integral part of the development process, an evaluation process must be established for accessing and studying promising new technologies prior to their use in actual products. This process of evaluating new technologies, prior to their use, is often called qualification testing or off-line maturing.

TABLE 6.1 Integrated Test Plan

Identification of all validation and verification activities to be performed

Identification of what type of prototypes, modeling, simulations, tests, demonstrations, and trials are needed to validate and verify that product and design requirements have been met for all potential environments and uses (e.g., performance, reliability, producibility, maintainability, safety, environmental, and useful life)

Identification of specific information to be gathered from each test

Identification of the data that must be collected from the environment (e.g., thermal, mechanical shock, and vibration)

Evaluation of where combined testing can be cost effective

Identification of required resources (personnel, equipment, facilities, and length of time) and schedule to perform and support testing

Identification of the quantity of test prototypes, trials, and demonstrations necessary to achieve desired confidence level

Identification of sequence and schedule of testing

Methods for analyzing and resolving problems quickly

A multi-phased bottom-up (i.e., testing at lowest hardware and/or software levels and progressing up toward the entire system) test program can reveal the appropriateness of moving from one phase to another in concert with the maturing of the designed product. Although products have become more sophisticated, test requirements are too often specified from previous projects with little consideration of changes in technology, duplication of test efforts, or the elimination of older tests that are no longer needed. Too many people simply modify existing test plans and requirements that were developed for a previous product. Attempts have also been made to "standardize" test requirements. In many instances, these standard requirements have shown little relation to the actual operational environment, resulting in costly changes to systems after production. To identify and correct deficiencies prior to production, an effective integrated test strategy and program must be implemented. Some of the many different types of tests are shown in Table 6.2.

When a program is behind schedule, management is often tempted to meet program milestones by reducing or canceling various tests or evaluations. As shown in the refrigeration example at the first of the chapter, it is almost always costly in the end. Days gained in the early program stages become lost weeks or months in the schedule, as costly redesigns are required to correct flaws found in later test and evaluations.

All tests should be organized to improve, not merely to pass or fail, the product. An effective feedback loop ensures that all detected failures and unusual results require analysis and corrective action. A planned program requires that all available test data be reported in a consistent format and analyzed to determine reliability growth and technical risk. A key function to be performed is the determination of a uniform reporting format that will ensure the collection of life and reliability data during all subsequent testing. The plan assures maximum feedback for analysis.

6.4   DESIGN REVIEWS

Design reviews should begin early in the design process to verify that the proper analyses and trade-offs are being made. Early deign reviews will focus on design drawings and approaches. As prototypes and test results become available these can be reviewed.

A successful design review concentrates on a detailed analysis of the technical aspects and risks of the design. The best review is conducted by non-project, impartial, objective senior technical experts. The technical competence of the reviewers must be high in order to ensure positive contributions and sound action. The review is very detailed about all technical risks.

TABLE 6.2 Types of Developmental Tests

Test	Test Strategy
Customer prototypes	Initially identify customer needs and then validate that the design meets them
Design prototypes	Evaluate design features, requirements, approaches and identify problems
Modeling and Simulation	Use models and simulation to study the design
Reliability growth	Test to failure; analyze each failure in terms of improving reliability through redesign
Environmental limit	Assure that the design performs at the extreme limits and conditions of the operating envelope
Life and accelerated test	Evaluate the useful life of a system's critical parts for problem identification, root cause analysis and improvement
Qualification tests	Assure that the part, software etc. meet all requirements, and that the vendor uses proven methods to ensure the highest levels quality.
Environmental stress screening	Identify parts with latent defects to ensure high levels of quality in manufacturing
Human engineering operation and repair tests	Evaluate all human interfaces by using actual users, maintenance procedures, and support equipment in a user environment
Built-in test and diagnostics	Evaluate the capability level and quality of built-in test, self-diagnostics, and self maintenance
Producibility and manufacturing qualification	Evaluate the producibility of the design and qualify all manufacturing processes
Packaging	Evaluate the ability of the packaging to protect the product

The design review process is especially critical for producibility, to ensure timely identification of potential manufacturing problems and their solutions. It is an efficient way to evaluate the maturity of the design, find areas needing correction, spot technical risks, look at the producibility and design margins, and evaluate the manufacturing and quality aspects of the design and potential vendors. As shown in the refrigerator example, the independence and competence of the reviewers are essential.

6.5   PROTOTYPING, DESIGN MODELING, AND SIMULATION

Prototyping, design modeling, and simulation are valuable methods for evaluating and testing new ideas, technologies, products, services and support systems. For instance, prototyping reduces the need for many tests and evaluations. They also provide a communication medium for information, integration and collaboration. Product development uses desktop solid-modeling CAD systems, rapid-prototyping tools, and virtual-reality environments to design and evaluate products, components, parts, and software user interfaces. Prototypes can be physical, electrical, software models, or simple physical mock-ups. These can be made of insulation foam, cardboard, etc., or computer simulations using solid modeling and animation. Rapid prototyping describes any technology that can produce prototypes in a very short time. They can shorten product development time by evaluating design requirements quickly and allowing more design iterations, thereby improving the reliability of the final designs (R&D, 1996).

New rapid prototyping tools include stereo lithography where a CAD-generated image is directly converted into a solid plastic model. Currently, the limitations for stereo lithography include a limited part envelope, an inability to form large thin sections, low thermal stability of acryl ate parts and limits on the types of materials that can be used. Later in product development, production prototypes can be used.

Prototypes allow the design teams to better visualize designs and more prototypes can be used to fine-tune the design, documentation, and service. "People often have difficulty envisioning a complex three dimensional object from various two dimensional representations, but human pattern recognition capabilities come into play when one can touch, feel, rotate, and view a real, physical, three dimensional object from different angles" (Jacobs, 1996).

Prototypes can serve as test samples to study consumer preferences to provide valuable design and marketing information. Prototypes can also provide form and fit tests of components. Production planning can determine requirements for tools and fixtures. Models can help to design packaging and dunnage that hold parts for shipping.

Using many prototypes, the successful Hewlett Packard Desk Jet printer was completed in just nine months, even though it required many inventions and innovations (Bowen et al, 1994). Bowen noted that, "the team started with breadboards to confirm the technology; studied a series of rapidly constructed prototypes to check out form, fit and function; evolved prototypes that were tested for customer reaction and served as the basis for final tool design. Prototypes provided a common basis for everyone's work, unifying the contributions of marketing, manufacturing and design. Smart prototyping was instrumental in the speedy realization of a breakthrough project (Peters, 1995)."

Design Validations and Verification Using Modeling and Simulation

Testing of many products is becoming too expensive and time consuming. To reduce these costs and still verify the product's design, many companies are using design modeling, simulation, and virtual reality to compliment the typical test program. This is especially true in the semiconductor, aircraft, and aircraft engine industries. Complex models are developed which simulate the product under different conditions. Solid CAD systems greatly enhance the ability to verify a design. As noted by Jacobs (1996), "solid" models are much easier to visualize than wire frame-drawings, surface models, or multiple blueprints. The designer can catch errors that would have been missed with drawings. The more complex the part, the easier it is to miss something (Jacobs, 1996).

In one virtual reality project, users can load proposed car-trunk designs with virtual luggage to determine the ease of loading a particular trunk configuration and the amount of storage space available under different conditions (R&D, 1996).

6.6 DESIGN FOR TEST

Testability is a design characteristic that measures how quickly, effectively, and cost efficiently problems can be identified, isolated, and diagnosed in a product. Unlike production testing, validation testing focuses on characterizing all outputs and specifications. The test information should be sufficient to allow effective troubleshooting and "what if" analyses that facilitate quick decision-making. Key design parameters for testability are accessibility, controllability, observability and compatibility. For example, does the design allow the test probes easy access the various test points and connectors on a printed circuit board; or can a software tester easily locate and access different areas of software code? Can varying input signals identify their affect on the output signals? The design should be developed so that existing resources can be used with few modifications. Updating existing test software and fixtures instead of buying new ones saves thousands of dollars and weeks of time. Design parameters include density, technologies used, connectors, accessibility, hole sizes, and dimensions. Testability will be discussed in a later chapter.

6.7   TEST, ANALYZE, AND FIX METHODOLOGY

The most effective approach for obtaining a highly reliable product is to ensure reliability growth and design maturity through a structured test, analyze, and fix (TAAF) methodology. The TAAF methodology identifies failure sources in the design by subjecting hardware and software to increasing levels of stress. Each failure is then analyzed and the design is fixed (i.e., improved), so that the failure will not reoccur. Simulated user environments are used to develop parameters for the accelerated tests. These tests are usually performed at the subsystem level, with emphasis on those subsystems with low predicted reliability. Improvements in the reliability of these subsystems then have a major impact on overall system reliability. Subsystems that cannot meet the reliability requirements are then subjected to more extensive reliability development testing or a major redesign effort. These tests should be integrated with other tests to avoid duplication and to make effective use of limited test resources.

Reliability growth testing typically begins with early engineering models and is especially important for new products. The environment for the reliability growth test is often more stringent than that expected during normal use, to ensure that any critical problems are identified as early as possible.

In planning the TAAF test program, the predicted reliability should be greater than the required level. Tracking of reliability growth begins with initial laboratory testing and extends through operational field testing. Measures for different levels of risk can also be established, as shown in Figure 6.1. The levels shown are the recommended levels of the U.S. Department of Defense Directive 4245.7-M (DOD, 1985).

In addition to the results of the reliability development tests, other development and operational tests are used to assess system reliability. Sufficient hardware must be dedicated to this function to realize a fast growth curve.

The reliability of a new program can be planned by estimating a starting point (usually 10-20% of the predicted value for a complex system) and an estimated reliability growth rate. Using the estimated values, the total test time needed to reach the specified MTBF requirements can then be estimated.

6.7.1   Reliability Growth

Reliability growth derives from the premise that as long as a successful reliability improvement effort continues on the design, the design's reliability will improve. As mentioned earlier, one major competitive goal is to increase reliability using reliability growth. The concept of reliability growth is based on a model developed by J. T. Duane.

Reliability growth is measured by an increase in a reliability index such as the MTBF, approximately proportional to the square root of the cumulative test time. When plotted on log-log paper, with MTBF as the ordinate and cumulative test time as the abscissa, reliability growth approximates a straight line. The slope of the line is a measure of the growth rate. Although the

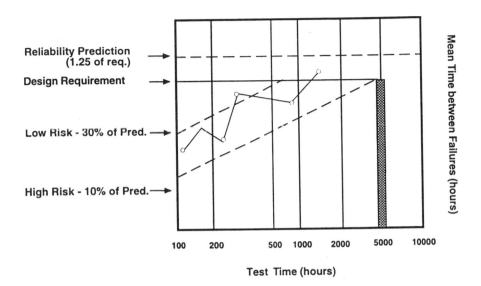

FIGURE 6.1 Reliability growth planning model.

Duane Curve has focused on hardware; the concept for software is also true. Growth rates have typically varied from 0.1 to 0.6. The rate varies based on how aggressive, well planned, and fully executed the test program is.

Reliability growth requires TAAF's iterative design improvement process. A product is tested to identify failure sources, and further design effort is spent to correct the identified problems. The steps are to 1.) Investigate all possible causes, 2.) Identify and correct the problem and 3.) Prevent any reoccurrence of this or a similar problem. The rate at which reliability grows during this process is dependent on how rapidly the sources of failure are detected, and how well the redesign effort solves the identified problems without introducing new problems.

A technical parameter that can be determined from a reliability growth curve is the instantaneous reliability for the design. The instantaneous or current reliability is the system reliability that would be obtained if the test program were stopped at that point in time. Periodic reliability growth assessments can then be made and compared with the planned reliability growth values.

Practical Suggestions for Reliability Growth Testing

Using years of experience, Martin Meth (1994/93) published an excellent list of suggestions for reliability testing. Some of his rules include:

- "Duane Reliability Growth Curve" is a rule of thumb. There is no underlying mathematics related to physical phenomena that would explain the way the Duane growth curve works.
- On the average, the effectiveness of design corrections to improve reliability will be 70%. (i.e. 30% of the time the correction does not fix the problem or adds another problem to the design! This is especially true for software.)
- Predicted reliability should exceed the design requirements by at least 25%.
- Thermal cycling is the most successful method for uncovering unanticipated hardware failure modes.

6.8   SOFTWARE TEST AND EVALUATION

In a typical software development project, approximately 20-30% of the elapsed time is devoted to the integration and test phase. Module-level testing during the code and debug phase can take a significant portion of the programming time. As systems become more complex, the software becomes extremely difficult to test adequately. No single technique exists to thoroughly test software. The best approach is to use several test methods continuously throughout design and coding. Documentation is extremely important for effective testing.

There is no way to test all possible software paths for a complex system involving immense logic complexity. Some of these paths are eventually exercised after the system is produced and some legitimate problems occur because they were not tested specifically. Many past studies have illustrated how the cost of correcting a software error multiplies if the problem is not found early in the design effort. The goal is to find errors by executing the program(s) in a test or simulated environment. General testing rules are described in Table 6.3 (Myers, 1976).

The test effort should demonstrate that the product satisfies the customer, meets the design requirements, performs properly under all conditions, and satisfies the interface requirements of hardware and other software components

The product is validated using the scenarios from the early design phases (see chapter 3). Since each scenario provides a complete description and can be immediately prototyped, this allows user validation to begin very early in the development process. The validation process also uses scenarios for documentation, training, and user manual purposes. Feedback is gathered from users as they learn to use the system. Problem issues are efficiently defined within the scenario task context and task ontology. Problems can be automatically traced back to the appropriate design area. Analysis of the system by users may result in new tasks or scenarios, which can be described in user

TABLE 6.3 Testing Rules

1. A good test is one that discovers an undiscovered error.
2. The biggest problem in testing is knowing when enough testing has been done.
3. A programmer cannot adequately test his or her own program.
4. Every test plan must describe the expected results.
5. Avoid unpredictable testing.
6. Develop tests for unexpected conditions as well as expected conditions.
7. Thoroughly inspect all test results.
8. As the number of detected errors increases, the probability that more errors exist also increases.
9. Assign the most creative programmers to the test group.
10. Testability must be designed into a program. Do not alter the program to make tests.

Source: Adapted from Myers, 1976.

terms. The system matures iteratively through continuous user feedback and development iterations.

The task and subtask descriptions from the task analysis provide input for designing the validation testing process. Task descriptions, specialized domain models, application architecture, and the reference requirements have all contributed information for the application requirements. Task links and application requirements map back to each functional requirement. This mapping is as used as an audit trail. The task descriptions and the domain models undergo a requirements analysis validation and verification. The requirements are generalized versions of the user and expert definitions of what the system can do and how well it can do it. The requirements do not contain or specify quantitative values or states. It is merely the highest level of abstraction for defining the system's capabilities. Instantiated components and newly constructed components are verified independently and as part of the system.

6.9 ENVIRONMENT, ACCELERATED LIFE, AND HALT TESTING

Each design should be proven correct by complete operational and environmental testing. This should include part, subsystem, and system testing. Maximum stress tests are performed when practical. If this is done early, future failures can be identified and avoided. Testing includes worst-case operating conditions, including operation at its maximum and minimum specified limits. Although there may be some degradation of performance at these operating extremes, the design should remain within specified limits. Test stresses can include input signals, temperature, rain, shock, vibration, salt, and fog. All failures and unusual results are analyzed thoroughly and not credited to "random failures or circumstances." All failures have a cause, and this cause must be found before any corrective action, such as a design change can be taken. Design changes can include redesign, selecting different parts, or selecting new vendors.

Life tests determine the predicted life of a part/product and information on how the product will fail or wear out. It consists of taking a random sample of parts/products from a lot and conducting tests under particular use and environmental conditions. There are three types of testing to be discussed: life, accelerated life, and highly accelerated (HALT) testing. Testing starts at the lower part level and progresses up to module and then to product level.

Traditional life testing tests sample parts in a reasonable environment for the life of the design. When parts/products are very reliable over a long period of time, complete life testing may not be very feasible since many years of testing under actual operating conditions would be required. Owing to this length of time, the parts may be obsolete by the time their reliability has been measured.

An accelerated life test is usually used to resolve this problem of time. Accelerated life tests predict a product's life by putting the product under environmental conditions that are far more severe than those normally encountered in practice. This reduces the time required for the test since the higher levels of environmental stress cause the parts to fail more quickly. To properly conduct and analyze an accelerated life test, one needs to understand the relationships between use and environmental conditions and their effects on the physics of failure of the part. Models are often used that have been previously derived from the knowledge of these physical properties. These include the power rule model, the Arrhenius reaction rate model, the Eyring model for a single stress, and the generalized Eyring model, which can be found in references on testing. These models account for the relationships between the failure rate of a part and the stress levels under a specific range of values of the stress. Success of the procedure depends not only on a proper selection of the model, but also on how accurately the model emulates the actual part. For example, the power rule model is preferred for paper capacitors, whereas the Arrhenius and Eyring models seem to be the most favored for such electronic components as semiconductors.

There are two important assumptions made when conducting and analyzing accelerated life test results. The first assumption is that the severity of the stress levels does not change the type of lifetime distribution, but does have an influence on the values of the parameters associated with the lifetime distribution. The second assumption is that the relationship between the stress level and the design parameters that results in failures is known and valid for certain ranges of the stress level.

The third test type is the highly accelerated stress test (HALT) where stresses are applied in steps well beyond the expected environmental limits until the part/product fails. The purpose is to detect any inherent design or manufacturing flaws. The process includes test, analyze, and fix (TAAF) to identify the root cause of each failure. With the failure information the team can improve the design or manufacturing process. The design team must identify which stresses are appropriate for testing such as vibration, temperature, cycling, current, etc.

All three tests are effective engineering tools, especially when normal life tests are not feasible. The key to success is corrective actions. Because of high levels of environmental stress, the team must ensure that unrealistic failure modes are not introduced. The part's physical properties and an analysis of the failure modes should indicate whether new failure modes have been initiated by the test itself. For example, increasing temperature beyond a certain level may change a material's characteristics, which would negate any information secured from the test.

6.10   QUALIFYING PARTS, TECHNOLOGIES, AND VENDORS

The purpose of qualifying new parts, technologies, software and vendors is to verify that a part or vendor selected for a design is suitable even for the most severe conditions encountered in normal use. Acceptance can be performed on a statistical sample to ensure that the design, workmanship and materials used conform to acceptable minimum standards.

During the early design and development phase, it is particularly important to evaluate whether vendors and subcontractors have the capability and qualifications to manufacture parts that will meet all requirements. Requirements as a minimum include documentation, performance, cost, quality, quality control procedures, reliability, schedule, packaging, technical risks, and management control. It is also important that critical items are identified and test plans developed for them, even though the actual assemblies may not be available for test until later.

The number of items to be tested and the type of environment used depend on experience, technical judgment, cost, schedule, and history of the vendor. The use of HALT or stress-to-failure testing can be very cost effective in ensuring an achieved safety margin or figure of merit (FOM) for any part tested. This type of testing requires an accurate realistic environmental analysis of the expected operational environment for the equipment being tested. The following are keys to successfully qualifying new parts:

- Identify the critical parts and vendors to be tested based on technological risk and other factors.
- Start testing as early as possible to identify and solve problems.
- Establish part specifications with qualification requirements in mind.
- Develop test procedures to describe the important parts of each test, and state explicit pass and fail conditions.

Beyond the typical tests for performance requirements, some tests that are commonly performed for electronic parts are seal test, thermal shock, vibration, moisture resistance, and acceleration tests. A brief summary of each of these tests is provided to give the reader a sense of the qualification process.

6.10.1   Seal Test

The purpose of this test is to determine the effectiveness or hermeticity of a part's seal. There are two general types of seal test:

Gross leak: a check for visible bubbles when the part is immersed and pressurized in a bath of fluorocarbon detector fluid. The detector fluid vaporizes at 125°C and, when trapped inside the part, emits a trail of bubbles.

Fine leak: a part is placed in a pressure vessel that is evacuated and back-filled with a tracer gas, such as helium or krypton. It is then pressurized to 60 psi for 2 hours. The test is done by a machine that evacuates a test chamber and checks for leakage of tracer gas.

Life testing helps to determine the effect of extended exposure for seals. In one test a small split was observed in the encapsulation after the 2000-hr operational life test at 130°C. The split was a result of surface tension caused by shrinkage during exposure to elevated temperatures. The design was changed to a molded package with increased thickness.

6.10.2   Thermal Shock Test

In the thermal shock test, the sample is subjected to repeated cycles of temperature extremes at high transition rates. This test helps to determine the part's ability to withstand exposure to temperature extremes and fluctuations. A common test may involve 10 cycles between -55 and +150°C. Common failure modes include the cracking and delaminating of the finish, opening of the seals and case seams, leakage of the filling materials, and electrical failures caused by mechanical damage.

6.10.3   Vibration Test

Vibration tests determine the effects of sinusoidal or random vibrations on a product. A common sinusoidal test for electronics is 20 g for 30 min. on each axis, with the frequency varying between 20 and 2000 Hz. A common failure for this test is broken solder fillets on the circuit board. In one example, the test revealed that the design allowed excessive movement around the center insert.

6.10.4   Moisture Resistance Test

A moisture resistance test accelerates the damaging effects of heat and humidity on a part. Temperature cycling between 25 and 71°C produces a "breathing action" that pulls moisture into the sample. In the vibration and low temperature subcycle of the test, the temperature is reduced to -10°C and the sample is vibrated at a 0.03-inch excursion for 1 min. on each axis. This accelerates the deterioration caused by moisture inside the sample. Common failure modes include physical distortion, decomposition of organic materials, leaching out of constituents of materials, changes in electrical properties, and corrosion of metals. For example, chip resistors can become detached from their leads after one cycle. To reduce the effects of stresses during thermal cycling, a method of providing strain relief was later placed on the leads during assembly.

6.10.5   Acceleration Test

The acceleration test is used to determine the effects of constant acceleration on electronic components or parts. It is useful for indicating structural and mechanical weaknesses and can also be used to determine the mechanical limits of the package or lead system. For example, a common test for small electronic parts is 70 g for 1 min. in each axis. A common test for large parts is 40 g for 1 min. in each axis.

6.11   PRODUCTION AND FIELD TESTING

Production testing includes all quality control tests including incoming testing and environmental stress testing. As the production cycle begins, incoming parts are evaluated to eliminate manufacturing defects and early part failures.

One special type of production test is called environmental stress screening (ESS) The goal of ESS is to test parts under stress conditions in order to transform latent defects into part failures that can be detected in testing. For electronic products, screening and conditioning are usually performed utilizing temperature cycling and vibration conditions. Testing for most commercial and consumer products are usually performed at the vendor's site and under less stringent conditions, because their environment is less severe. For commercial and consumer products, screening and conditioning criteria is based on an overall knowledge of the product. Owing to the cost-competitive nature of the commercial and consumer markets, product screening and conditioning are continuously tailored to maximize effectiveness. Initially, all products or new parts may be 100% screened and conditioned. As the product matures, the testing requirements (number and level) are reduced.

6.11.1   Part Testing To Improve Vendor Quality

As reported in the Best Manufacturing Practices Newsletter (Solheim, 1996), a project tested over 450,000 components during board assembly and system level testing. There were 50 components that failed and had to be replaced. The distribution of failures were integrated circuits (IC) -- 4%, resistors -- 20%, transistors -- 18%, capacitors and diodes -- 4% each. These figures related to the following parts per million (PPM): IC - 375 PPM, resistors -- 50 PPM, transistors --385 PPM, capacitors -- 20 PPM, and diodes -- 100 PPM. Within the 450,000 components, there were 646 different part types of which 30 part types were responsible for the 50 component failures. Pareto data analysis showed that one transistor part number was involved in 56% of the replacements.

This data is used to target those components and vendors responsible for the majority of the component rework. The data is further analyzed to determine which component vendor is responsible for which failures. This data is then used to develop an index. This index is combined with the vendor's quoted price to determine the total projected cost; thereby determining which vendor will be awarded future contracts. (Solheim, 1994)

6.11.2   User And Field Testing

The test to be discussed here is often the most important test, user testing. User tests are extremely important in determining how well the product will satisfy the customer and actually perform in the real-world environment. Customer opinions and perceptions are the final measure of success and quality. These are especially important in service oriented products. Tests are conducted with typical users under actual use conditions. These tests can also evaluate parameters such as aesthetics and identify problems that could not have been identified in any other tests. Since users (i.e. humans) will often take very unexpected actions, how the product or service responds to this non-predicted action is important.

6.12   NOTEBOOK COMPUTER TEST AND EVALUATION

The test program for the notebook computer would focus on technical risks such as new technologies, processes, software, and vendors. User tests would solicit feedback for parameters such as aesthetics, weight and size.

6.13   SUMMARY

A goal of every test and design review should be to identify areas for design improvement. Test and evaluation is a designer's tool for identifying and correcting problems. All tests should therefore be coordinated in an integrated and systematic approach to develop as much design information as possible. By combining and organizing the results from various tests, considerably more information is available without the costs of additional testing. This chapter has reviewed the key design practices and techniques in test and evaluation for improving reliability and producibility.

6.14   REVIEW QUESTIONS

1.   How would you distinguish prototyping from rapid prototyping and from virtual reality? What are the advantages of each?
2.   Describe how computerized design verification will continue to grow in the future.
3.   What is reliability growth and how is it accomplished?

4. What are life testing and accelerated life testing?
5. Discuss the trade-off between test and evaluation cost and product reliability growth.

6.15   SUGGESTED READINGS

1. T. F. O'Boyle, "Chilling Tale: GE's Woes with a New Refrigerator Show the Risks of Introducing Big Product Changes." Wall Street Journal.

6.16   REFERENCES

1. H.K. Bowen, K. Clark, C. Holloeray, and S. Wheelwright, The Perpetual Enterprise Machine: Seven Keys to Corporate Renewal, 1994.
2. Department of Defense (DOD), Transition from Development to Production. Directive 4245.7M, Washington, D.C., September 1985.
3. Jacobs, Software Computers etc., Automotive Production, p. 50, April 1996.
4. M.A. Meth, "Practical Rules for Reliability Test Programs", Reliability Analysis Center Journal, Vol 2, No. 3, 1994, (reprint) and International Test and Evaluation Journal, Vol 14, No. 4, December 1993.
5. G. J. Myers, Software Reliability Principles and Practices, Wiley, N.Y. 1976
6. T.F. O'Boyle, Chilling Tale: GE's Woes with a New Refrigerator Show the Risks of Introducing Big Product Changes, Wall Street Journal, 1983.
7. T. Peters, Do It Now, Stupid!, Forbes ASAP, August 28, 1995, p 170-172.
8. R&D, Research Speeds Development, Research and Development p. 14 March, 1996.
9. B. Solheim, Approach to Achieving 100 Parts per Million Program, Navy BMP Survey, March 1994.

Chapter 7

MANUFACTURING: STRATEGIES, PLANNING AND METHODOLOGIES

Manufacturing Can Ruin the Best Design

Manufacturing can mess up a good design with poor quality, missed schedules, and cost overruns. Today's manufacturing is facing increasingly formidable challenges caused by global competition, new technologies and electronic commerce. Manufacturing must quickly manufacture high-quality, low-cost, customized products. This is driving major changes in manufacturing strategies, methodologies and technologies. For example, lean manufacturing wants to accomplish this with less inventory, material movement, floor space, and variability. The entire product development team must understand and assist manufacturing to ensure that a design can be efficiently and quickly produced.

Best Practices

- Manufacturing Strategy
- Manufacturing Planning
- Producibility
- Process Development
- Manufacturing Qualification, Verification and Prototyping
- Design Release and Production Readiness
- Design to the Methodologies and Technologies of Manufacturing

BRACE FOR NEW MANUFACTURING STRATEGIES

Over the years, manufacturing has tried many strategies to improve lead-time, quality, cost and technical risk. One new strategy to significantly reduce lead-time for faster customer response is called by many names including make on demand, mass customization, flexibility, agility, robustness or nimbleness. Can manufacturing produce individualized/customized products quickly?

As reported by Stewar in Forbes Magazine (1992), "The theory behind flexibility is simple. If you and I are competing and I can read the market quicker, manufacture many different products on the same line, switch from one to another instantly and at low cost, make as much profit on short runs as on long ones, and bring out new offerings faster than you, or do most of these things, then I win." Product designers and manufacturing must react quickly to changes in customer needs. "Their focus: more and better product features, mass customization, flexible factories, expanded customer service, and rapid outpourings of new products (Stewar, 1992)".

Another major strategy is outsourcing and vendor partnerships. Having other companies produce more of the parts and services is one method to reduce costs and free resources for other activities. Vendors become partners in the product's development. Information sharing with vendors allows problems to be jointly solved. For example, Dell computer minimizes manufacturing by only assembling the computer. Other companies make all parts. Build and ship to order allows Dell to provide a quality product at a very competitive price without the costs of extensive manufacturing facilities or large amounts of inventory.

Other major strategies include lean manufacturing, electronic commerce, rapid prototyping, six sigma quality, mistake proofing, enterprise resource planning (ERP), Internet purchasing etc. As a result, design and manufacturing are starting a new era that will require even greater amounts of teamwork and producibility. The designer must understand and accommodate these changes as well as manufacturing's capabilities to ensure that a design can be efficiently produced

In the past, verifying that a product can be easily manufactured generally occurred too late, or was limited to a simple review by manufacturing or the vendor. This condition results in a design that too often requires:

- Unfamiliar, high-risk manufacturing techniques
- Very tight requirements and tolerances relative to the capability of the selected manufacturing process
- Poor choice in vendors that cause schedule and quality problems

When producibility is considered throughout the design process, easier to manufacture designs will result. This chapter will review the various strategies, functions, and methodologies of manufacturing and how design can help.

7.1 IMPORTANT DEFINITIONS

Manufacturing strategies are the vision and framework for accomplishing long-term corporate goals. This framework helps to focus manufacturing goals and provides plans for integrating the necessary functions and resources into a coordinated effort to improve production.

Manufacturing planning is the roadmap that identifies the approach and tasks for all critical paths between design, production, and the tasks necessary to ensure a successful transition from design to manufacturing. Since manufacturing planning continues in effect throughout an entire program, it becomes the heart of the front-end production effort and the road map for the establishment of all production specifications. As the design develops, the comprehensiveness and thoroughness of the plan will increase.

Producibility is a discipline directed toward achieving design requirements that are compatible with the capabilities and realities of manufacturing. More specifically, producibility is a measure of the relative ease of manufacturing a product in terms of cost, quality, lead-time, and technical risk. Design for producibility is often called by other names, including manufacturability, design for manufacturing, design for automation, design for robotics, and design for production. Regardless of the terms used, designing for producibility is the philosophy of designing a product so that it can be produced in an extremely efficient and quick manner with the highest levels of quality. This is accomplished through an awareness of how design decisions affect the production process, including the capabilities and limitations of specific production equipment.

The four basic manufacturing metrics used in industry are:

1. **Cost** (cost per unit produced, productivity, inventory costs, facility costs, labor costs)
2. **Quality** (yield, non-conformances or defects per unit produced)
3. **Manufacturing and vendor lead times** (time to produce a product)
4. **Technical risks** (number of new processes, technologies, requirements and vendors)

7.2 BEST PRACTICES FOR MANUFACTURING

Product development teams must have manufacturing capabilities in mind. In the past, not enough of the team's attention was placed upon the processes of manufacturing, inspection, test, and repair. The best practices for incorporating manufacturing considerations are as follows:

*   **Manufacturing's strategies and the company's business environment** are considered when developing a product's design.

- **Manufacturing planning** delivers a roadmap that identifies in detail all tasks necessary for successful production, and focuses on the specifics of the required manufacturing process.
- **Producibility techniques** are used by designers to develop design requirements that are reasonable.
- **Process development** is performed concurrently with the design when new or unique manufacturing processes or requirements are needed.
- **Manufacturing process qualification and verification** uses prototypes to ensure that all manufacturing processes and procedures are capable and verified before production begins.
- **Design release and production readiness** are based on technical issues and are thoroughly documented for production.
- **Design to the many methodologies and technologies of manufacturing** to ensure they are compatible with available manufacturing capabilities.

7.3   MANUFACTURING'S STRATEGIES AND THE COMPANY'S BUSINESS ENVIRONMENT

Most successful companies were initially formed around an innovative or superior product design. This trend resulted in a situation were many companies considered design and marketing the company's most important functions and, therefore, received the most attention and resources. When the United States had superior manufacturing capabilities in the 1950s and 1960s, management could neglect manufacturing and still be successful. Manufacturing was treated as a service organization and evaluated in the negative terms of poor quality, low productivity, and high wage rates. During this time, manufacturing was not expected to make a positive contribution to a company's success. Japanese success in manufacturing higher quality, lower cost products show the error in this judgment. Unfortunately, large capital investment alone cannot immediately correct problems caused by years of neglect. Improving a company's manufacturing capabilities is a difficult long-term process that requires considerable reserves of both expertise and capital.

One of the most popular new strategies in manufacturing is called "lean manufacturing". Perfected by Taichi Ohno, this strategy focuses mainly on the elimination of waste in all areas with a focus on inventory, work-in-process, material handling, cost of quality, labor costs, set-up time, lead-time and worker skills. There are several principles to lean thinking including eliminating waste, standardize work, produce zero defects, and institute one-piece flow. The method focuses on the Value Stream that is defined as the specific network of activities required designing, ordering, and providing a specific product, from concept to delivery to the customer (Womack, 1996) This flow should have no excess steps, stoppages, scrap or backflows. It also emphasizes the complete elimination of muda (waste, unnecessary tasks, etc.) so that only activities that create value are in the value stream. Timely responses are critical for this strategy to be effective. This provides a way to support value-creating action in the best sequence. Lean thinking provides a way to do more and more with less and less: inventory, human effort, equipment, time, space, while coming closer and closer to providing the customer with exactly what they want

Manufacturing strategies are the vision and framework for accomplishing long-term corporate goals. The vision establishes the company's goals for manufacturing. The framework helps to focus efforts on meeting manufacturing goals by planning for integrating the necessary functions and resources into a coordinated effort. Communication of this strategy sets the right climate for the teamwork and long-term planning that are necessary to improve manufacturing capabilities. The strategy should be well known throughout the company, with regularly scheduled reviews to monitor progress toward the goals.

Without a well-defined manufacturing strategy position, companies can too often look for short-term solutions that may prove detrimental in the long run. Long-range strategic plans allow sufficient emphasis to be placed on identifying and anticipating manufacturing technologies of the future. In this manner, manufacturing is prepared for new technologies with the expertise and equipment early enough to stay ahead of competitors.

A manufacturing strategy addresses the following concerns:

- Are future manufacturing technologies and requirements identified and essential expertise acquired early in development efforts?
- Is the manufacturing strategy compatible with long-range corporate objectives and factory modernization initiatives?
- Is there a long-term commitment for continuously improving manufacturing and vendor capability?
- Do the manufacturing, vendors and design functions interactively develop both product and manufacturing process designs?
- Are important vendors identified and long term partnerships established?
- Are the "make or buy" decision criteria/parameters for outsourcing established for determining whether to outsource or manufacture within the company?

7.4   MANUFACTURING PLANNING

The manufacturing plan coordinates the various production planning elements, such as production readiness and qualification. Without the benefit of thorough manufacturing planning, major problems will occur when a product is first produced. As shown in Table 7.1, this results in high rework and scrap rates, low quality, missed schedules, poor communication, cost overruns, and degraded product performance.

TABLE 7.1 Results of Inadequate Production Planning

Inadequate Planning	Manufacturing Problems
Producibility and manufacturing not involved during development	Unproducible design resulting in high production costs requiring a large number of design changes
Producibility and manufacturing issues not addressed in design reviews	Manufacturing planning may be seriously flawed owing to lack of information
Insufficient time allocated and no prototypes available for manufacturing qualification prior to production start-up	Major production start-up problems due to invalidated production processes
Critical factors of manufacturing process not identified or reviewed prior to production	Process, tooling, methods, and procedure problems occur in production start-up
Inadequate design documentation and test requirements	Quality problems caused by inadequate instructions
Vendors are not qualified or ready	Inconsistent quality and schedules from vendors

The manufacturing plan identifies the approach and details all tasks necessary for accomplishing manufacturing's strategies. This includes all critical paths between design and production and the tasks necessary to assure a successful transition from design to manufacturing. Since the manufacturing plan continues in effect throughout an entire program, it is the heart of the front-end production effort and the road map for the establishment of all production specifications. As the design develops, the plan will become more comprehensive and thorough. Although no standard plan exists that is adequate for all products, all manufacturing plans are concerned with meeting the cost, schedule, quality, performance, and environmental goals established for the product.

A manufacturing plan may consist of a single milestone chart that shows the interrelationships of the many facets of production, or a series of milestone charts with each having detailed subplans. To be most effective, the manufacturing plan should first be addressed during early product development and updated as the product design and customer needs change. This early manufacturing involvement is essential to assure that the product design is compatible with the overall production capabilities of the many fabrication, assembly, test, and procurement organizations that will ultimately be required to produce the product. An example of a manufacturing plan outline is shown in Table 7.2.

TABLE 7.2 Manufacturing Plan Outline

Product definition and requirements planning:
    Product requirements and configuration
    Procured technologies and vendors
Product schedule and quantities:
    Product development and release schedule
    Production schedule and quantities
Product procurement and supply chain approach:
    Design guidelines and standards
    Make parts in house or buy from vendors (i.e. outsourcing)
    Vendor benchmarking and selection
    New technologies/vendors/services required and qualification plan
    Quality control
    Supply chain requirements for shipping, packaging, and environmental issues
Manufacturing processes and prototypes:
    Processes required and capabilities (precision and quality)
    Qualification plan including prototypes
Manufacturing functional plan:
    Product cost (breakdown)
    Processes and equipment utilized
    New process development
    Make or buy criteria and decisions
    Methods, training, skills required
    Manufacturing capacity and facilities
Test functional plan:
    Comprehensive test plan

7.5   PROCESS DEVELOPMENT

As noted earlier, producibility is a discipline directed toward achieving design requirements that are compatible with the capabilities and realities of manufacturing. All design requirements are reviewed to ensure that manufacturing can meet them at a reasonable cost, with high quality and low technical risk.

The steps for product development to ensure effective manufacturing are to:

1. **Identify the company's strategies** and detail how they affect the design process and product design.
2. **Comprehensively study the current business environment, competitors and new technologies** and incorporate this knowledge into the design process and product design.
3. **Develop a detailed manufacturing plan** that includes key vendors and global partners

4. **Identify manufacturing and vendor processes** that need to be developed or improved
5. **Establish a comprehensive producibility program** that includes design guidelines for manufacturing's capabilities, methodologies and processes and process qualification

Effective communication and involvement between designers, vendors, and manufacturing starts early in the design process. For the lean thinking strategy, Value and the Value stream are defined with the key emphasis placed in product and process simplification. Often, new processes must be developed or existing processes must be significantly improved. Just as design must continually use new technologies to stay competitive, manufacturing must also develop new processes and technologies to stay competitive and support new design technologies. Having the design based on existing manufacturing processes insures a maximum degree of standardization in product designs and manufacturing processes. Process development is similar to product development except that the "product" is a new manufacturing process, method, equipment, or technology. The methodologies of product development are used in process development. This process is illustrated in Figure 7.1.

7.6   PROCESS QUALIFICATION AND VERIFICATION

Before production begins, an important task is the qualification of vendors, methods, tools, software and processes. This process is used to assure that manufacturing is ready. Each manufacturing process and vendor is reviewed and verified with respect to its adequacy to support the program objectives. Prototypes are used to test each manufacturing process for performance parameters such as tolerances, cost, cycle time, and quality. Documentation such as production procedures and test requirements are verified. The process begins with the manufacturing plan. The process of qualification and verification requires the coordinated efforts of all areas of the program team.

7.7   DESIGN RELEASE AND PRODUCTION READINESS

One of the major milestones of any program is when the design of the product is stopped and the design is released to manufacturing. The major concern of design release is whether the design is "mature" or "production ready". Many companies have guidelines that control the point in time when the design is released to manufacturing. This point can, however, vary quite a bit by a company's size and the design's complexity, technical risks, and schedule requirements.

One shortcoming arises when a design is released according to predetermined schedule requirements rather than its technical progress (i.e., assessment of the design's maturity). Setting unrealistic completion dates in order to increase sales or please upper management is the major cause of this problem. This situation has people scrambling to meet unrealistic deadlines, which cause them to deviate from procedures, thus increasing the probability of major problems.

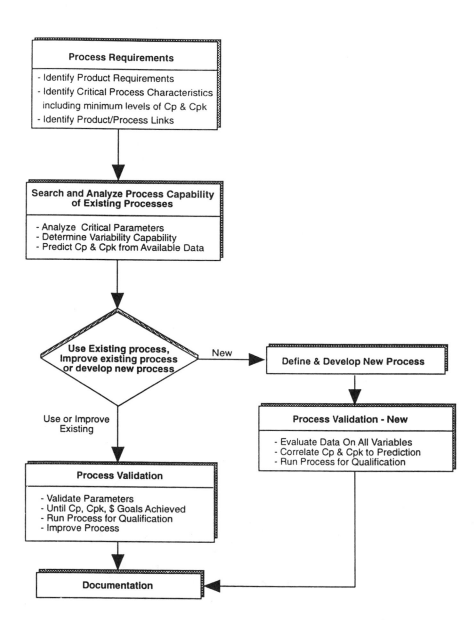

FIGURE 7.1 Manufacturing process development.

Technical progress and the risks involved with initiating the manufacture of the product are the criteria that should determine the design's release date. When the program involves the production of hardware or software that has been developed and manufactured before, a historical database can be used to schedule a design's release date. When the program involves a product that is entirely new, however, management must closely monitor the technical progress of the design and manufacturing process.

Understanding the technical risks involved allows management to schedule realistic delivery and design release dates. A checklist for evaluating the production readiness for fabricated parts includes:

- Is the part producible (i.e. does it meet Cp, Cpk, cost, and schedule requirements)?
- Are design and test drawings complete and comprehensive?
- Is the tooling package complete and in place?
- Are all shipping, packaging, and environmental issues documented?
- Have purchased materials and parts been analyzed to insure the best is used? (i.e., cost, vendor, producibility, environmental, etc.)
- Have alternative processes or vendors been evaluated?
- Do any scheduling obstacles exist?
- Has all software been tested?
- Were prototypes used to verify processes?
- Have experienced personnel been solicited for input into the documentation and producibility of the design?

7.8   DESIGN TO THE METHODOLOGIES AND TECHNOLOGIES OF MANUFACTURING

A successful product development team understands current manufacturing methodologies and technologies, how to design for them, and where the required expertise resides. The state of the art in manufacturing technology is changing so rapidly that everyone must continually stay abreast of the changes to insure the producibility of each design. This is extended to everyone in the supply chain (i.e. designers, vendors, partners, logistics, etc.). In this section, several manufacturing methodologies and technologies are discussed with emphasis on producibility i.e. how the design can influence its successful implementation. The areas for further discussion are listed below.

- Electronic commerce and web purchasing
- Computer aided design and manufacturing
- Production documentation and procedures
- Computer-aided process planning

- Ergonomics, human engineering, mistake proofing, and Poka Yoke
- Material requirements and enterprise resource planning
- Inventory control and just in time
- Production systems control
- Quality control
- Group technology and cellular manufacturing
- Automation and robotics
- Test and inspection
- Vendor partnerships and supply chain management

7.8.1 Electronic Commerce and Web Purchasing

Electronic Commerce (EC) is the paperless exchange via computer networks of engineering and business information using e-mail, Electronic Funds Transfer (EFT), and Electronic Data Interchange (EDI). Electronic Commerce is dramatically changing the procurement process and it will eventually impact all suppliers and vendors. The advantages for are to reduce lead time, information transfer paperwork, provide better configuration control among multiple entities and get lower prices due to more competition based on more companies bidding on work. The Internet allows new and more worldwide suppliers to bid on contracts. Requests for bid can be sent to thousands of vendors saving time and money. One company reports savings of over 20%. The automotive companies have already started their own web supply systems, which will become a multi-billion dollar marketplace.

For example, Internet file transfer of design CAD data has already become a part of the printed circuit board production cycle at many companies. The factory electronically receives the data and then immediately manufactures the printed circuit board. Some companies solicit price quotes and bids from vendors. Another example of EDI is where electronic transactions are combined with manufacturing inventory and planning information to have parts directly delivered to the factory floor. As technology and the Internet evolve, file transfer methods will continue to improve.

7.8.2 Computer-Aided Design and Manufacturing

The most common problem in manufacturing is ineffective communication between design and the various production departments. Previously, most communication was accomplished through drawings, design reviews, and company design guides. The computer aided design and computer network environments are beginning to provide a means of improving communication between production and design. The connected factory allows manufacturing to continuously communicate with design and evaluate design choices or criteria. This is especially important when design and manufacturing are located in different areas. Suggestions on how design can help manufacturing are summarized in Table 7.3.

TABLE 7.3 Design Helping Manufacturing

Problems/Factors between Manufacturing and Design	Design Solutions
Software compatibility	Selection of design and manufacturing software that can be easily interfaced with Web and each other's systems including ontology.
Communication access	Network bulletin boards, E-mail, easy access to design and manufacturing databases, CAD terminals in manufacturing etc.
Communication quality	Accurate documentation
Understanding and visualizing parts	Visual simulation, solid modeling

A complete computerized approach for communicating between design databases and manufacturing databases is not currently viable, except in some applications, because it requires a greater level of software integration than currently exists in industry. Future computer-aided drafting, design, analysis, and manufacturing tools will enable the electronic transfer of data for rapid development of problem solutions. Data standards will allow seamless and efficient electronic interchange of business and technical information. Its future success is highly dependent on how well company procedures and design practices can be adapted to this concept.

In order to computerize producibility analyses, a complete understanding of the manufacturing process is necessary. Decision processes must be translated into logical representations and then developed into models that emulate the decision-making process. The software can make these decisions only when all the information required for consideration is available in a computer-readable format. This points to the need to have accurate design and manufacturing data. A challenge in future years will be for the manufacturing and product design functions to work together to define requirements and develop common integrated databases.

7.8.3   Production Documentation and Procedures

Correct production procedures, such as assembly and manufacturing instructions, are essential to assure that manufacturing methods are consistent and meet all design requirements. A lean manufacturing system must have a detailed understanding of the manufacturing operations such as the product types, methods, and processes. Instructions must be thorough and descriptive, agree with the drawing, and contain all necessary processes and tooling references required for product compliance and consistency. Design documentation provides the basis for production procedures. If the design documentation is incomplete, inaccurate, or confusing, the production procedures are more likely to be wrong. Additionally, a system for controlling changes to these instructions is needed to ensure configuration control procedures are existent and adequate. This task is especially critical when new technologies or unique processes are required. Suggestions on design's role in helping manufacturing are:

Problems in Production Procedures	Design Solutions
Procedures do not convey correct information to production workers	Designers provide complete and effective documentation in a timely manner
	Designers review manufacturing procedures for technical correctness

7.8.4   Computer-Aided Process Planning

Process planning is where product designs are transformed into a step-by-step manufacturing plan. A process plan includes a list of the manufacturing processes from the raw materials and vendor parts to the finished product. Most companies depend on experienced manufacturing personnel to create process plans based on their specific manufacturing expertise. The process plans are either handwritten on paper or are simply keyed into a computer file. In recent years, computer-aided process planning has increased the overall productivity and quality of the process planning function. The full benefits of computer-aided process planning are realized when the manufacturing's capability database is comprehensive, accurate and available. Information should include cost, availability, precision, error rates and types, and functional properties.

Computer-aided process planning, as it exists today, utilizes either a group technology approach or a generative approach. The group technology approach involves designing a general process plan for a specific part family or redesigning an existing process plan to include a new addition to the part family. The generative approach utilizes decision logic processing to generate a unique process plan based on specific product design parameters.

When group technology in design is implemented, the designer retrieves an existing design that is most similar in manufacturing characteristics to the new part to be designed. The new design is then based, as much as possible, on the existing design's manufacturing processes. This approach will help to insure a maximum degree of standardization in product designs and manufacturing processes. How design can help process planning is:

Problems for Process Planning	Design Solution
New or unique requirements and processes or new process sequences	Design requirements are compatible with proven manufacturing processes and methods

7.8.5   Ergonomics, Mistake Proofing, And Poka Yoke

Most manufacturing, assembly, inspection, and test operations are performed manually. Human elements will continue to dominant production in the future, except in the cases of high-volume, simple to automate, or products having requirements that require automation. Manufacturing must ensure that the workplace is designed properly; test ensures that sufficient test capabilities are provided; and quality ensures that proper controls are in place. The methodology for properly designing a workplace for humans is called human engineering, ergonomics, or human factors.

An important design method for improving quality by reducing human errors is called mistake proofing or Poka Yoke. The Japanese popularized this technique. The concept is that a manufacturing process and/or product is designed so that an untrained person can correctly manufacture the product in only one-way. Specific design considerations include the provisions of guide pins to ensure proper alignment and go/no-go part fits. Suggestions on design's role in helping manufacturing to reduce human error are shown below. This technique will be discussed in greater detail in a later chapter on simplification.

Human Problems	Design Solutions
Operator errors, poor quality, and fatigue	Mistake proofing and manufacturing workplace design
Many operator injuries especially back strains and carpel tunnel problems	Human engineering, mistake proofing, and poka yoke guidelines to minimize the effects of manufacturing variability
Production operators can perform tasks in more than one way resulting in many human errors	Mistake proofing and poka yoke design only allows parts to be assembled or manufactured in one way

7.8.6   Material Requirements and Enterprise Resource Planning

Planning, scheduling, producing, purchasing, and controlling all the parts and materials used in manufacturing including vendors and suppliers is an extremely complex process. Due to the tremendous number of parts and materials used by most companies, formal planning and control systems are utilized. The process is referred to as material requirements planning (MRP) and enterprise resource planning (ERP). This is a key component for a lean thinking strategy, because it focuses on eliminating all waste in terms of lead-time, material, necessary storage space, money, unnecessary processes, and inventories.

An enormous amount of design and manufacturing data is required for MRP. Data from each part is stored in databases. The first is the item master file, which contains information about the part, such as description, source (e.g., made or purchased), unit of measure, lot size, cost per unit and lead-time stock balance. Another is the product structure file, which contains the computerized version of the engineering drawing parts list with additional manufacturing information.

Programmed logic evaluates the schedule and decides when parts will be needed. The program utilizes available inventory and offsets the release of orders by the necessary lead-time. A total of seven parameters for each part number are used to develop the plans, and each must be continuously updated:

- Gross quantity
- Gross need date
- Available inventory
- Open order quantity

- Open order due date
- Order quantity
- Order due date

ERP expands this information to include all aspects of the entire enterprise such as management. The design team plays a major role in the success or failure of MRP and ERP. Since both are involved with the precision scheduling of thousands of different products and parts, its success depends on the quality of the information in the system. Product delivery schedules should be correct and up-to-date if the resulting schedules are to be of any use. Engineering drawings must be released in time to support manufacturing lead times. Since the lowest levels of the product structure and long lead items are usually manufactured or purchased first, the design team should concentrate its efforts on designing and releasing these drawings first. Because of the large number of parts in the system, last-minute design changes or additions create major problems and usually result in schedule delays.

Another design decision that directly affects MRP/ERP is the selection of materials, components, and vendors. Since the logic attempts to use available inventory as the first source of supply, selecting standard parts allows the system to use available inventory wherever possible. Even when the inventory level of a standard part is too low, the system will already have historical information such as purchasing specifications for determining realistic lead times and identifying multiple suppliers.

In contrast, when new parts are selected, specifications must be developed and lead times are only estimates. Choosing parts from unknown vendors also introduces scheduling risks into the process. A summary of the designer's role is:

Manufacturing Delivery Problems	Design Solutions
Incorrect delivery schedules	Correct, up to date schedules
	Release drawings on schedule
	Minimize the number of design changes
	Minimize the number of vendors
Risks of delivery delays	Use proven and preferred vendors
	Release drawings on schedule
New parts or vendors not in database	Use standard parts or vendors

7.8.7   Inventory Control and Just In Time

One of the major goals in all companies is to reduce inventory because it is so costly to maintain. Just in time (JIT) is a phrase that refers to several manufacturing techniques that have proved to be very successful. Other terms are Kan-Ban, zero inventory, demand scheduling, and pull through scheduling. These processes represent the state of the art in the reduction and control of inventory.

Just in time is an inventory philosophy that minimizes the storage of parts, work in process and finished products by manufacturing a product or ordering parts only when it is needed, not before. This reduces inventory and work-in-process levels, but the improved quality effect is perhaps the biggest benefit offered. As work-in-process and inventory levels are reduced, quality problems are identified and corrected earlier. This results in improved quality. The responsibility for quality is placed on each person. This is true from the raw material vendor up to the final shipping clerk. With reduced inventories between operations, the feedback on quality problems is instantaneous and results in a production line stoppage. Production problems quickly surface and must be solved.

Anything so promising and simple must have a catch. In the case of JIT, implementing this type of production system is difficult. Companies that have experimented with JIT concepts have had various degrees of success. Some have tried to install the system into existing production operations and some have tried to utilize Japanese-trained managers. The most successful have set up new operations with trained managers and production workers. When Hewlett-Packard implemented JIT in their Fort Collins Systems Division, reductions of 75% in work in process and 15% in space requirements were found (Editor, 1985).

The designer can greatly reduce a company's inventory costs and assist in implementing just in time. The major problems that usually result in the inventory costs are shown below. The designer can play a great role in reducing these costs by putting the suggested design solutions to use.

Problems Resulting in High Inventory Cost	Design Solutions
Large number of different parts and raw materials	Part and material standardization and minimize the number of vendors
Large number of parts	Part reductions
Storage of in-process parts and assemblies	Part families and standard designs to utilize benefits of group technology
Storage of partially completed parts and assemblies because of lack of parts or design changes	Verified and mature design not requiring design changes during production Realistic schedules, design releases, documentation, and deliveries

7.8.8   Production Systems Control

Production systems control is the process of tracking and controlling a part or product through manufacturing. The system collects manufacturing, labor, and material information as the product is manufactured. The design team has little impact on this system unless the method of data collection requires specific design features, such as bar codes on the product.

Information Problems	Design solution
Capturing data as system flows through the plant	Provide data collection information on the product such as bar codes or machine vision readable codes

7.8.9   Quality Control

Total quality management (TQM), total quality control (TQC), statistical quality control (SQC), and zero defects are methods that have caused a remarkable turnaround in quality. Designing a product right from its conception plays a major role in quality because it can reduce the cost of quality prevention, detection and appraisal. IBM has estimated that 30% of its product's manufacturing cost (i.e., the total cost of quality prevention, detection, and appraisal) arises directly from not doing it right the first time. Significant quality and manufacturability of design, the pursuit of zero defects, and the systematic stress testing of products during design and manufacturing can all contribute to lowering cost (Garvin, 1983).

Total quality management is much more than a slogan; it is a systematic process that involves integration of design, manufacturing, production workers, vendors and the traditional quality control groups in the manufacturing process. Under this concept, all groups are responsible for the quality of the manufactured product.

One company's TQM's focus is to (DOD, 1989):

- Emphasize continuous improvement of processes, not compliance to standards
- Motivate to improve from within, rather than wait for complaints/demands from users
- Involve all functions, not just the quality organization
- Satisfy the customer, not merely conform to requirements
- Use guides and target values as goals to improve on, not standards to conform
- Understand the effects of variation on processes and their implications for process improvement

Another method to improve quality is to meet international standards of quality such as ISO 9000 certification. In 1980, ISO set up international technical committees to address the issue of quality standards. This was prompted by demands on industry to justify their quality procedures and methods to national and international customers. The ISO 9000 standards set forth basic rules for quality systems, from concept to implementation, whatever the product or service. Compliance with these standards should ensure that a supplier has the quality procedures in place to produce the required goods or services.

The implementation of problem-solving techniques establishes a mentality that allows continual improvement in the quality of products and processes. One of the major changes deals with the operators on the production floor. Each operator becomes responsible for the quality of the part he or she produces, and the operator personally performs the required inspections. As a result, defective parts are identified immediately and corrective actions can be taken. Test and operators must be properly trained and they must have the necessary equipment. Necessary equipment may include templates, gauges, special machines, and special lighting.

Quality manufacturing needs active participation by the design department to help manufacturing to quickly resolve quality problems. Minor design modifications or better communication can often significantly improve quality. Some reasons for poor quality that are directly impacted by the design function are shown.

Production of a high-quality product requires a systematic procedure for resolving all quality problems found on the production line. The effective feedback of quality information to management, production, and design is critical if a company expects a continuous improvement in quality. Feedback alone, however, is not enough. The reasons for each quality problem must be identified and then addressed. This determined method of resolving quality problems is often called enforced problem solving. This method is only successful if the various design and manufacturing groups participate fully.

Factors Causing Poor Quality	Design Solutions
No communication with designers on manufacturing problems	Effective total quality control program quickly identifies problems, notifies design, and corrective actions are taken immediately.
Inadequate procedures	Complete and accurate design documentation. Designers review technical aspects of manufacturing procedures.
Untested manufacturing procedures and processes	Utilize standard manufacturing and test processes. For new technologies, evaluate and qualify the new manufacturing process.
Engineering change notices	Only mature and verified designs are released to manufacturing to minimize the risk of needing design changes later.
Designs that are difficult to produce and test	Design for producibility, repairability, and testability.

The first step in this process is to gather quality information from the manufacturing processes. After the raw data has been compiled, the design team must examine this information to identify problem areas. Some of the methods commonly used for examining quality data include:

- Process flowcharts
- Pareto charts
- Histograms
- Failure modes and effects analysis
- Scatter diagrams
- Cause-effect diagrams
- Ishikawa charts

- Failure analysis laboratory
- Quality analysis by outside groups
- Experimental design and tests
- Taguchi methods
- Random samples

For example, Pareto analysis classifies all problems according to their cause or reason. These are then summarized into tables showing the number of occurrences for each cause. The team places their highest priority on eliminating the causes for the highest occurrences. The Pareto principle is to "Focus on the important few problem causes." An example of a Pareto analysis of a notebook computer is shown in Table 7.4. Although this methodology requires extensive effort, no other quality technique is as effective as enforced problem solving.

TABLE 7.4 Pareto Analysis of Assembly Errors for a Notebook Computer

Human errors by type

     50% improper human performance of task

     25% omission of a part or task

     25% wrong part assembled

Human errors by part type

     25% labels and nameplates

     16% screws and fasteners

     16% keys and locks

     16% packs of documentation and software

7.8.10   Group Technology and Cellular Manufacturing

Group technology is a technique in which similar parts are identified, grouped together, and manufactured in a common production line environment. The purpose of group technology is to capitalize on similarities in manufacturing and design. These parts with common manufacturing characteristics are called part families. For example, a plant producing a large number of different products may be able to group the majority of these products into a few distinct families with common manufacturing process characteristics. Therefore, each product of a given part family would be produced similarly to every other product of the family. This results in higher production rates and greater manufacturing efficiencies.

There are three general methods of grouping products into families:

1. In visual inspection, knowledgeable engineers identify the families. This is the least sophisticated and the least expensive method.
2. Formal classification and coding systems are based on design and production data. This method is the most complicated and time consuming.
3. Production flow analysis uses historical production information to identify similarities in production flow.

After the families are identified, special production lines called cellular manufacturing or manufacturing cells are then developed to produce each family. Since the new cellular line treats each member of the family the same, many of the advantages of mass production can be implemented. The common production flow provides the basis for automation, quality improvement, and reduced levels of inventory.

The design plays a major role in the success or failure of group technology. Designers must be familiar with the different part families and their production lines to ensure that the design parameters are compatible with group technology. Group technology generally constrains the designer as to what processes can be used, tolerances, material selection, and part sizes. To ensure compatibility, many companies have instituted design reviews in their design process. Although certain design limitations may be imposed by group technology, the benefits of reduced lead times and improved quality should more than offset this inconvenience. A list of design solutions is shown below.

Problems in Group Technology	Design Solutions
Parts/Assemblies do not fit the manufacturing process or vendor capabilities of the cell	Design to the cell's manufacturing capabilities such as part size, materials used, tolerances, surface finish, hole sizes, etc.

7.8.11   Automation and Robotics

Automation is a major method for improving productivity and quality. Low labor costs and high repeatability for quality improvement make automation a major goal for manufacturing. Identifying automation opportunities early in product design is important. Designing a product so that it is easy to automate is difficult and must be integrated early in the design process. In addition, automation equipment tends to require long periods of time for design and development, which are sometimes longer than developing the product itself.

Fixed automation occurs in high-volume products when the automation is specifically designed for a unique product or manufacturing function. This approach is used in industries with products that are not changed very often. These systems are usually hardware oriented and require considerable cost and design effort when process changes must be made. When automation is designed for frequent modifications due to different products, the flexibility is usually provided by software. Robots are a special type of flexible automation that are becoming a major part of the manufacturing environment.

Producibility design guidelines are used to ensure that a design is compatible with the process being used. General manufacturing guidelines for robots and automation are discussed in a later Chapter, Producibility Design Guidelines.

7.8.12   Test and Inspection

The current emphasis on quality and reliability and the current competitive state of the international market have resulted in both greater visibility and increased responsibility for test and inspection. What tests need to be performed and what data to record and display are major decisions. A common trap is to assume that ingenuity in the design of manufacturing test and inspection equipment can compensate for design deficiencies. Many design techniques can significantly improve the testability of a product. Major trends in testing and inspection are:

1. Higher levels of product complexity and new technologies to be tested
2. Increased levels of automation used in test and inspection tasks
3. More comprehensive inspection and testing requirements
4. Lower cost and higher quality per tested and inspected function

7.8.13 Vendor Partnerships and Supply Chain Management

Manufacturers usually rely upon many other companies (i.e., outsourcing to vendors) to provide many of the major and minor components of their products. Many electronic companies make fewer than 10% of the parts that go into their product. The author worked on one project where purchased part costs constituted 95% of the product's total manufacturing cost! As a result, a company's success is highly dependent upon vendors, subcontractors, and suppliers. Vendors are likewise dependent upon the company for their continued business and growth. The key to a successful relationship is communication and teamwork. The goal is to change vendor relationships from an adversarial, cutthroat win or lose battle to a cooperative, win-win proposition. This is becoming increasingly harder with the use of more and more overseas vendors. Although cost may one reason for choosing an outside vendor, the primary reason is that in-house capability and expertise does not exist to produce the product.

Vendors can be recognized as an extension of in-house capability and treated as partners. This allows the company to gain access to world-class capabilities, share risks, and free up resources for other activities. Partnerships allow companies to share development costs, technical risk, logistics, and expertise. Effective and shared communication is essential for an effective partnership relationship with a vendor, supplier, or subcontractor, regardless of where the vendor is located. Information on product forecasts, sales, cost issues, inventory, etc. should be shared. The amount and type of communication will vary depending on the location and historical relationship between the contractors. When this relationship has been healthy, the flow of information will be frequent and more informal. When the relationship becomes antagonistic, the communications will, by necessity, be formal and less frequent. Although the free flow of information can be very effective in producing results, the lack of formalized systems to document contractual agreements may later result in misunderstandings.

Another form of communication is to provide continuous feedback on the overall program's progress to the vendor and subcontractor. The company should keep the vendor and subcontractor aware of long-range business projections to allow them to prepare for significant production rate increases or decreases.

The initial selection and qualification of a vendor, as well as the decision to establish one source or multiple sources for a particular item, is dependent upon many considerations. As a minimum, vendors should be committed to being the "best in class", willing to work in partnership, and have a working knowledge of statistical quality control. A strong positive correlation has been found between vendor partnering and effective product development. Considering market share, sales, and qualitative success criteria, 80 best-in-class companies were identified (Management Roundtable, 1996). Vendors who frequently partnership were more often identified as best in class. Some features of these best-in-class companies are listed below (Management Roundtable, 1996).

- The best companies were twice as likely to have a supplier permanently on site to collaborate on development "often" or "always".
- The best were twice as likely to have suppliers share some strategic planning and product planning with them "often" or "always".
- 35 percent of the best-in-class companies said that suppliers "often" or "always" give input on product plans, versus 15 percent for the rest.

The company's goal is to keep the number of vendors and subcontractors to a minimum. This practice reduces the resource management needed to maintain control and provides greater opportunity for cost improvement. Often, however, it is essential that additional suppliers be established for an item to assure sufficient production capacity, quality conformance, and pricing competition. Design solutions for vendors are:

Manufacturing Problems	Design Solution
Problems with	Use qualified or preferred vendors with vendors proven history
	Use proven designs and technologies
	Selection is based on multiple criteria
	Fewer vendors with longer-term contracts
	Share information with vendors
	Provide immediate feedback to vendors and jointly solve problems and improvement

Although vendor selection is unique in each individual case, the following factors need to be carefully examined in evaluating vendors:

- Commitment to be the best with a history of continuous improvement
- Company objectives and business viability
- Technical knowledge and viability
- Willingness to work in partnerships
- Level of manufacturing technology
- Current production and future capacity
- Capability of increasing production levels

- History of meeting delivery schedules
- Quality control procedures and qualifications such as ISO 9000
- Manufacturing process capability measures (Cp and Cpk)
- Cost analysis
- Technical risk analysis

7.9   SUMMARY

A challenge for today's industry is how it can adapt to the changing worldwide production environment. To the design team, this requires a continuing awareness of how to design for manufacturing. Each team member must understand the various manufacturing elements, how to design for them, and where the required expertise resides. In this chapter, these elements and techniques were discussed with emphasis on their influence on the design team.

7.10   REVIEW QUESTIONS

1.   List a number of well-known companies and describe their perceived manufacturing strategy. How well are they accomplishing their strategies?
2.   How can design help manufacturing for the different manufacturing methods and technologies?

7.11   REFERENCES

1.   DOD, Total Quality Management: A Guide for Implementation, DOD 5000.51-6, March, 1989.
2.   Editor, Certificate of Merit Awards, Assembly Engineering, June: 43 1985.
3.   Garvin, Quality on the Line Harvard Business Review, September: 65 - 75 1983.
4.   Management Roundtable, Preliminary Results from 1995-96, Best Practices Survey,.p. 9, 1996.
5.   T.A. Stewar, Brace for Japan's How New Strategy, Forbes, p. 63-77, September 21, 1992.
6.   James P. Womack and Daniel T. Jones, Lean Thinking, Simon & Schuster, p.16-28, September, 1996.

Chapter 8

SUPPLY CHAIN: LOGISTICS, PACKAGING, AND THE ENVIRONMENT

Support after Manufacturing

For many companies, supply chain and environmental costs surpass all other direct costs. Issues include packaging, shipping, service, environment, government regulations, customer rules, warehouses, local agents, partners, repair centers and global considerations. Product design must minimize these support costs while fully preserving the integrity and innovation of the product and having a minimum negative impact on the environment. Logistics or supply chain planning must be flexible since shipping, environmental and government parameters are constantly changing

Best Practices

- Supply Chain, Logistics And Environmental Trade-off Analysis
- Design For Logistics and Supply Chain
- Design For Service and Maintenance
- Design For Disassembly (DFD)
- Packaging Design
- Design For The Environment (DFE)
- ISO 14000

DESIGN FOR DISASSEMBLY

One new supply chain and environmental trend in product development is called design for disassembly (DFD). As described by Bylinsky in Fortune Magazine (1995), the goal of DFD is to conceive, develop, and build a product with a long term view of how its components can be effectively and efficiently repaired, refurbished, reused or disposed of safely in an environmentally friendly manner at the end of the product's life.

Bylinsky describes how disassembling old computers began a few years ago to retrieve precious metals like gold and platinum. Boards are sold to chip retrievers who resell chips to such users as toy manufacturers. "Computer makers that can reduce the number of parts and the time it takes to disassemble a PC will profit when the product, like a sort of silicon salmon, returns to its place of origin (Bylinsky, 1995)".

Two examples of DFD were identified by Bylinsky. One is Siemens Nixdorf's personal computer, the green PC41. It contains 29 assembly pieces versus 87 in its older model. The new computer also has only two cable connections, versus 13 in the old one. The new PC41 is assembled in seven minutes and can be taken apart in four. The older computer takes 33 minutes to put together and 18 minutes to take apart. This lower disassembly time reduces the cost of recycling (Bylinsky, 1995).

In another example, Kodak redesigned its disposable cameras to be recycled. "In the recycling center, the covers and lenses are removed; plastic parts are ground into pellets, and molded into new camera parts. The camera's interior, moving parts, and electronics are tested and reused up to ten times. By weight, 87% of a camera is reused or recycled (Bylinsky, 1995)".

Design for disassembly is part of extended product responsibility often called EPR. Returnable auto batteries, deposits on bottles and cans and returnable shipping containers are examples of considering a product's affect on the environment.

Logistics, supply chain and environmental issues are becoming more important in product development. For example, electronic commerce is drastically changing business methods. Orders can be placed in seconds but shipping can still take days. Using global vendors causes shipping to take weeks or more making scheduling even more difficult. Global packaging design must consideration harsher environments. For global products, providing logistic services in foreign countries is a major cost driver. Important logistic concerns include distribution, shipping, packaging, repair and support costs. Important environmental concerns include reduced generation, recycling, reuse, and effective disposal. This chapter will focus on design's role in supply chain, packaging, and the environment.

8.1   IMPORTANT DEFINITIONS

Supply chain is the "complete flow" of the product and includes all of the companies with a collective interest in a product's success, from suppliers to manufacturers to distributors. It is includes vendors, and their suppliers, manufacturing, sales, customers, repair, customer service, and disposal. It includes all information flow, processes and transactions with vendors and customers. Today's business climate is concerned more with developing equity relationships and forming joint ventures around the success of the entire supply chain rather than an individual company's gains or losses. A key to success is for everyone on the supply chain to have the latest and best information from everyone else.

Logistics is a discipline that reduces life cycle installation and support costs by planning and controlling the flow and storage of material, parts, products, and information from conception to disposal. Direct expenditures on transportation represent 17 percent of the U.S. gross domestic product (GDP) (Leake, 1995). Direct expenditures for industrial logistics are more than 10 percent of GDP. This forces companies to reassess their logistics methods and approaches in product design. For many companies, logistic and environmental costs surpass all other direct costs. Companies now realize the potential of streamlining their logistics, transportation, distribution, and environmental efforts. The metric for logistic trade-off analysis is cost and leadtime. Cost includes packaging, shipping, support-equipment design, technical publications, maintenance plans and procedures, spares provisioning, field repair services, training classes, facility engineering, and disposal. Leadtime can be the time needed to ship the product, service response, repair, etc.

Packaging design's purpose is to reduce shipping costs, increase shipping protection, provide necessary information, minimize the environmental impact and be safe. Packaging design is often a minor issue in the product design process and indeed it is often a last minute design task. Packaging design, however, can have a big impact on the logistic, reliability, environmental, and cost aspects of a product. A product's integrity (i.e., reliability) may be compromised upon delivery unless the package is able to properly protect the product during distribution and storage.

Environmental design's goal is to minimize a product's effect and cost on all aspects of the environment. Design goals include reuse, recycling, remanufacturability, disassembly, ease of disposal, use of recycled materials, using environmentally friendly manufacturing processes and selecting vendors with good environmental histories. Short-term environmental discussions include compliance to regulations and laws; whereas long-term decisions include environmental liability and anticipation of global environmental concerns. Environmental design is unique since governmental law determines most of the requirements and these can change quickly and radically. Design must anticipate environmental laws that will exist in the future. Environmental trade-off analyses traditionally use cost metrics including the materials, processes used, the wastes produced and the final disposal of the product. Other metrics are energy used, pollution produced, and amount of waste materials.

8.2   BEST PRACTICES FOR SUPPLY CHAIN, PACKAGING, AND ENVIRONMENT

Since the terms logistics and supply chain are used so interchangeable, this Chapter will use both terms. The best practices for supply chain are:

- **Supply chain and environmental considerations** are part of all trade-off analysis and incorporated early into the design of the product, it's manufacturing processes, packaging, vendor selection and other product related items.
- **Design methods** include:
  - Design For Supply Chain and Logistics
  - Customer Service and Maintenance
  - Design For Disassembly (DFD)
  - Packaging Design
  - Design For The Environment (DFE)
  - ISO 14000

8.3   DESIGN FOR SUPPLY CHAIN AND LOGISTICS

The purpose of design for supply chain is to ensure that the product design is a cost-effective, fully supportable system throughout a product's life. This is accomplished by designing the supply chain system concurrently with the product. As defined by Byrne (1992), quality in supply chain means meeting the company's cost goals and meeting customer requirements. Customer requirements and expectations including the following logistic design parameters:

- On time delivery
- Ease of inquiry, order placement and order transmission
- Timely communications about delivery
- Accurate, complete, undamaged orders and error-free paperwork
- Responsive post-sales support such as technical information, repair and warranty
- Commitment to environmental concerns including packaging and disposal

The contract supply chain industry is expected to more than triple in size over the next five years to $50 billion in annual revenue (Bigness, 1995). "This is a lot of money for an industry that is based upon moving things around." The world economy depends on the traditional and complex science of supply chain. Parts and supplies must be shipped into manufacturing plants on time. Finished products must then be distributed efficiently to customers with needed support (i.e., repairs and warranty) and disposal of the product and manufacturing wastes when no longer in use. Until recently, most companies handled both incoming and outgoing logistics. "This is rapidly changing as logistic costs become a larger portion of a product's total cost" (Bigness, 1995). Many companies are now using third party companies to handle all of their logistics.

The Internet and electronic commerce are also changing supply chain methods. Orders can be placed in seconds but the "physical delivery" of the product can take days. Delivery will become an important performance measure. Logistic design must consider how orders are placed, suppliers notified, entire orders are grouped together and packaged and then how the complete order is shipped to the customer. For products or parts with long lead-times, such as overseas products, effective methods for predicting order size for future demand must be developed. A key design concern is to ensure scalability of the supply chain process in order to meet changing levels of demand.

The foundation for the supply chain planning process is developed from design, reliability, maintenance, vendor and environmental data. One approach for analyzing this data is the logistics support analysis procedure. Logistic support analysis is a formalized technique used to include maintenance and supportability features into the design, identify quantitative and qualitative logistics resource requirements, and influence other design aspects such as requirements, packaging, and vendor selection. The primary objective of the process is to identify logistic design constraints and support risks and to ensure their consideration into the design. Additional objectives are the identification of required support resources and the coordination and integration of the efforts of the logistic disciplines in developing quantitative and qualitative logistic resource requirements. Like any design analysis, it is an iterative process to optimize system support criteria during the design process.

Logistics is a systems analysis discipline that covers the entire life cycle of the product, i.e., cradle to grave. The steps are to:

1. Identify customer, key supplier and shipping methods requirements and design/logistic capabilities.
2. Identify all viable design and support system alternatives and the risks associated with each
3. Perform extensive trade-off analyses to identify the "best" combination of product design and logistics approach

Alternative designs and support approaches are evaluated quantitatively and qualitatively, relative to their impact on operational readiness and availability. For the notebook computer, logistic trade-offs can include:

- Outsource logistic planning and repair activities?
- Whether to have a toll free technical phone support center.
- Number and locations of factory authorized repair centers with the types of repairs to be performed and test equipment needed.
- Level of built-in test (BIT) and their associated impact on warranty costs and reliability.
- Best methods for shipping, packaging, and disposing of the products. Use third party logistic company? Should retailers or warehouses install certain parts to increase packing density?

As shown in Figure 8.1, the process uses the results of many different analyses performed by various disciplines for evaluating the effects of alternative designs on support requirements and costs. Some logistic design considerations are listed in Table 8.1.

A determination must be made as to which support system satisfies the need with the best balance between cost, schedule, performance, and supportability. New or critical support resources must be identified. Global considerations include government regulations, customer rules, bonded warehouses, international shipping, and role of local agents, partners and repair centers.

Technology is one method for improving logistics while lowering costs. Some trucks in the USA have in-dash computers that incorporate voice activated speech recognition, navigational (mapping) systems, global positioning system (GPS), electronic logbooks, and telecommunications. The receiving system keeps track of all of the trucks every few minutes automatically. The intelligent system can monitor trucker performance and provide shipment data to the customer. In addition, for safety the system will recognize if the truck is too far off course or has been idle too long. When this occurs, the appropriate personnel can be contacted.

Supply chain includes everyone from suppliers to retailers. Outbound logistics is the distribution of the product and its support resources out to the user. Packaging, shipping, warehousing, and order processing are some of these tasks. Environmentally sound logistic practices encourage fewer shipments, higher space utilization, less modal transfers (e.g., truck to airplane), and shorter movements using environmentally friendly methods. A key to success is for everyone on the supply chain to have the latest and best information from everyone else. Forecasted schedules will be more accurate. This allows each partner to perform his or her tasks exactly when needed. The right part is

TABLE 8.1 Logistic and Supply Chain Design Considerations

Logistics Area	Design Considerations (Design for ...)
Facilities and support resources	Compatible with existing resources and minimizes their use
Shipping and transportation	Most timely, effective and efficient method and robust to shipping damage
Packaging and handling	Provide cost effective protection, Compatible with storage, shipping, and recycling requirements. Minimize space and weight
Installation	Easy or no installation, no required tools, test equipment, or special labor skills required
Training	No training required or compatible with existing skills and repair equipment, fixtures, software, and environmental, health and safety
Technical documentation	Design documentation is accurate, easy to read, complete and concise.
Warranty and repair	No maintenance or self maintaining if possible, easy to use phone or internet help service. spare parts, test equipment, and services are available for the future. Parts that will wear out are identified and easily accessible
Environmental issues	Plan for reuse and recycling. Select vendors with good environmental histories.

delivered at the right time and not before. For example, the supplier can have access to the customer's actual sales in real time.

Synchronized supply chain is where the entire supply chain (e.g. all vendors, factories, shippers, distributors, etc.) is modeled so that material flow and other constraints can be simultaneously simulated and optimized. This information can be used to cut out unnecessary steps (lean logistics), identify bottlenecks, and allow constant changes to the schedule.

Disposal and recycling for some countries are required. Logistic design must deal with disposal of waste and the reverse flow of collecting and transporting recyclable materials. This last part of logistics or supply chain is called reverse or inbound logistics. In this stage, materials and parts in the supply chain system are gathered and processed. This includes packaging materials and worn out products. It typically has many environmental issues and problems.

For example, Toyota Motor Manufacturing's USA Kentucky plant uses returnable plastic containers to ship most of its materials and parts from suppliers. The plastic containers and pallets meet specifications for maximum cube space use for truck trailers and also reduce the environmental impact on local landfills. The container system works well with the plant's just in time (JIT)

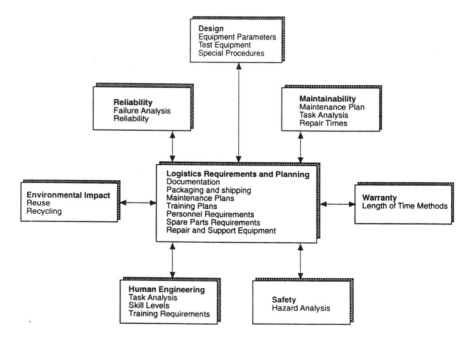

FIGURE 8.1 Logistics.

operation.

The process of exporting products for international trade can be a challenge. Typical questions that must be answered include (adapted from Patel, 1996; Onkrisit, 1993; Gordon, 1993; Pegels, 1987):

1. How do the goods move to the point of export? Are there cost, volume, or weight trade-offs? Should they be insured? How many products per shipment are effective?
2. Is an export or import license(s) or other government documentation required?
3. Who will handle preparation of the documents for the export shipment?
4. What packaging is needed?
5. What product, instructions, or packaging modifications will be performed at the export location? What resources are needed?
6. How will repair and warranty tasks be performed?
7. Can the firm show conformity to ISO 9000 quality standards and ISO 14000 environmental standards?

The product development team evaluates their particular application in order to answer these questions.

8.4 DESIGN FOR CUSTOMER SERVICE AND MAINTENANCE

Ongoing product support (also called customer service) must be provided to the customer to answer questions or resolve problems that may arise. Often the customer may not be able to properly identify the problem over the telephone. Good documentation is critical to this effort. Flowcharts of repair actions and lists of common failure mode and then corrective actions. When changes or updates are needed, the user or team needs enough documentation to be able to evaluate the feasibility of the requested changes. Global products will need special documentation so that customer service can respond to the unique aspects of different country's product uses, culture, climate, repair methods etc.

After the product is installed, the customer may find something that needs to be revised or improved. Changes in technology, hardware, or other software products may also require design changes. Since many software products are used for several years, upgrades or revisions are expected by the customer. In industry, legacy systems is a term for older software systems that require large amounts of resources to keep them performing properly. In many companies, the operation expenses for legacy systems are the largest budget item for computers. When changes are to be made, the software designers must ensure that the software discipline is followed while making those changes.

Software modification and extending its useful life is a major task that occurs after software is in operation. Costs of maintaining older "legacy" software products are consuming ever-increasing percentages of already tight departmental budgets. Pressman (1992) gives some depressing news: maintenance costs accounted for roughly 60% of the typical information systems department's software budget. Older languages often require a higher percentage of maintenance programmers than development programmers. Moreover, "the cost to maintain one line of source code may be 20 to 40 times the cost of the initial development of that line" (Pressman, 1992). Of the time spent performing maintenance, more of the developer's time is spent trying to understand the program or reverse engineering rather than designing new code. This shows the importance of good documentation.

8.5   DESIGN FOR DISASSEMBLY

Design for disassembly is a design discipline that ensures that a product can be taken apart quickly, efficiently, safely, in an environmentally friendly manner, and with a minimum amount of human and equipment resources. If a product cannot be disassembled quickly or cheaply enough, the environmental savings will be lost. For designers, it is very similar to design for assembly and repairability but in reverse. Some design considerations for disassembly include (adapted from Henstock, 1988):

- Disassembly should not affect or damage other products
- Part placement and accessibility methods should consider disassembly
- Fastening and joining methods should be easy to disassemble
  For example, screws are good but adhesives and welding are not
- Removal and recovery of liquid components and wear out components must be considered
- All plastic parts should be of the same recyclable material where possible
- Plastic parts should be identified by material type
- Easy disassembly should not encourage vandalism, theft, or cause potential safety problems
- Repeated recycling should not cause quality or safety problems

Software and methodologies are available to help designers such as products called DIANATM by POGO International (1996).

8.6   PACKAGING DESIGN

There are many different methods for packaging a product. For shipping, the outside box can be cardboard, wood, plastic, metal etc. Inside support can be foam, foam peanuts, cardboard dividers etc. For consumer products, labeling and styling become very important. The product's box (i.e. the box that the product is sold to the user in) may also be cardboard but many companies are going to shrink-fit plastic. Business to business parts can use returnable containers. Software products can be purchased in boxes or can be downloaded from the Internet. Internet packaging design includes how the customer can access the software to be downloaded, the number of customer options, security issues, customer help for downloading and what software should be included for downloading. Designing the best package is difficult. Formal guidelines should be developed for suppliers who define the requirements for their material packaging. The guidelines should include provisions to ensure compatibility with receiving methods, quick unloading, easy inspection and verification, efficient movement to storage or production, and direct use on manufacturing floor or repair use (Sclke, 1990).

There are six major purposes of packaging that are:

1. Cost
2. Protection
3. Communication and labeling
4. Convenience
5. Environmental considerations
6. Government and customer regulations

Package design must also meet governmental regulations and practice responsible environmental stewardship. The design of packaging must address several stages in the product life cycle including receiving of incoming parts, materials, and packaging materials, movement of parts and materials through out the plant, distribution or shipment of finished goods, unpacking, reuse or disposal of packaging materials, and distribution and shipment of repair and warranty parts

The best solution is to use no package if possible. This is only possible with a non-fragile item. A bag is used to package products whenever possible to minimize space. It may be a paper envelope with minimum dimensions with bubble wrap dunnage for added protection or it may be a clear plastic bag to protect the part from scratches until it is combined with other parts for final shipment. Shrink-wrap plastic is also popular.

8.6.1   Cost

Cost is usually the most important parameter in designing packaging. The customer buys the product for many different reasons but rarely is packaging a consideration. To lower cost, the design team tries to pack more product(s) in a smaller package, use cost effective methods, and minimize weight. When the product is designed for these parameters great savings can be realized. The best example of design for packaging is stackable products. Products such as plastic chairs that can be stacked one upon the other results in a very large saving in shipment size.

Package standardization reduces the diversity of packaging types to take advantage of purchasing quantity discounts and allows standardization of transportation, material handling, and storage. Standard packages must be traded off against package designs that are custom fit to the product to reduce void fillers and dunnage. Custom designed packages increase packaging density which requires less space, freight cost, and handling during transportation and improves utilization of storage facilities.

Domestic freight costs are calculated based primarily on weight with an occasional excess size charge. International freight costs are calculated by package dimensions and weight. International freight cost reductions are directly proportional to reduced package volume. For example, air freight is based on the greater value of either the weight or volume shipped, with rate breakpoints and minimums. Ocean freight is based on the greater of cubic meters or weight shipped, by the type of cargo. Airborne freight and air small package handling (e.g. Federal Express) have different and unique parameters that must be considered.

Package dimensioning should optimize material handling and consider the use of automatic movement and storage systems. These systems restrict packages to certain dimensions and weights. Automatic guided vehicle systems (AGVS) require that packages must be secure, stable, and sized within certain dimensions. Robots require precise part location and orientation. Automated electronic component placement machines have special package requirements such as tubes, pallets, or taped rolls. The packaging dimensions should also be compatible with other mechanization such as part feeding and material conveyor systems.

8.6.2   Protection

A product must be protected from conditions that may damage or degrade it. Packaging provides a barrier between the product and the environment. The product's fragility is defined according to the handling requirements in each stage of its distribution cycle. Obviously, the packaging should protect the product from being scratched, crushed, punctured, overheated, compressed, decompression, etc. Testing may be needed.

Some products have areas that can withstand much higher loads than other areas. Weaker areas may require isolation. For example, a circuit board with connectors will require the protrusions to be isolated. Delicate parts can damage each other by scratching, bending or breaking when placed in a bin unprotected. Some may be damaged if carried by an untrained material handler. The packaging should not be more protective than required, as excessive packaging designs can be very costly in terms of material, freight, space, and disposal.

The type of package will vary depending on the method of shipping used (i.e., mail, truck, air, or ship). Packaging should be sufficient but not excessive for protection during shipping. Labor costs will increase by requiring too much wrapping and unwrapping at progressive steps. The shipping and receiving groups can work together to optimize the packaging. Final packaging is only performed when the object will not be combined with other parts.

When a package is loaded onto a company truck for movement to another building, the company can control the handling methods used. This is not the case for external shipments. External shipping conditions vary dramatically. It is not uncommon for boxes to be dropped, scraped, crushed, and stacked to a height of ten feet. The package should ensure that the product will be protected during normal and rough handling, but not necessarily from gross mishandling. Internationally shipped cargo needs extra strong structured packing and containerization or unitization in order to have a better measure of protection.

External environmental conditions such as humidity and temperature have an effect on the product and the packaging itself. This is especially true for food and bio-medical products. The conditions to which the packaged product may be exposed should be defined and considered in the packaging material selection and design. Many products are subject to condensation in the hold of a ship. One method of eliminating moisture is shrink wrapping (i.e., sealing merchandise in a plastic film). Water proofing can include waterproof inner liners, moisture absorbing agents, or the coating of metal parts with a preservative or rust inhibitor. Desiccants (i.e., moisture-absorbing material), moisture-barriers, vapor-barrier paper or plastic wraps, sheets, and shrouds can also protect products from water leakage or condensation damage. Palletized shipments should be able to withstand being placed on a wet floor.

Boxes with appropriate dunnage should be used when the product is too fragile, large, or heavy for adequate protection in a bag. A box without an integral pallet may be shipped and stored with any side up regardless of printed instructions on the package.

There are many possibilities for dunnage including recycled cardboard pellets, polyethylene foam, polyether foam, polyester foam, polyurethane foam, convoluted foam, cellulose, single face corrugate, instapak foam, bubble wrap, quilted paper, etc. Whenever possible, recyclable dunnage is used. Paper based dunnage such as quilted paper or corrugate paper is often used because they have less environmental impact. Boxes should be made with recyclable materials and easy to break down for reuse or recycling.

Products should be adequately protected against theft. Methods for discouraging theft include shrink wrapping, sealing tape, and strapping. Patterned sealing tapes will quickly reveal any sign of tampering.

8.6.3   Communication and Labeling

Major issues in the communication function of packaging include information about the product, contents in the box, and methods for handling. A common type of information printed on the package is for marketing purposes. This is to help sell the product, such as the company name, product name, pictures, logos, styling, and advertising. This is often the box that the product will be sold in. For example, Barbie dolls are placed in their final box prior to being shipped to the store. Several product boxes are then placed in cardboard shipping boxes, which are then placed in shipping containers for overseas shipment. One marketing goal is to "catch the customer's eyes." Attractiveness and styling of the package can directly affect sales.

Designers must consider international requirements for shipping such as language and culture. For example, in Canada the French language on the label must be of at least equal prominence with the English language used on the same label.

Packaging communicates information about the contents of the package. The name of the product, ingredients or components and a list of hazardous materials are required by law for shipment in the United States. The quantity, size, and model or type of product should be identified for ease of receiving, inspection, transportation, and warehousing. Other countries will have additional information requirements. Products such as electronic products that are often stolen may be labeled in a code rather than the actual product name.

The weight and fragility of the contents are important information for anyone who may handle the package. Label information is used to convey instructions on proper orientation, lift points, weight, center of gravity, opening instructions, etc. Warning information is used to protect the product as well as the user. Labeling the side of a package with orientation instructions (i.e., This Side Up) does not guarantee it will be placed properly. Designing the package with an integral pallet will define the package bottom and increase the probability of proper orientation. The information must be readable for the various countries that the product will be sold and for the various reading skills of the people there. Multiple languages may be required. In many countries the users or workers may not be able to read. This makes the use of figures very useful. International symbols can be used to convey information. Personnel using different languages may handle the package. Labels may also contain safety issues such as antidotes in the case of human contact and clean up specifications for an emergency response team in the event of a spill.

8.6.4   Convenience

Design issues for convenience include safety, easy handling, and dual-purpose use of packaging. For example, features designed into a package can improve opening, storing, transporting, handling and disposing of the contents and/or packaging itself. Handles or lift points can be designed into the package for ease of lifting and to reduce accidents. Containers can be designed to nest in order to facilitate stacking and to minimize shifting during transportation. When possible, packages should be able to be opened without tools such as knives. Packages can fold up or collapse when empty to use less space during storage and transport and then reopen with a pull for reuse.

Designing for dual-purpose use is to design packages for uses other than product protection. One example is when the package can be reused for shipping other products or materials back to the vendor. Another example is where the product can be partially used the package should be suitable for storage after opening (e.g., paint, ink or glue). This reduces the loss and disposal of unused product. Another example, a piece of equipment uses a 75 pound power supply that is installed into the side at approximately eight inches above the floor. In production, a hoist is used to unbox the power supplies and install them, because the weight and posture required makes a manual lift an ergonomic risk. In order to change out the power supply in the field, a package was designed to double as a fixture for handling. The top and bottom of the package are identical and are used to position the power supply at the correct height off the floor. The supply can then be slid into the equipment without having to manually lift it out of a box and then hold it in the air eight inches above the ground for precision placement. The middle section of the package is a sleeve to complete the protective packaging unit for shipment. When a replacement power supply is sent to the field, the field service engineer removes the top of the package, inverts it, positions it next to the equipment, and slides the bad power supply out onto the fixture. The sleeve is then removed from the replacement part that is positioned next to the equipment, before sliding the replacement part into place. The sleeve is then replaced on the package top, which is now the bottom, and the bottom is used as the top.

8.6.5   Environmental Considerations

For many products, the product's packaging causes more environmental problems than the product itself. It is a source of litter, ground water pollution, depletion of resources, air pollution, and the greatest portion of solid waste. In 1984, 54% of the nearly 45 million tons of packaging used in the U.S. was paper based (Selke, 1990).

Many design aspects of packaging can maximize performance, cost, and environmental issues all at the same time. For example, a package design that minimizes the use of materials in turn minimizes weight and density.

An adapted list of Selke's (1990) environmental package design guidelines include:

- Eliminate the use of toxic constituents in packaging materials.
- Use reusable packaging materials. Design packaging that can be easily returned and used again. If replacement parts are sent to a customer, supply the return freight costs to have the customer return the discarded parts or the empty package.
- Use single materials whenever possible in package design. Multi-material packages are less suitable for recycling because the materials are difficult to identify and separate.

- Use previously recycled and recyclable materials in the packaging and product design.
- Encourage recycling and reuse by requiring the user to return the packing box or returning the defective part.

8.6.6   Government and Special Customer Requirements

Many customers and governments have specialized packaging requirements. This can include protection, security, size, disposal, environmental, or labeling requirements.

In order to protect the environment from the effects of hazardous materials, most countries have published special packaging, labeling and handling requirements. In the U.S., the Department of Transportation regulates all shipments that include hazardous materials such as hazardous chemicals, radiological, biohazards, etc.

8.7   DESIGN FOR THE ENVIRONMENT

A rational balance between economics (i.e. making a profit) and environmental responsibility (i.e. saving our environment) is a difficult task in product development. The difficulty of the problem includes diversity of the environmental issues involved, lack of scientific knowledge on many issues, government laws and regulations, differences between countries and unknowns in directly relating design decisions to environmental results

The philosophy behind design for the environment (DFE) is to consider the complete product life cycle when designing a product. The design team must consider environmental issues. The past point is especially difficult to predict.

The hierarchy of design steps for DFE as defined by Wentz, (1989) is to:

1. Eliminate all waste, if not then
2. Reduce, if not then
3. Recycle, if not then
4. Reuse and recovery, if not then
5. Treatment, if not then
6. Disposal

DFE design decisions include the types of materials that are used in the manufacture, packaging, and use of the product. Considering a material's recyclability and reusability capabilities; the materials long term impact on the environment; and the amount of energy (and efficiency) required for the product's manufacture and assembly. Designing for easy disassembly for manufacturing; including remanufacturing characteristics. Consideration of the products disposal characteristics; will it disintegrate in a landfill? Complex tradeoffs between different environmental design alternatives. For example, is the use of paper packaging better than plastic? Paper packaging causes trees to be cut but it is degradable so the effects on landfills are much less than styrofoam packaging which uses petroleum, more energy, and is non-degradable so it is filling up our landfills.

Various environmental related analyses such as materials balance, thermodynamics, and environmental economics can be used. Theories of material and energy balances can also include problems of pollution, emissions, and waste, not only in the processes of producing energy, but also in the processes of using or consuming energy.

8.7.1   Life Cycle Stages

The development team must consider all environmental life cycles stages in a typical manufactured product (Huggett, 1995). These are:

Stage 1.   -   **Premanufacturing.** Includes suppliers of raw materials and parts, which generally use virgin resources.
Stage 2.   -   **Manufacturing operations.** Includes the energy use, wastes, and scrap of the manufacturing processes. It is usually the best understood and most evolved stage in product development.
Stage 3.   -   **Packaging and shipping.** Includes packaging materials, packing density, and different methods for shipping, and reuse or disposal of packaging materials.
Stage 4.   -   **Customer use.** Not directly controlled by the manufacturer, but is influenced by how products are designed, manufacturer maintenance, and regulatory requirements instituted by a government
Stage 5.   -   **Disposal or remanufacture termination.** When a product is no longer satisfactory because of obsolescence, component degradation, or changed business. The product can then be refurbished or discarded.

Many products impact the environment at more than one stage. For automobiles, the greatest impact results from the combustion of gasoline and the release of tailpipe emission during the driving cycle. There are other aspects of the product that affect the environment, such as the use of oil and other lubricants, discarding of tires and the ultimate disposal of the vehicle.

8.7.2   Environmentally Conscious Business Practices

There are many environmentally conscious business practices. One practice is design for the environment. For design, the first and most important consideration is elimination. Eliminating an environmental problem is environmentally easier and cheaper than most other techniques.

Source reduction, the second practice of design, is doing the same things with fewer resources. Design for source reduction includes eliminating or minimizing packaging, consolidating freight and performing tasks closer to the ultimate consumer. For example, some automotive companies use closed compartments when shipping in order to eliminate the need of spraying a protective sealer on the car. Smaller packaging and rearranged pallet patterns can reduce materials usage, increase space utilization in the warehouse, truck trailers and shipping containers, and reduce the amount of handling required. This results in less packaging waste, fewer shipping vehicles, fewer accidents, and easier handling in warehouses. Another method is to manufacture or add variety (i.e., options) to the product at a location that is closer to the consumer to reduce shipping costs and environmental problems.

Designing for reuse and recyclability are different in terms of degree. The relationships between recycling and reuse can be defined by the amount of treatment required. Minimal treatment of a material is more closely associated with reuse of a product, while a material that requires some amount of treatment can be considered to have undergone recycling. The Federal EPA considers recycling to be a waste management practice outside the control of the plant.

IBM announced in 1999 the first personal computer made of 100% recycled resin for all major plastic parts. A saving of up to 20% is expected for some parts. Another example is that most semiconductor factories now recycle their water.

Design for disassembly focuses on designing a product so that it may be taken apart easily. Technological and design characteristics may include various libraries and materials of alternative adhesives and connection devices that can be effectively used to form and disassemble products.

Design for remanufacturing refers to the design of a product with respect to repair, rework, or refurbishment of its components. In a typical refurbishment process, worn-out components and equipment are grouped together, disassembled, cleaned, refurbished where needed and then reassembled. For example, a remanufactured gasoline engine requires 33% less labor and 50% less energy. (Hormozi, 1999)

Designing for disposal requires the selection of environmentally friendly materials and parts. Eventually, the wastes of a product may end up in a landfill or some disposal location. The issues of a product's biodegradability and toxicity play a large role. For some products, designing the product for controlled incineration may be preferred.

8.7.3   Recycling Automobiles

An article by the Crain News Services highlighted several interesting facts about recycling automobiles (1996). "Eleven million vehicles are scrapped annually in the United States. Nearly 95 percent of them are 75 percent recycled (some newer cars such, as the Chrysler Contour and Cirrus are up to 80 percent recyclable). That's more than aluminum cans (68 percent) and newspapers (45 percent). Auto recycling saves an estimated 85 million barrels of oil annually that would otherwise be used in the manufacture of replacement parts".

The author explains that when a car is recycled the fluids are drained first. Some are refined for reuse, and others are discarded. The antifreeze is checked for viability and the good stuff is resold Ford estimates that the time to remove the recyclable parts of any vehicle must be reduced to 15 minutes to ensure efficient recycling. The automobile must be designed accordingly.

The recycling for parts works on three levels; direct reuse (i.e. available in a car junk yard), rebuilt for reuse, and material recovery. The first, direct reuse, is called "pure recycling, a part for a part." Working fuel pumps are removed and sold. The second level of reuse is a part that can be sold to a company and rebuilt for resale. The third level is the materials recovery level, in which parts are sold to be melted down and made into something else.

In the third level of material recovery, magnets first separate out ferrous material. Large fans blow the lightweight materials (i.e., fluff) away from the reusable, non-ferrous material including aluminum, copper and zinc. The company has two piles of scrap for sale, ferrous and non-ferrous materials. A third pile contains non-saleable, non-reusable materials (consisting of plastics, rubber, glass, foam, fabric and adhesives). A company pays to send this material to a landfill. Non-reusable materials represent about 25 percent of the car's weight (usually about 500 pounds), and makes up two percent of all waste put into landfill each year. The Vehicle Recycling Development Center (VRDC) is dedicated to finding ways to decrease the 25 percent of non-recyclable content. Be able to drain 95 percent in 20 minutes. The VRDC wants to standardize drain plugs and fluid caps so that the same draining equipment is used for all models.

The automakers have already taken a step by using a label system, endorsed by the Society of Automotive Engineers, which marks the material type for each plastic part. This allows recycling companies to know what type of plastic it is such as. PP for polypropylene, PC for polycarbonate, PA for nylon, and so on (Crain News Service, 1996).

8.7.4 Environmental Analysis

Life cycle costing is a major tool used in design for the environment. Environmental cost considerations include the materials, processes and energy used, the wastes produced and the final disposal of the product.

A suitable assessment system for environmentally responsible products (ERPs) should have the following characteristics (Fabrycky, 1991):

- Enable direct comparisons among different products
- Usable and consistent across different assessment teams
- Encompass all stages of product life cycles and all relevant environmental concerns

- Simple enough to permit relatively quick and inexpensive assessments

Another analysis tool is inventory analysis, which is the identification and quantification of all energy and resource uses and their environmental effects on natural resources. It is the process of gathering inventory data along the production chain of each product. The data are brought into uniform units, such as cost, so that they can be added into total numbers for all components and life-cycle processes. This analysis is performed by identifying the process waste for each stage of the product life cycle then measuring or estimating its quantity. Impact analysis is the assessment of the consequences that wastes have on the environment. It evaluates an array of alternatives and identifies the activities with greater and lesser environmental consequences.

8.7.5   Environmental Considerations For Waste Disposal

A company should minimize the environmental impact of its products and packaging from conception to disposal. According to Sony's vice-president for Environment, Safety, and Health, "Everything we make eventually ends up as waste. We can make it a liability or turn it into a profit." Some programs implemented by Sony include a stand-by power mode for televisions that use 50% less power than the industry average, lithium ion rechargeable batteries that eliminate the use of toxic substances, and a program to recycle electronics products that is the largest one ever attempted in the U.S.

This discussion will focus on waste disposal. Major issues in environmentally friendly product disposal and package design include:

- Minimize waste generation
- Assimilation of waste into the environment
- Resource recovery of energy, materials, and parts

The most effective way to reduce solid waste is to reduce the rate of waste generation. The Environmental Protection Agency (EPA) defines waste reduction as the prevention of waste at its source, either by redesigning products or changing societal patterns of consumption waste generation.

There are several ways to reduce waste generation.

- **Reduce the quantity of material used.** Reducing the package size, weight or levels of wrapping.
- **Increase the average lifetime of products.** Reducing the number of products being discarded and the number of replacements, the amount of material being used is reduced.

- **Substitute reusable products and packages for disposables.** Containers that can be refilled require slightly more material in some cases in order to survive more cycles but the refill containers require much less material.
- **Reducing consumption.** Educate consumers how to use your product more efficiently. Remind them to turn down their hot water heaters when on vacation, accurately measure laundry detergent when using your new more concentrated product, and to lubricate equipment for long lasting use. Sell your product based on its long lasting and efficiency specifications. Sell maintenance agreements that emphasize product efficiency and longevity benefits.

Assimilation of Waste into the Environment

Landfills in many countries are quickly running out of space. Because of the increasing regulations governing landfills, it is getting harder to find new sites causing disposal prices to rise. If solid waste must be landfilled, it should be as degradable as possible. Biodegradable is where the waste will be assimilated into the environment naturally from the action of living organisms, photodegradable (from the action of light), hydrodegradable (from the action of water) or thermal-degradable (from the action of heat) (Selke, 1990).

It is certainly better for waste to be degradable than inert. The product design concept of depending on degradability, however, is difficult. There is little chance of knowing exactly how long the degradation will take to occur. It may occur prematurely depending on the external environmental conditions of transportation and storage. It can also take much longer depending upon the conditions in the landfill. Degradation of any material is difficult in a "dry" landfill where it has little exposure to some of the required agents such as light and water.

Reuse of Energy and/or Material

Reuse, recycling, and incineration are all forms of resource recovery. Recycling allows recovery of material to either produce more of the same product such as aluminum cans or to produce an alternative product such as lower grade paper products made from some portion of recycled paper fiber. Recycling is useless if there is no market for the recovered material. It is essential that manufacturers of products and packaging use materials that have been recycled to help generate the market.

Reusability can be a good option in circumstances where the packaging remains in the manufacturer or vendor control. One example of resource recovering is discussed for a circuit board manufacturer. In order to transport circuit boards, they designed a standard size box with convoluted foam inside to hold most boards and protect them from movement and damage. When closed, a reusable cardboard sleeve slides over the box similar to a matchbox cover. When a board is sent to the field as a replacement the field service engineer returns the bad board to the plant in the replacement board's box by placing a return address label over the original address label on the sleeve. Labeling and sealing of packages often is a problem when the box is reused over and over so the sleeve can be replaced whenever if it becomes too damaged from repeated sealing.

Recovery of energy can be accomplished through incineration of combustible solid wastes. Unless the waste is primarily combustible, the residue may be toxic and must still be buried. Incineration also causes air pollution requiring the addition of expensive control equipment.

8.8   ISO 14000

ISO 14000 is designed to allow companies to identify vendors and other companies that are environmentally friendly. The purpose is to compliment a country's current environment laws and regulations. The goal of the standard is to help a company develop management or process standards so a company can consistently meet their environmental obligations on all fronts: regulatory, customer, community, employee, and stockholder. The original scope is not intended to duplicate or replace a country's regulatory system.

ISO 14001 focuses on the specification of systems and guidance for use, which form the core of the ISO 14000 standards. The major elements of the proposed standards within this category include the setting of environmental policy planning, implementation and operation, checking and corrective action, and management review. Firms will have to show that they have environmental control programs to be accepted in commerce.

8.9   SUMMARY

The entire product development team must minimize support costs while fully preserving the integrity of the product and having a minimum negative impact on the environment. A good design will add to the competitive advantage of a company. It will minimize the cost of distribution by reducing the weight and volume being stored and shipped.

8.10   REVIEW QUESTIONS

1. What is the design team's role in supply chain planning?
2. State the goal of environmental design. What are the financial benefits and costs of environmental design?
3. What are the major purposes of packaging?
4. What are some of the many parameters in the selection of package type?

·8.11   REFERENCES

1. J. Bigness, In Toady's Economy, There is Big Money to be Made in Logistics, Wall Street Journal, September 6, 1995.
2. G. Bylinsky, Manufacturing for Reuse, Fortune, p. 103, February 6, 1995.
3. P.M. Byrne, Global Logistics: Improve the Customer Service Cycle, Transportation and Distribution, 33(6): p.66-67, 1992.
4. Crain News Service, The Last Road, Shredder, Recycling Await Million Vehicles Annually, Dallas Morning News, p. 6D, February 24, 1996.
5. W. Fabrycky, and B. Blanchard, Life Cycle Cost and Economic Analysis, Englewood Cliffs, New Jersey, 1991.
6. J.S. Gordon, Profitable Exporting - A Complete Guide to Marketing Your Products Abroad, John Wiley and Sons, Inc., 1993.
7. M.E. Henstock, Design for Recyclability, Institute of Metals, London, England, 1988.
8. Hormozi, Make It Again, IIE Solutions, April 1999.
9. R. Huggett, Environmental Science & Technology, March 1995.
10. W.W. Leake, IIE Supports 13, IIE Solutions, p. 8, September 1995.
11. S. Onkrisit and J.J. Shaw, International Marketing Analysis and Strategy, Macmillan Publishing Company, 1993.
12. Patel, Design for International Use, Unpublished Student Report, The University of Texas at Arlington, 1996.
13. C. Pegels, Management and Industry in China, Praeger Publishers, 1987.
14. POGO International, DIANA, Disassembly Analyses Software, College Station, TX 1996.
15. R.S. Pressman, Software Engineering, McGraw-Hill, New York, 1992.
16. J. Sarkis, G. Nehman, and J. Priest, A Systematic Evaluation Model for Environmentally Conscious Business Practices and Strategy, IEEE Symposium on Electronics and the Environment, Dallas, TX, 1996.
17. S.E.M. Selke, Packaging and the Environment, Technomic Publishing Company, Inc., 1990.
18. Wentz, Hazardous Waste Management, McGraw Hill, New York, NY 1989.

Chapter 9

DESIGN FOR PEOPLE: ERGONOMICS, REPAIRABILITY, SAFETY, AND PRODUCT LIABILITY

Consider the User and Support Personnel in the Design

The performance and reliability of any product depends upon the effectiveness of both the product design and the human user. The design must consider all types of people who might directly or indirectly interact with the product including users, spectators, production workers, maintenance, transportation, and support personnel. Designing for people maximizes human performance and minimizes human errors. This means fewer, faster and simpler tasks. Advertisements using such terms as "user friendly", "human engineered", "ergonomics", and "easy to maintain" demonstrate that designing for people is a critical design parameter.

Best Practices

- Simple And Effective Human Interfaces
- Functional Task Allocation Analysis.
- Task Analysis, Failure Mode Analysis, Maintenance Analysis, Safety Analysis, And Product Liability Analysis
- Design Guidelines
- Mistake Proofing and Simplification
- Standardized/Common Tasks And Human Interface Designs
- Prototype Testing Of Human Tasks
- Effective Design Documentation
- Safety Hazard And Product Liability Analyses

DESIGN MUST CONSIDER FORESEEABLE OCCURRENCES

Anyone who has played golf knows that when hitting the ball, many times the golf ball will not go where the player wants it to go. Occasionally, even golf professionals will hit a wild shot. When a golfer's wild or erratic shot hits someone else; the golfer is often not held responsible by the U.S. court system. Most people understand that the golfer did not intentionally hit the person with the golf ball. The following questions were developed from an idea in a Wall Street Journal article.

If someone is hurt, or property is damaged, however, who should pay for the damages? Who is responsible?

The United States court system's answer to these questions is sometimes the designer and operators of the golf course. The designer is responsible for planning for wild and erratic golf shots. The designer must minimize any chance or probability that a golf ball may hit a person.

The design team must anticipate and consider all foreseeable occurrences caused by the capabilities (i.e. limitations) of persons or the environment that will directly or indirectly interface with the product. This is true whether designing a golf course, calculator, toy, medical equipment, or a jet aircraft. Since everyone including the designer knows that people hit wild shots; it is the designer's responsibility to take this "wildness" into account when designing a golf course. This includes worst-case values and normal distributions of shots. An additional consideration is that the person that is hit by the golf ball may not even be playing the game of golf, such as a spectator or someone walking by the course. The design team must also consider this potential situation in the design of a golf course.

A successful design is one that is safe to use, within the capabilities of its users, and easily maintained under operational conditions. In short, products must be designed for people. The purpose of designing for people is to optimize performance by reducing the potential for human error, ensuring that users can efficiently perform all tasks required by the design and minimizing any undesirable effects of the design on people. Just as the effects of the environment must be considered, the design must also consider all people who will possibly directly or indirectly interact with the product, including users, spectators, production, inspection, maintenance, transportation, and support personnel. Advertisements using such terms as "user friendly", "human engineered", "ergonomics", and "easy to maintain" demonstrate that designing for people is a critical design parameter.

This chapter provides an overview of the design disciplines of ergonomics, human engineering, repairability, maintainability, product safety and product liability.

9.1 IMPORTANT DEFINITIONS

Ergonomics, human engineering or human factors are the systematic application of knowledge about a human's capabilities in the design of a product. The idea is to fit the product to the human. Human engineering is a design discipline that seeks to ensure compatibility between a design and its user's capabilities and limitations. The objective is to achieve maximum human efficiency (and hence, acceptable systems performance) in fabrication, operation, and maintenance. Common measures include task time (minutes necessary to complete a task), number of tasks and human error rate (number of human errors in a specified period of time or based on the number of error opportunities).

Repairability, serviceability or maintainability is the design discipline concerned with the ability of a product or service to be repaired/maintained throughout its intended useful life span with minimal expenditures of money and effort. Either term is often used as a performance measure of a design expressed as the probability that a product failure can be repaired within a given period of time. Almost all repairs require human involvement. Designing for repair is similar to ergonomics but highlights the design issues involved with maintenance tasks, such as repair concepts, accessibility, procedures, and equipment used. The design objective is to maximize the probability of successful maintenance actions through designing easy-to-use equipment and procedures. Common measures include mean time to repair (MTTR), life cycle cost to repair, and availability.

Design for safety is the design discipline that reduces the number and the severity of potential hazards by identifying and preventing potential hazards. Its purpose is to prevent accidents, not to react to accidents. Whereas the techniques of reliability are to decrease the number of failures, the techniques of system safety reduce the number of potential hazards and the severity level of a hazard when failures occur. Correcting hazardous designs prior to a product's failure can save many human lives and millions of dollars in potential product liability and product recalls. Measures include the number of accidents reported, severity of accidents and the number of potential hazards identified but not resolved.

Design for product liability is the design discipline that minimizes the potential liability of a product by identifying all risks, designing to avoid these risks, and maintaining all quality records. The key purpose is to ensure that the design is safe for all "foreseeable" uses, users, environments, and events. It must consider both today's and future requirements and all of the countries that the product may be used in. Measures include product liability costs and number of liability lawsuits.

9.2   BEST PRACTICES TO DESIGN FOR PEOPLE

The performance and reliability of any product depends upon the effectiveness of both the product design and the human user. Regardless of the hardware and software performance characteristics of a design, some designs result in better human performance and are safer than others. The best practices to successfully design for people are as follows:

- **Simple and effective human interface** is developed to improve human performance, reduce human errors and increase number of potential users by using the design techniques of ergonomics, repairability, maintainability, safety, and product liability.
- **Functional task allocation analysis** maximizes system performance by effectively dividing performance, control, and maintenance tasks between the product and personnel.
- **Task analysis, failure modes, maintenance analysis, safety analysis, and product liability analysis** are used to analyze human requirements, identify and correct potential problems and predict human performance.
- **Design guidelines and analysis promote mistake proofing with simplified, standardized/common tasks and human interface designs** to improve personal performance.
- **Prototype testing of human tasks including repair and manufacturing** is used to identify design improvements, potential human errors, and hazards.
- **Effective design documentation** is critical for user, manufacturing, and repair instructions.
- **Safety hazard and product liability analyses** identify and correct all potential hazards including all foreseeable situations.

The overall design goal is to ensure effective and safe human performance. Since poor performance and human error are major causes of product failure, understanding and designing to human capabilities are critical.

Ergonomics or human engineering started during World War II, when personnel could not properly operate some of the developed complex systems. It is also called ergonomics and human factors. Quick solutions, such as personnel selection and additional training, could no longer solve this problem. Researchers began to develop design principles and guidelines for equipment that a large number of people could easily use. The steps for ergonomics are to:

1. Identify all of the people that will be affected or have contact with the product for all stages of use including operators, manufacturing, support, disposal, etc.

2. Identify all human parameters and conditions of use including physical, mental, cultural and environment
3. Perform functional task allocation to determine which tasks should be performed by humans and which by the system
4. Perform task analysis to determine human task requirements and identify and correct potential problems
5. Use design guidelines and analyses such as mistake proofing to improve human performance and reduce errors
6. Use prototypes and testing for all of the above tasks to optimize human performance

Ergonomic analyses can be performed by the design team, an ergonomist, or by a human engineering specialist working with the design team. In the absence of a human engineering specialist, the design team will need to review some of the references listed at the end of the chapter before performing the analyses.

9.3 HUMAN INTERFACE

Human performance is often measured by time such as response/reaction time, accuracy such as errors, or accessibility such as total number of people that can use the product. Time is dependent on the number of tasks, options/alternatives, complexity of the sensory input, complexity of the physical or mental task, and the number of outputs. Accuracy is dependent on the number of alternatives, complexity of the task, compatability between stimuli and response, the interface control used, speed required, and feedback. Accessibility is dependent on the level of physical and mental capabilities, skills, and training that is needed to use the product. For consumer products, it directly affects the number of potential customers. Designing for disabilities such as manual dexterity and carpal tunnel syndrome can increase sales. Designing for other cultures can also increase the potential population of users. Improving the human interface such as simplifying the human task, reducing the number of human tasks, or having the software perform more of the task are several solutions.

A key is to make the human interfaces simple. Most people do not want complex products. In an interesting article Rising Heat: New Thermostat Designed by Brilliant Morons by Virginia Postrel (1999), she describes the dilemma of products that are too smart for the user. Her new programmable thermostat comes with 56 different temperature defaults. Instead of easily setting for one temperature like simple thermostats, this model requires programming. How many programmable VCRs, stereos, and microwaves do you ever use in the program mode? The problem is not the technology but the way it is applied. You should provide a simple design for the user and make the technology and design adapt.

9.3.1   Types Of Human Error

Human error can be defined as any deviation from required performance. Meister (1964) and DeGreene (1970) defined error as one of the following:

- Incorrect performance of a required action
- Failure to perform a required action
- Out-of-sequence performance of a required action
- Performance of an unrequired action
- Failure to perform a required action within an allotted time period

Errors may be classified in terms of cause, consequences, and the stage of system development in which they occur. For example, the design, environment, manufacturing, maintenance, operator, or documentation can cause error.

Design Induced Human Error

Design characteristics that cause or encourage human error are often classified as design induced. Improper functional task allocation or a failure to follow design guidelines for human capabilities usually causes design-induced errors. Functional task allocation is used to determine which functions or tasks will be performed by the product and which by the personnel. Errors result when humans are assigned functions they cannot perform or when a product performs a function that might better be performed by a human. Design errors are likely to occur in the following situations where the team:

- Does not completely analyze product and system requirements for the human element in all areas.
- Relies exclusively on his or her experience rather than designing for actual users and maintenance personnel.
- Assumes that the user, production worker, and maintenance personnel are all highly skilled and motivated.

Environmentally Induced Human Error

The environment in which the product is being used can obviously have a detrimental effect on human performance. The external environment (i.e., temperature, snow, rain, and so on) or the internal environment (i.e., level of training, motivation) can cause environmentally induced human errors. The design team must be completely aware of the types of environment in which their product will be used in order to minimize any detrimental effects.

Manufacturing Induced Human Errors

Manufacturing, production, or workmanship errors are errors that occur in production and can be defined as internal to the production worker (i.e., motivation or lack of training) or external. Over 90% of quality defects in assembly are caused by workers. Reducing these errors is a major goal of producibility and mistake proofing (i.e. poka yoke). External causes listed by Meister and Rabideau, as cited in Meister (1971), include the following:

- Inadequate human factor design of product, machinery, tools, test equipment, and documentation
- Inadequate methods for handling, transporting, storing, or inspecting equipment
- Inadequate job planning (poor or nonexistent procedures)
- Poor supervision or training
- Inadequate work space or poor work layout
- Unsatisfactory environmental conditions (such as lighting, temperature, and acoustic noise)

Maintenance and Installation Induced Human Errors

Maintenance-induced errors (including installation errors) occur during installation and routine maintenance actions. They can occur throughout a system's life cycle and often increase, as the system becomes older. An example of a maintenance-induced error was found on the U.S. Army's Hummer vehicle, which replaced the jeep-style vehicle. The Army found that several polyurethane components used to muffle the Hummer's diesel engine decomposed when they come in contact with either diesel fuel or windshield washer fluid. Since diesel fuel and washer fluid are used in daily maintenance actions, design changes were required. The final design solution was to coat the defective parts in neoprene, a tough plastic. This design change resulted in an additional cost of $13 million for the project.

Installation errors occur when setting up and initially using a product. They are especially a problem for software products. Everyone has encountered some problems when installing a new computer software program onto their personal computer. Many of these problems are due to hard to understand or incorrect documentation.

User-induced Human Error

User induced error is the kind of human error with which most people are familiar. In many products, human error may be frequent but without serious consequences. For example, many software failures are caused by human error but these do not usually cause serious life threatening or costly problems. In major systems, such as aircraft or military systems, human error may be less frequent, but the consequences may be serious or even fatal. As with manufacturing errors, the causes for user error may be internal (i.e., motivation, etc.) or external (i.e., inadequate human engineering, documentation, training, etc.). Designing products to human engineering criteria can reduce these external causes.

To Make Skies Safer, Focus On Pilots

For example, improving the safety of the airplanes is a major concern. Statistically speaking, however, only 9.7% of the fatal accidents in the last 10 years have been caused by the airplane itself, and 74% of the hull losses were due to actions by the cockpit crew or flaws in maintenance (Grey, 1999). Grey noted that nine out of 10 people who have died in U.S. air carrier accidents would have died even if the planes themselves were perfect. In order to improve air travel safety, then, the biggest gains are to be made by reducing human errors. More attention should be given to design and training, especially in dealing with situations.

9.4   FUNCTIONAL TASK ALLOCATION

Functional task allocation is the process of apportioning (i.e. dividing) system performance functions among humans, product, or some combination of the two. It is usually performed early in the design process during requirement definition or conceptual design. Failures occur, when a functional task is assigned to a human, but the necessary information is not provided to the user. An example is the failure to include an easy to understand display to warn someone of an emergency condition.

Failures can also occur when the hardware or software of the product automatically performs functions. An example is software products that automatically perform tasks, which the operator does not want. The task is to decide whether a function will best be performed in a manual, semiautomatic, or automatic manner.

The steps are to:

1.   Identify all tasks/functions that need to be performed
2.   Evaluate all possible combinations of the user or product performing the task
3.   Flow chart the tasks to promote simplicity and identify improvements
4.   Allocate tasks based on the first three steps

A critical but common mistake occurs when the design team attempts to automate every function that can be automated. Only functions that cannot be automated through hardware or software are then left for the human operator.

This design approach results in excessive complexity, schedule delays, and assignment of tasks in a manner that the human cannot perform properly. Appropriate functions for humans and products are occasionally provided in handbooks, but they are only guidelines for evaluating specific functions in real-world design.

9.4.1 Functional Allocation is a Critical Design Task

Could a minor change in design have prevented an airliner crash? An article by W. M. Carley in the Wall Street Journal (1996) shows some of the difficult trade-off that can occur in functional allocation. Initial press reports focused on how pilot errors apparently caused a navigational problem that then resulted in a crash.

On some jet designs, the spoilers automatically retract during an emergency climb. "What has gone almost unnoticed is that this plane, with a minor design difference, might have flown out of the tight spot safely. Because the jet had been readying to land, small flaps called spoilers or speed brakes had been extended from the tops of its wings - literally spoiling the airplane's lift and hindering its ability to climb" (Carley, 1996). If retracted, the plane "may" have made it safely over the trees.

Should airplanes leave more of the flying to the pilot or more guided by computers that automatically take control in emergencies? The two major aircraft companies have answered this question differently. Differences in task allocation are often caused by a difference in design philosophy. Both companies have excellent reasons for their different philosophies. Company A's design relies more on a pilot's judgment. They believe that the pilot, not a computer, should have complete control in emergencies. Company B's design on the other hand, relies more on computerized features in its airliner design which limit the control options available to a pilot. For example, Company B's jets have a computer system that senses when a plane is approaching a defined level of stall the spoilers automatically retract, increasing climb. These automatic systems have reduced some types of pilot errors but added new types of errors directly caused by the automated systems. What if an airplane crashes because the automated flight control software does not function properly and the pilot cannot override the system in time? The author notes that several accidents "may" have been caused by the pilots' difficulties in coping with the increasing level of automation. As this Wall Street Journal story shows, dividing system tasks between humans and machines is a very difficult job.

9.5 TASK ANALYSIS AND FAILURE MODES ANALYSIS

Task analysis (sometimes called task-equipment analysis) is a design technique that evaluates specific task requirements for an operator with respect to an operator's capabilities. The analysis identifies each human activity or task required and then compares the demand of the tasks with human capabilities and the resources available. The goal is to identify and correct potential problems. The steps are to simply:

1. List all human tasks
2. For each task, describe the task requirements in terms of response time, physical, sensory, mental, training, and environment. This includes a description and analysis of all interfaces, tools, and equipment that the person may need.
3. Identify and correct potential problems and failure modes through design changes where possible.

A failure modes analysis (FMEA) identifies the human failures that could be introduced by all people involved with the product. It is no different than a traditional FMEA except that human failure modes are analyzed. Simulation and virtual prototyping of the human should be considered.

The level of detail for the task analysis and FMEA are based on the design information required and the importance of each task. Design team must look at their designs critically to see that the task requirements do not exceed human capabilities. A variety of formats have been used to organize task analysis data. An example of one format is shown in Figure 9.1. In addition to the operator's tasks, task analysis can also be performed for maintenance and support personnel.

9.6   DESIGN GUIDELINES

The third major design activity involves the use of guidelines to provide proven criteria for designing to human capabilities, mistake proofing the design, and helping to standardize common tasks. They can range from simple checklists to specific design criteria. Checklists are considered general guidelines for initial data in the analysis. Examples of specific design criteria for U.S. Department of Defense equipment controls are shown in Figure 9.2. The steps are to:

1. Identify all types of users that will interface with the product.
2. Identify existing guidelines
3. Perform analyses of human performance
4. Develop additional data when needed through testing

Human data to be used in the design effort is dependent on who will be using the product and under what conditions it will be used. This is especially important for products that will be sold internationally. The physical, cultural, and mental characteristics of the potential users are a key design consideration for any product or system. A major source of human engineering physical information is called anthropometry, the study of human body measurements.

There are over 300 anthropometric measures of the human body, which are based on actual studies that scientifically measured a sample of people. The

Function: Operate Aircraft Landing System

Task: Enter Airports Coordinates

Subtask	Action Stimulus	Required Action	Feedback	Potential Errors	Time		Skill Level
					Allow able	Neces sary	
Enter Coordinates	Prompt on display	Depress buttons on keyboard	Values shown on display	a. Wrong coordinates b. Wrong button is pushed	1 min.	1 min.	Qualified pilot

FIGURE 9.1 Task analysis

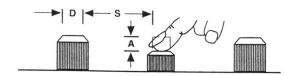

	Dimensions		Resistance		
	Diameter (D)				
	Fingertip	Thumb or Palm	Single Finger	Different Fingers	Thumb or Palm
Minimum	9.5 mm (3/8 in.)	19 mm (3/4 in.)	2.8 N (10 oz.)	1.4 N (5 oz.)	2.6 N (10 oz.)
Maximum	25 mm (1 in.)		11 N (40 oz.)	5.6 N (20 oz.)	23 N (80 oz.)

				Separation (S)	
	Displacement (A)			Single Finger	Single Finger Sequential
Minimum	2 mm (5/64 in.)		Minimum Preferred	13 mm (1/2 in.)	6 mm (1/4 in.)
Maximum	6 mm (1/4 in.)			50 mm (2 in.)	13 mm (1/2 in.)

FIGURE 9.2   Recommended separation between adjacent controls.

data is presented either as an average value or as the percentile of people within a certain value. Percentiles are most often used to determine a range, ensuring that a specified percentage of people would be capable of using the product. A good example of this is the reasons for forward-backward adjustments of automobile seats. If only the average human size value was used to design seats, many people would be unable to drive a car comfortably or safely. Although it is usually not possible to design for all extreme values (maximum or minimum), the design should compensate for the largest population possible including physically challenged users and other cultures. A typical design goal is to design for 90 to 95 percent of the population.

Human performance requirements for the notebook computer could include:

- Keyboard size and activation pressure
- Display resolution, size, and brightness
- Device locations
- Mouse/trackball location, size, and resistance
- Labeling and colors
- Adaptability for human physical limitations or disabilities
- Ease of initial setup by the user
- How to implement self-diagnostic capability

Imrhan (1996) discusses a variety of anthropometric, musculoskeletal and visual issues relating to stress, strain and performance for using a computer. Design principles with respect to the physical dimensions and illumination and the effects of poor designs are treated in detail (Imrhan, 1996).

9.7   PROTOTYPING AND TESTING

Prototype evaluation is a very effective method for optimizing the human-machine interface. Research testing allows different designs to be evaluated for customer preferences, identify design problems, and potential human errors. Prototyping is especially important in designing software products and products for persons with physical limitations.

One company uses human research to design more comfortable automobile seats and compares them to rival products. Researchers developed a pressure-sensing mat that determines how a motorist's weight is distributed onto the seat. After analyzing data on hundreds of test subjects, the company has mapped nine critical pressure zones on each car seat. This data is then directly used in the seat's design. Rather than designing many different seats and then testing them, it uses human engineering research to get it right the first time.

9.7.1   Software User Simulation and Prototyping

For many products, software development plays a very important role. Prototyping is often used for design requirements, analysis, and for testing the final design. McDonnell Douglas in St. Louis (BMP, 1996) uses software tools to accomplish the following design analyses:

- Rapid prototyping for cockpit displays with built-in flight simulation
- Human factors analysis of human body fit and function
- Analysis of the visual impact of components in a pilot's visual path
- Modeling and analysis of human-machine integration requirements
- Simulation and analysis of the thermal behavior of the human body
- Assessment of the performance of the ejection system
- Simulation of combat maneuver G forces and resultant human responses

For the final user interface, software creates a buffer between the end user and the hardware components of the computer (i.e., registers, memories, etc.) and the software tools used by the programmer to make the application run (Dumas, 1988). An average of 48% of the coding time for software is devoted to the user interface, and 50% of the implementation time is devoted to implementing the user interfaces (Myers, 1995).

The seven principles of a good user software interface are (Dumas, 1988):

1. Put the user in control
2. Address the user's level of skill and experience
3. Be consistent
4. Protect the user from the inner workings of the hardware and software
5. Provide on-line documentation
6. Minimize the burden on the user's memory
7. Follow the principles of good graphics design

Two important decisions during the design phase of the software user interface are in the presentation and hardware (Stubbings, 1996). The presentation decision involves choosing the interaction objects that the user will interface with. Interaction objects are visible objects on the screen which the user will manipulate and view to enter information (Larson, 1992). Examples are menus, command boxes, icons, and cursors. The hardware decision involves determining what devices will physically be used to input the information to complete the job. Examples are a keyboard, mouse, touch-sensitive screen, joystick, and a trackball.

9.8 DOCUMENTATION FOR USERS

Effective design documentation is critical for user, manufacturing, customer service and repair instructions. Instructions should be easy to understand and correct to minimize human errors. The keys to good documentation were discussed in the chapter on Early Design. For the notebook computer example, some examples of warnings that are printed in one instruction manual include:

- Using the computer keyboard incorrectly can result in discomfort and possible injury. If your hand wrists and/or arms bother you while typing, discontinue using the computer and rest. If discomfort persists, consult a physician.
- Don't spill liquids into the computer. If you spill a liquid into the keyboard, turn the computer off, unplug it from the AC power source, and let it dry completely before turning it on again.
- To prevent possible overheating of the CPU, do not block the fan.

9.9 REPAIRABILITY AND MAINTAINABILITY

With unparalleled increases in equipment and software complexity, locating repair personnel with the high levels of education and training necessary to repair today's equipment in many different countries is a major concern in industry. As a result of these concerns, concerted efforts are being made to reduce the complexity of the maintenance function by designing for repair. The meaning of repairability and maintainability are very similar and often used interchangeably. Repairablity is the design discipline concerned with the ability of a product to be easily and effectively repaired at the production facility. Maintainability is the design discipline concerned with the ability of the product to be satisfactorily maintained throughout its intended useful life span with minimal expenditures of money and effort.

Both are an extension of human engineering so the steps are similar. 1.) Describe all support task requirements including failure modes, support equipment, environments, etc. 2.) Simplify the design and repair process, identify and 3.) Correct all potential problems through redesign of product or repair process. When possible, self-diagnostics, built-in test and self-maintenance are used to improve the repair process.

The purpose of maintainability is to design products so that they may be easily maintained and kept in a serviceable condition after the product is sold. To accomplish this goal, the system should be easily and quickly maintained by:

- User or technician with minimum skill levels, the operators themselves, or the product automatically diagnoses a problem or repairs itself (self maintenance or self diagnostics)
- Minimal number of special training, tools, support equipment, software, and technical documentation
- Reducing scheduled or preventive maintenance requirements

Maintenance tasks consist of two general types of tasks (Moss, 1985):

- **Corrective** tasks are those performed in order to correct a problem or to restore the product to acceptable operating conditions after a failure has occurred.
- **Preventive** tasks are performed in order to defer or prevent the occurrence of a future anticipated failure. Preventive maintenance thus serves to extend actual product performance and reliability beyond that expected without the corrective action.

9.9.1   Maintainability for Automobile Engines

Designing for maintainability in the automotive industry is shown in the many design improvements that have been incorporated. Improvements include:

- Speed of repair by replacing complete items rather than repairing the failed item
- Longer time between oil changes
- Longer life spark plug design
- More reliable alternator design (modified bearing housings for longer life; permanent lubrication feature)
- Increased life battery design
- Electronics that can monitor and optimize system performance and directly communicate with external test equipment

Very soon computers and sensors in the car will be capable to perform self-maintenance. Systems will monitor performance, identify problems immediately when they occur, communicate through satellites to the car company, identify corrective actions for the problem, and then either fix the problem automatically or tell the driver to take the car in for maintenance.

9.9.2   Repair and Maintenance Concept/Plan

The repair and maintenance concept or plan for a product is a description, flowchart and timeline of the way in which the repair actions will be conducted. The concept is established early in the design phase and continually updated to provide guidance for the development team. Initially, management or the customer defines the overall concept. The repair and maintenance concept later develops into a more formalized plan.

The maintenance plan must consider the user and their capabilities. The owner of a notebook computer with a problem needs information to decide whether to attempt to fix it themselves, call an information number, or take it to a service and repair site. A high level of built-in test capability and the ability for the computer to communicate directly with the service center over the Internet are design actions that can greatly assist the user. Self-maintenance is also helpful. Figure 9.3 shows a typical maintenance concept for a notebook computer.

9.9.3   Repairability Design Guidelines

Designing for repairability and maintainability means simplification, standardization, and inclusion of those features that can be expected to assist the technician. The design and its' support systems must therefore consider the technician's capabilities. A goal is to design equipment that can be operated and repaired effectively by the least experienced personnel with little or no outside assistance. A "typical" technician is assumed to possess minimal levels of education, training, and motivation. For international products, the "typical" technician will vary from country to country. When designing for maintenance, the familiar Murphy's law - If it is possible to do it wrong, someone will surely do it - is too often found to be true. Many of the human engineering design considerations relating to operators are applicable to repair personnel.

Since the design will ultimately fail, the first step is to identify all failure modes including those caused by test equipment, operators and repair personnel. In electronic equipment and software, fault detection and location often incur larger costs than the cost of the actual repair (i.e., replacement of modules, printed circuit cards or software). The impact of design for repair and testability (increasing the capability to detect and locate equipment faults) is felt in both production and logistics costs, which often represent 85% of the total life cycle costs. Reliability will also be improved by averting subsequent failures caused by repair or maintenance-induced human errors. Listed in Table 9.1 are some design considerations that will increase the repairability of a system (Imrhan, 1992).

Company Unveils Diagnostic Software

One computer company is developing an automated technical help program that includes software capable of independently diagnosing and repairing problems. (Goldstein, 1999) This software is based on a database of common potential problems and fixes. For example, if a customer can't get a printer to work, the computer automatically reviews the settings being used with

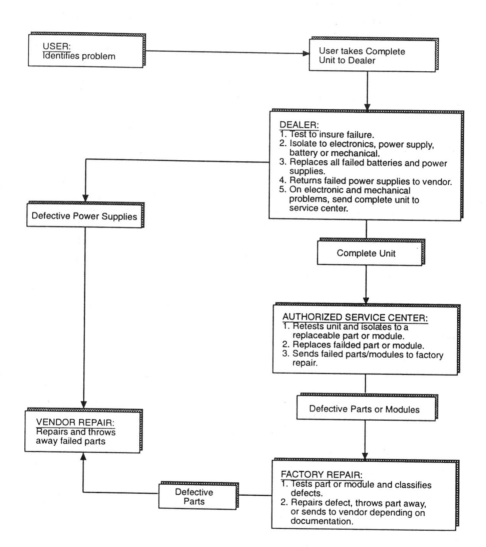

FIGURE 9.3 Maintenance concept for a personal computer.

TABLE 9.1 Design Consideration Examples

1. Locating and orienting parts to aid physical access

- Locate maintenance controls in front of operator, within his/her immediate view and reach.
- Locate lines such as wires, cables, piping and tubing so that the time taken to retrieve, to setup and to use them can be minimized.
- Locate units so that, when they are being removed, they will follow straight line or slightly curved paths instead of sharp turns.
- Orient parts (seats, pins, etc.) in order to minimize time for repair, removal or insertion; and to avoid constrained body posture.
- Position components and sub-assemblies so that they can be reached (removed and replaced) easily and quickly. Avoid locations that are difficult to reach, such as under seats, inside recesses, behind other components, and so on.
- Locate the most frequently failing (or most critical) components so that they are the easiest to access.
- Mount parts in an orderly way on a flat surface, rather than one on top of the other, to make them more accessible.

2. Task simplification and mistake proofing

- Design to promote quick and positive identification of the malfunctioning part or unit such as built-in diagnostic software.
- Design to minimize the number and complexity of maintenance tasks.
- Minimize the number and types of tools, training, skill level and test equipment required for maintenance, in order to avoid supply problems, frequent tool changes and the use of wrong tools.
- Standardize equipment (fastener sizes, threads, connectors, etc.) in order to minimize the number and type of tools needed for maintenance, and to eliminate supply logistics problems.
- Avoid using equipment parts that require special tools.
- Use interchangeable components within and across equipment to reduce supply problems except where components are not functionally interchangeable.
- Use a minimum number of fasteners necessary to maintain equipment integrity and personnel safety.
- Use snap-in retainers, latches, spring-loaded hinges, etc. To facilitate the removal and replacement of parts.

the machine looking for possible conflicts. One advantage to this system is that a log of all diagnostic work done by the computer is kept throughout the process so the user doesn't have to repeatedly explain the problem and the company gets summary data on user problems. This shows that processors are getting smarter about fixing themselves. In the future, many products will have self-diagnosis and self-maintenance capabilities. This will include software products that can upgrade themselves when necessary and refrigerators that can call for repair service before they fail.

9.9.4   Maintainability Analysis and Demonstrations

Maintainability analysis evaluates a product's repair tasks and estimates repair times. Important data necessary in this analysis are the concept, tool and test equipment requirements, failure rates, limited life items, and repair time predictions.

The design team can use repair time predictions for trade-off studies to evaluate design, test, logistics, and warranty approaches. Predictions for all actions can be summarized, plotted, and statistically presented. For example, this illustrates that 50% of all actions can be performed in 25 minutes or less. In this way, predictions help determine requirements, predict repair costs, compliance, and identify maintenance design features requiring corrective action. Predictions are either performed using random sampling or based on reliability-failure rate predictions. A principal use of the failure rate is to "weigh" the repair times for various categories of repair activities and thereby provide an estimate of their contribution to the total maintenance time.

An example of a format for documenting repair time predictions using reliability failure rates is shown in Figure 9.4. Note that the elements of a prediction are failure rates of units and repair time estimates of units. Repair time estimates are derived either from experience or from documented tables found in various textbooks concerning maintainability.

Repair demonstrations are another method of ensuring that the equipment can be repaired according to specifications. These are actual demonstrations of actions performed by technicians on the designed equipment. Repair times that might be evaluated include the following:

- Test time for built in test and problem detection
- Setup and test time for test equipment
- Fault detection time and fault isolation time
- Access time to the failed unit/software
- Replacement time for the unit/software
- Test time to verify repair

Inserting known faults into the equipment, one at a time usually starts a demonstration. Thus, many faults can be demonstrated and an average time can be determined. The number of faults allocated to each part of a design is usually based on the part's failure rate.

Subsystem	Failure rate	Average Time (hours) to Perform Corrective Maintenance Tasks							Avg. Time to Repair	Avg. Time Failure Rate
		Locali-zation	Isola-tion	Disas-sembly	Inter-change	Reassem-bly	Align-ment	Check out		
Transmitter	1400	.02	.25	.10	N/A	.10	N/A	.15	.62	868
Receiver	650	.02	.10	.15	N/A	.15	.10	.15	.67	435
Processor	550	.02	.25	.10	N/A	.10	N/A	.20	.67	368
Display	990	.02	.05	.05	N/A	.10	.10	.20	.52	515
Total	3500									2168

$$\frac{\text{Avg. Time}}{\text{to Repair}} = \frac{\Sigma\,(\,\text{Avg. Time x Failure Rate}\,)}{\Sigma\,\text{Failure Rate}} = \frac{2168}{3500} = .619\ \text{hours}$$

FIGURE 9.4 Repair time predictions.

9.10 DESIGN FOR SAFETY AND PRODUCT LIABILITY

Reducing the number and severity of the accidents caused by product failures requires the design team to anticipate and resolve potential hazards. Many, however, are uncomfortable with the negative viewpoint required for effective safety and product liability analyses. A design and support system is normally viewed in the positive sense of performing an intended function while operated by trained personnel in the appropriate environment. Safety and product liability analyses, however, require that the design be viewed in the most critical and negative light, under both appropriate and inappropriate conditions of use. System failures and product misuse are assumed in safety and product liability analysis. This negative viewpoint is required to identify all potential hazards rather than concentrating only on more probable hazards.

Evaluation of all hazard possibilities is not a reflection on the quality of the design. This is especially true for evaluating those events that are not a part of the normal function, such as product misuse, unique environments, and unusual failures. Two key steps for reducing the potential hazards and potential liability of a product are as follows:

1. **Implement a hazard analysis approach** to identify and correct all potential hazards, including these caused by misuse and abuse.
2. **Incorporate** design strategies and guidelines that minimize the hazard potential.
3. **Verify** safety requirements.
4. **Document** safety activities.

Safety techniques are used to help protect both the manufacturer and the customer. Although successful accomplishment of these techniques may often seem unproductive during the design phase, the real gain is the reduction in potential accidents, product recalls, and liability costs.

9.10.1 Hazard Analysis

A hazard is generally any condition or situation (real or potential) capable of injuring people or damaging the product itself, adjacent property or the environment. It is also important to recognize that product and property damage is also a safety consideration. It is similar to a FMEA except that the focus is on safety. The steps in an effective hazard analysis are as follows:

1. Identify major hazards.
2. Identify the reasons and factors that can cause the hazard.
3. Evaluate and identify all potential effects of the hazard.
4. Categorize the identified hazard as catastrophic, critical, marginal, or negligible.
5. Implement design changes that minimize the number and level of hazards.

Hazards are often classified according to the severity of the detrimental effects that could result. For example, the Department of Defense, in MIL-STD 882, System Safety Program Requirements, classifies hazard severity as follows:

Severity	Category	Accident definition
Catastrophic	I	Death or total system loss
Critical	II	Severe injury, severe occupational illness, or major system damage
Marginal	III	Minor injury, minor occupational illness, or minor system damage
Negligible	IV	Less than minor injury, occupational illness, or system damage

Accidents are generally considered random events with a certain probability distribution. For the purpose of safety analysis, the determination of the probability of an accident is usually simplified by assigning a qualitative value or by using reliability data. The MIL-STD-882 qualitative probability definitions are as follows:

Level of probability	Description
Frequent	Likely to occur; frequently experienced
Probable	Will occur several times in life of an item
Occasional	Likely to occur sometime; several times in life of an item; occur in life of an item; expected to occur
Improbable	So unlikely it can be assumed occurrence may not be experienced

The overall risk of a hazard (i.e., accident) can then be quantified as the combination of the hazard severity and its probability. A numerical combination of severity and probability is called the real hazard index (RHI). This index allows design decisions based on the acceptance of certain levels of risk or the prioritization of which hazards should receive the most attention. After the hazards are classified, each hazard can then be evaluated for the most appropriate design strategy.

The first step is hazard identification. The purpose of this task is to identify all "foreseeable" situations that could involve the product and result in a hazard. Controversy in this step is often caused by the perception that any hazard constitutes a flaw in a design. This is definitely not the case, since many hazards are unavoidable. Identifying every possible hazard at every level of the design can represent a significant effort. If not careful, designers become so intimately familiar with their design that they cannot perceive potential hazards that would be readily apparent to someone not having detailed knowledge of the design or specialized skills. Consequently, other designers and outside specialists are often used to identify hazards. Special training may also be required to accomplish this task efficiently and accurately.

In addition to hazards resulting from the design's primary functions, there are hazards resulting from activities that support the primary function, such as maintenance, production, and other products. In some cases these hazards represent more risk than those inherent in the primary function itself. For example, design considerations for the maintenance of electrical equipment are usually more difficult to incorporate than designing for the user. This is usually the result of the necessity to disassemble the product to perform the repair. In a disassembled state, hazards that are normally controlled and contained such as high voltages will be exposed. These potential repair hazards must be considered and controlled by the designer.

The means of controlling hazards is dependent on the nature of the hazard. Design features that are necessary for the functional performance of the design are often "inherent" hazards (e.g., fire is an inherent hazard of a kitchen match or electrical shock of an electric appliance). Inherent hazards cannot be designed out. Rather, the hazard is to be identified and controlled. Another hazard type can also result from the way an element is used in a design. This is an "implementation" hazard. This kind of hazard may be corrected by a design

TABLE 9.2 Hazard Analysis

System	Aircraft landing system
Subsystem	Transmitter
Hazard	Overheating caused by thermal radiation prior to flight
Effects	Operational: electronics overheating resulting in failure prior to takeoff Maintenance: possible burns if maintenance personnel touches unit
Hazard category	Marginal
Corrective actions	Built-in test evaluates potential failures caused by overheating Unit automatically turns off when inside temperature exceeds 15°C Maintenance personnel instructed to wear gloves when touching unit Maintenance procedures include warnings of high temperature

change without affecting the function of the product. Another class of hazards result from the inappropriate application of an element in a design. This is an "application" hazard. This kind of hazard may also be corrected without affecting function. Knowledge of the nature of hazards allows elimination of all but inherent hazards in the product.

A general outline for a typical hazard analysis is illustrated in Table 9.2. Since hundreds of potential hazards for a product may be identified, a systematic method of documentation is usually used. Typical safety analyses include preliminary hazard analysis, system hazard analysis, and operational hazard analysis. Certain types of hazard analyses are more beneficial during specific periods in the development of a product or system. As the program's life cycle progresses, the hazard analyses are then updated, revised, and expanded to ensure that all hazards are identified and subsequently minimized and controlled.

A preliminary hazard analysis is usually accomplished early in the requirements definition and conceptual design phases. This permits the early development of design and procedural safety plans for controlling the major hazards.

Other hazard analyses are a continuation and expansion of the preliminary hazard analysis. As the design of the system becomes more specific, the hazard analysis can also be more specific. The results can determine what safety requirements are needed to minimize and control the hazards to an acceptable level. Other analyses, such as fault tree, sneak circuit, and failure modes and effects are also used for safety analysis.

The efficient application of these analyses requires knowledge such as analysis methodology, federal regulations, typical failure modes, industry standards, and so on. Additional knowledge in specialties, such as user disabilities, explosives, lasers, and radiation, may be necessary, depending on the product line. In addition, some specialized unique techniques may be required, such as hazardous waste handling, explosive hazard classifications, and biomedical incident reports.

9.10.2   Weakness of Hazard Analysis

In System Safety Engineering and Management, Roland and Moriarty (1983) identified some common problems in performing several types of hazard analysis, limited experience of the designer, bias of personnel, unforeseen failure modes, and incomplete operational data base In addition, Hammer (1972) has pointed out that too much effort is usually concentrated on hardware problems and not enough on human errors. During years of work in industry, the authors have found many hazards caused by poorly designed software and the product's interface with other products.

9.10.3   Design Strategies for Safety

Several design choices are available when a potential hazard is identified. The following is a prioritized list of four design strategies:

1.   Design to eliminate or minimize hazard.
2.   Attach safety devices and protection.
3.   Provide warning labels and devices.
4.   Develop special procedures or training.

Level 1 Design to Eliminate or Minimize Hazard

The best strategy is to design a product that is inherently safe. To utilize this strategy, hazards are identified that have available design options to eliminate or minimize the hazards. This strategy requires design tradeoffs between safety and other requirements. It is not always possible to satisfy the product specification and eliminate all hazards. A completely safe knife will not cut. Electric appliances designed for households use hazardous 115 V AC as the power source. This choice of power source, however, presents the unavoidable possibility of fatal electric shock to the customer. Designing for a minimum hazard approach for the notebook computer would place the voltage step down transformer at the electrical wall socket and only run a reduced voltage and current to the product. For software products, the minimum hazard design approach could use a hardware or software approach to restrict access to any software areas, which could cause problems. This is often called a "firewall" to represent a wall that could stop access to an area of software code.

Level 2 Safety Devices

The next step is applied only when the design options for eliminating or minimizing the hazard are not possible. Safety devices are functions "added" to the design to control hazards. A very common type of safety device is a guard or barrier between some hazard and the user. For our appliance example the guard would eliminate all access to the 115V AC unless the guard was removed.

Safety devices must be carefully designed to be effective and convenient, since no additional hazards or reliability problems should be introduced because of a safety device itself. For example, a barrier on a high voltage terminal strip may need to be removable to allow repair. However, if the barrier is free to be removed, it is possible (and even likely) that sometime in the future it will not be replaced after it is removed. The safety capability of the barrier is improved if it is permanently hinged to allow access to repair points while preventing total removal. Further improvements in capability are gained if the configuration of the hinged barrier is such that the product cannot be reassembled unless the barrier is in its proper position. Software guards, firewalls, or barriers are safety devices for software products to limit access so that the user or other software products do not destroy or modify software code, which results in failures.

Level 3 Warning Labels and Devices

Warning labels and devices are often ineffective and are used only when safety devices cannot adequately protect the user. Labels are of little use when a person cannot read the message because of age, education, language or nationality. Liability suits have consistently demonstrated that warning labels do not protect a manufacturer from product liability. In fact, some people believe that warning labels signify that the company knew that the product was unsafe and, therefore, should have designed a safer product. Warnings do not replace the need for safety devices, although these labels may be necessary in conjunction with them. When absolutely required, the warning must be positive and unambiguous.

When considering the previously mentioned high-voltage barrier, it should be marked with a label to indicate the nature of the hazard exposed when it is moved. Labels are also used in software products to warn the consumers on how to protect the packaged media (i.e., diskette, CD, etc.) and methods to ensure proper operation of the software. Many labels have been developed for international use and are to be used when possible.

Level 4 Procedures and Training

Special safety procedures or training of personnel is the last resort for safety. Both are admissions that the design is potentially hazardous. Procedures are especially ineffective since they are apt to be ignored, lost, or forgotten. For example, how many people read a computer's instruction manual prior to turning on the computer and trying it? Over periods of time, the possibility increases of using incorrect or out-of-date procedures. The use of special training to prevent hazards is no better than the use of procedures, since it will provide protection only for those who have received the special training.

9.10.4   Safety Documentation

Since accidents and their resultant product liability suits will occur as long as a product remains in use, safety related design documentation is critical and must never be destroyed. Records to keep include design specifications, requirement interpretations, hazard analyses, and hazard control measures. These records are important since they substantiate the intent to design a safe product.

9.11   PRODUCT LIABILITY

Anyone injured because of a faulty product has a right of legal action against the designer, manufacturer, or seller of that product. The liability is called product (or products) liability. Customers, general public, or government on grounds of product liability, breach of contract, and breach of warranty can sue manufacturers. **Design for product liability is the design discipline that minimizes the potential liability of a product by identifying all risks, designing to avoid these risks, and maintaining all quality records.** Especially in the United States, product liability is a substantial financial risk of doing business. From both a humanitarian standpoint and from simple economics, it is important for designers to minimize this cost. Product liability cases usually arise under tort law and the law of contracts (i.e., warranty).

One method to avoid liability for product related injury is a contract clause known as "hold harmless". These are often called liability "waivers" or "disclaimers". The company attempts to absolve themselves from all liability for the malfunction of the product in writing. Customers agree to the clause as part of the purchase. An example of a disclaimer for a notebook computer manual is:

> This manual has been validated and reviewed for accuracy. Company A assumes no liability for damages incurred directly or indirectly from errors, omissions or discrepancies between the computer and the manual.

There are numerous cases defended on the basis of having such clauses in formal contracts. However, the validity of a disclaimer for most consumer products must be determined in court. The idea is well entrenched in law that one cannot contract himself out of liability to the public (Vaughn, 1974).

Negligence caused by manufacturing is difficult for someone to prove. Most people understand that manufacturing will have some defects. To prove manufacturing negligence, the individual must prove that the manufacturing process or quality control is negligent.

Negligence in product development, however, is much easier. A prominent question in product liability cases is whether the designer could have foreseen the problem. This "degree of foreseeability" places a large responsibility on the design group. The designer must foresee possible injury causing events throughout the product's life. "When we leave the realm of past and present and look at the future, we are leaving fact and turning to fantasy. What will happen to the product during its life and what will the product liability law become are somewhat difficult to surmise. Yet this is what we ask the designer to foresee" (Vaughn, 1974).

The California Supreme Court stated:

It has been pointed out that an injury to a bystander is often a perfectly foreseeable risk of the maker's enterprise, and the considerations for imposing such risks on the maker without regard to his fault do not stop with those who undertake to use the chattel.

If anything, bystanders should be entitled to greater protection than the consumer or user, where injury to bystanders from the defect is reasonably foreseeable. Consumers and users, at least, have the opportunity to inspect for defects and to limit their purchases to articles made by reputable manufacturers and sold by reputable retailers, whereas the bystander ordinarily has no such opportunities. In short, the bystander is in greater need of protection from defective products which are dangerous, and if any distinction should be made between bystanders and users, it should be made, contrary to the position of defendants, to extend greater liability in favor of the bystanders (Vaughn, 1974).

Product liability is also a growing concern for computer software and database developers. With the growing dependence on computers, economic and physical injuries can result from systems that do not perform adequately. The case of the aircraft crash in the section on functional allocation is a good example. What would happen if an airplane crashes because the automated flight control software does not function properly and the pilot cannot override the system in time? Potential liability from software failure can far outweigh the costs of completely testing the software or designing in more safeguards.

In summary, product liability analysis issues as listed by Burgunder (1995) are shown in Table 9.3.

The tasks to minimize product liability costs are similar to system safety. They are:

- Identify all product liability risks
- Design to eliminate or minimize risks

TABLE 9.3 Product Liability Issues for Manufacturing and Design

I. **Manufacturing**

 The product fails because of quality control procedures of the company or its vendors

II. **Design**

 A. Consumer expectation: The product fails to perform as safely as an ordinary consumer would expect.
 - Intended uses
 - Reasonably "foreseeable" unintended uses

 B. Risk benefits analysis: The risk of danger in the design outweighs the benefits in the design. Important factors are:
 - Gravity and likelihood of danger
 - Feasibility of alternative designs
 - Cost of alternative designs

 C. Failure to Warn Factors
 - Normal expectations by consumer
 - Complexity of product
 - Potential magnitude and likelihood of danger
 - Feasibility and effect of including a warning

Adapted from Burgunder, 1995

- Design to all applicable standards and knowledge of safety research
- Use stringent inspection and quality control procedures
- Maintain quality control and safety design records.

The wide variety of potential product faults and the difficulty of foreseeing all situations make the task of identifying all risks almost impossible.

Company Recalls Notebook Battery Charger

Even products that would appear to be safe can have problems. This story was reported in many United States newspapers. In cooperation with the U.S. Consumer Products Safety Commission, a company has launched a voluntary recall of about 3,200 external battery chargers for batteries used with their notebook computers. The chargers may have a defect involving a small electronic component that could create a potential fire hazard. Owners with part number XXX-YYY should stop using the chargers immediately and call or E-mail their name, address and phone number via the Internet.

9.12 SUMMARY

Designing for people is a critical parameter in the design process. Its purpose is to optimize system performance by reducing the potential for human error and ensuring that users can effectively and safely perform all tasks required by the design. This effort will result in a user-friendly design that can be easily maintained and supported after production. In addition, a system safety program will reduce the number of accidents and product liability costs. This chapter has provided an overview of the key practices of human engineering, repairability, maintainability, safety and product liability. This overview has, it is hoped, provided each member of the product development team with the importance of each discipline and their methodologies.

9.13   REVIEW QUESTIONS

1. In the golf case study, how would a designer take into account the "wildness" of golf shots?
2. List and explain the key techniques of human engineering.
3. Describe design concerns when a product will be used internationally.
4. How does the maintenance concept affect a product's design and vice versa?
5. Why is hazard analysis a difficult design task to perform successfully?
6. What potential safety and product liability risks exist for the following products?
    a. Notebook computer
    b. Automatic washing machine

9.14   SUGGESTED READING

1. S.N. Imrhan, Equipment Design for Maintenance: Part I - Guidelines for the Practitioner, International Journal of Industrial Ergonomics, Vol. 10, p. 35-40, 1992.

9.15   REFERENCES

1. Best Manufacturing Practices (BMP), Survey of Best Practices at McDonnell Douglas, www.bmpcoe.org, 1996.
2. L.B. Burgunder, Legal Aspects of Managing Technology, South-Western College Publishing, Cincinnati, Ohio, 1995.
3. W.M. Carley, Could a Minor Change in Design Have Saved Flight 965?, Wall Street Journal, p. 1, January 8, 1996.
4. B. DeGreene, System Psychology, McGraw-Hill, New York, 1970.
5. J.S. Dumas, Designing User Interfaces for Software, Englewood Cliffs, New Jersey, Prentice Hall, 1988.
6. A. Goldstein, Compaq To Unveil Diagnostic Software Dallas Morning News, September 28, 1999
7. J. Grey, To Make Skies Safer, Focus On Pilots, Wall Street Journal, November 3, 1999
8. Hammer, Handbook of System and Product Safety, Prentice-Hall, Englewood Cliffs, New Jersey, 1972.
9. S.N. Imrhan, Help! My Computer is Killing Me: Preventing Aches and Pains in the Computer Workplace, Taylor Publishing Co., Dallas, TX, 1996.

10. S.N. Imrhan, Equipment Design for Maintenance: Part I - Guidelines for the Practitioner, International Journal of Industrial Ergonomics, Vol. 10, p. 35-40, 1992.

11. J.A. Larson, Interactive Software: Tools for Building Interactive User Interfaces, Englewood Cliffs, New Jersey, Prentice Hall, 1992.

12. D. Meister, Human Factors: Theory and Practice, Wiley, New York, 1971.

13. D. Meister, Methods of Predicting Human Reliability in Man-Machine Systems, Human Factors, 6(6): 621-646, 1964.

14. MIL-STD-882, System Safety Program Requirements, U.S. Government Printing Office, Washington, D.C., 1984.

15. A. Moss, Designing for Minimal Maintenance Expense, Marcel Dekker, New York, 1985.

16. B.A. Myers, User Interface Software Tools, ACM Transactions on Computer-Human Interaction, Vol 2, No. 1, p. 64-67, March, 1995.

17. V. Postrel, Rising Heat: New Thermostat Designed by Brilliant Morons, Forbes ASAP, November 29, 1999.

18. E. Roland and B. Moriarty, System Safety Engineering and Management, Wiley, New York, 1983.

19. H. Stubbings, Student Report, 1998.

20. R.C. Vaughn, Quality Control, Iowa State University Press, Ames, Iowa, 1974.

Chapter 10

PRODUCIBILITY: STRATEGIES IN DESIGN FOR MANUFACTURING

Integral Part of the Design Process

Producibility is not a simple, one-time analysis performed near the end of the design process. Rather, a successful process requires a continuous journey of providing critical producibility information as needed to the design team and performing "many" different producibility analyses. This starts early in requirement definition and continues throughout a product's useful life. Success is fostered by an effective infrastructure that includes collaborative multi-discipline design teams with design, production, support areas, and vendors. Providing accurate, timely and quantified manufacturing and vendor information is essential for the team to make accurate producibility decisions.

Best Practices

- Producibility Infrastructure and Process
- Producibility Requirements Used for Optimizing Design and Manufacturing Decisions
- Consider Company's Business and Manufacturing Environment
- Knowledge and Lessons Learned Databases
- Competitive Benchmarking
- Process Capability Information
- Process Capability Studies and Design of Experiments
- Manufacturing Failure Mode Analysis (PFMEA)
- Producibility Analyses, Methods and Practices
- Design Reliability, Quality and Testability

Total Approach For Producibility

One of the most successful applications of producibility occurred in 1979. IBM Lexington began one of the largest automation projects in the world with an investment of nearly $350 million (Editor, 1985; Hegland, 1986; and Waterberry, 1986). Their strategy included:

1. New products designed specifically for automation
2. Early manufacturing involvement in the design process
3. Commonality of parts and modules
4. Continuous flow manufacturing

Design and manufacturing worked with each other from the beginning, thus promoting design for automation and manufacturing planning. This allowed the design development of the complex automation systems to be done in parallel with the product design. A good example of a producible design was the membrane keyboard, which has only 11 unique part numbers, contains no screws, and is assembled almost entirely by robots. The previous keyboard had 370 discrete part numbers, contained 65 screws, and was assembled by hand!

The size and complexity of the project suggested a modular approach. Rather than having project teams for each product, each new product was divided into five common areas or modules. This allowed the team members to concentrate on their specific areas and to minimize variations between different products. Each automated manufacturing area was then run as a separate unit with its own production manager.

All motors use the same end caps, bearings, terminators, and insulators. The only variable is the number of laminations in the stator stack. After assembly, robots automatically test and stack the motors. In final assembly, each unit is tested for printing quality. A selection of characters is printed, and a vision system examines the printed material, pixel by pixel, to determine that the shape, alignment, density, and skew of the characters are within the specifications.

Restricting itself to 60 suppliers, an 10-fold decrease from the past ensured continuous factory flow. Suppliers were committed to a just-in-time schedule and adhere to strict quality control requirements. As a result, 80% of the shipments from qualified vendors go directly to the shop floor without going through incoming inspection. Retail cost of a new versus previous model typewriter was reduced by an estimated 20%. The number of warranty repairs decreased by a factor of 4, indicating improved quality and reliability. The system can be reconfigured at a later time to produce another product at a cost of 10-15% of original capital. Thus, the design and factory was well positioned to respond to a changing marketplace.

This chapter reviews the strategies, processes and key practices of producibility.

10.1 IMPORTANT DEFINITIONS

Producibility is a discipline directed toward achieving design requirements that are compatible with available capabilities and realities of manufacturing. Other terms used interchangeably with producibility are manufacturability, design for manufacture, design for assembly, design for automation, design for robotics, design for production, and design for "X". Collaborative engineering, integrated product development and simultaneous engineering also imply producibility implementation. Regardless of the term used, producibility is the philosophy of designing a product so that it can be produced in an extremely efficient and quick manner with the highest levels of quality and minimal amount of technical risk. This is accomplished through an awareness of how design, management and manufacturing decisions affect the product development process, including the capabilities and limitations of new technologies, processes and vendors. Different companies use different methods and measures. The critical objective is to provide the "right information" to the design team at the "right time" to make informed decisions.

Producibility requirements and performance can be measured by the relative ease of manufacturing a product in terms of cost, lead-time, quality and technical risk. Some examples of producibility measures that are often used in design requirements and producibility analyses include:

1. Cost
   - Total manufacturing cost
   - Part and vendor cost
   - Direct labor cost to build a product
   - Complexity
     - Number of parts, parameters, features, etc.
     - Level of precision required
       (i.e., tolerances/manufacturing requirements)
     - Number of fasteners (assembly)
2. Schedule or lead time
   - Manufacturing and /or purchased part lead time
   - Total product lead time including ordering and shipping
3. Quality
   - Projected number of defects or yields, Cp and Cpk
   - Cost of quality (prevention, measurement, and warranties)
   - Variance of critical parameters
4. Technical risk
   - Number of new technologies, parts, vendors and processes
   - New or never before achieved levels of requirements

In 1987 while working on Motorola's Six Sigma Program, Dr. Mikel Harry, Lawson and others developed a more precise definition of producibility. They described **Producibility** as "the ability to define and characterize the various product and process elements that exert a large influence on the key product response parameters, and then optimize those parameters in such a manner that the critical product quality, reliability, and performance characteristics display:

- Robustness to random and systematic variations in the central tendency (μ) and variance (σ^2) of their physical elements,
- Maximize tolerances related to the "trivial many" elements and optimize tolerances for the "vital few",
- Minimize complexity in terms of product and process element count, and
- Optimize processing and assembly characteristics as measured by such indices as cost, lead time, etc. (Harry,1991)."

The key producibility recommendations of this definition for both design and manufacturing are:

- Concentrate on key design parameters (i.e. key characteristics)
- Develop design parameters that are "robust" to variation
- Loosen tolerances on "trivial many" non-critical parameters and develop optimum tolerances for the "vital few" critical parameters
- Minimize design complexity
- Optimize design parameters for manufacturing process and assembly

The producibility metric of a design with this definition is the predicted number of defects per million opportunities (dpmo) or the number of expected sigmas of quality. For a machined part design, producibility would be the projected number of defects per million opportunities expected based on a design's requirements (e.g. tolerance of .005) and the statistical distribution of the selected manufacturing process's variability. Note that the process capability will vary based on the factory, type of machine, material selected, brand name and age of the machine, operator skill, statistical confidence required etc.

10.2   BEST PRACTICES FOR PRODUCIBILITY

Producibility is often the difference between the success of a product or its failure. History shows many technically elegant designs that failed as products because their cost was higher than their perceived value or their quality was poor. Producibility needs to be an integral part of the design process. This is shown by:

1. **Establish an effective producibility process and infrastructure that includes:**
   - Producibility requirements are an integral part of the design process
   - Collaborative multidiscipline team design approach
   - Sufficient resources available
   - Information rich environment
2. **Producibility requirements are used for optimizing design, manufacturing, and support requirements** and are accurate, timely, up to date, easy to understand, etc.
3. **Tailor design to the company's current business and manufacturing environment that includes:**
   - Manufacturing and business strategies
   - Schedule and resource constraints
   - Production volume
   - Product mix
   - Availability and cost of capital
   - Global manufacturing and product development
   - Dependence on vendors and outsourcing
   - High overhead and parts cost relative to product cost
   - Rate of change
4. **Competitive benchmarking and prototyping** to identify the most competitive but reasonable requirements. In addition, innovative imitation (i.e. copying and improving the best ideas) develops more producible and supportable designs.
5. **Process capability information** includes limits, variability and technical risks.
6. **Process capability studies and design of experiments** when appropriate are conducted to identify process capabilities in order to select preferred low-cost, high-quality processes.
7. **Manufacturing failure mode analysis (PFMEA)** considers all potential manufacturing problems and then determines ways to minimize their occurrence or at least minimize their effects.
8. **Producibility guidelines, analyses, methods, and practices** evaluate designs, promote simplicity, predict/measure the level of producibility, identify problems and suggest areas for improvement.
9. **Several producibility methods** that have been successfully used in industry are implemented such as:
   - Best manufacturing practices program recommendations
   - McLeod's producibility assessment worksheet (PAW)
   - Mistake proofing and simplification
   - Boothroyd and Dewhurst design for assembly

- Robust design
- Taguchi methods
- Six sigma quality and producibility
- Failure mode analysis, Isakawa diagrams, and error budget analysis
- Design for quality manufacturing
- Formal vendor and manufacturing qualification of processes
10. **Design reliability, quality and testability** use proven design practices to improve a product's reliability and design quality.

Practices 1-6 are discussed in detail in this chapter. Practices 7-10 are discussed in later chapters in the book.

10.3   PRODUCIBILITY PROCESS

Design is a major cause of manufacturing problems. Articles on the success of Japanese manufacturing have mentioned a 40:30:30 percentage rule on the reasons for quality problems. This principle states that:

- 40% of all quality problems are the result of poor design
- 30% are the result of problems in manufacturing
- 30% are the result of nonconforming materials and parts supplied by outside vendors

Similarly a survey of companies in Europe showed that the top four major obstacles to automation were design related. these were design not assembly oriented, part-handling problems such as tangling or packaging, high degree of adapting and adjusting operations, and visual inspection required during assembly (Schraft, 1984).

The product development team must consider the many opportunities for manufacturing problems, which include:

- Electrical components
  - Every parameter
- Mechanical parts
  - Every dimension
- Software modules
  - Every line of code
- Manufacturing
  - Every time handled, every process, and every vendor
- Quality control
  - Every time tested, inspected, or handled
- Material handling, packaging, and shipping

- Every time handled and moved
- Every time packaged, stored, or shipped

Industry has too often concentrated solely on manufacturing solutions to problems. Lean manufacturing, quality control, just in time, statistical quality control, and other new manufacturing ideas will help, but more work must be done in addressing the design and logistics portion of the problem.

One answer is to "design in" producibility. How exactly does one proceed from an initial engineering idea or prototype to a highly producible design? Unfortunately, there is no simple answer to this question. As in design, producibility is both an analytic process and a creative science that draws on many disciplines. Producibility needs to be an integral element of the design process, with close coordination between the production and design. Producibility requirements and forming design teams with production, support areas, and vendor representative's foster integration of producibility factors in the design. The design is continuously evaluated to ensure that the producibility and supportability requirements are met. Production personnel participate in design development, and designers participate in production planning to ensure design compatibility with production. Vendors are included at every step to ensure high quality parts and on time delivery.

It is important that the team note the difference between early engineering designs and production designs (i.e., producible, easy to manufacture). Although an early engineering design may fulfill all performance requirements, a production design (i.e. producible) not only meet these requirements, but also minimizes production costs and technical risks and maximizes quality and reliability. This difference is probably best illustrated by the following examples. Although each example looks entirely different, similar producibility design techniques were used.

10.3.1   Producibility Analysis Examples

The following two examples from Texas Instruments (now Raytheon) provide specific examples of improvements to an engineering design that resulted from a producibility analysis. In both cases, the production design showed greater reliability, increased quality, and reduced production costs. Such benefits are common when producibility techniques are applied company wide.

Coldfinger Flange

The first example is the redesign of a coldfinger flange used to provide cooling in a cryostat or a sterling cycle cooler. The first step was to identify difficult to produce features of the flange design (i.e. technical risks). These were identified by a producibility analysis and either eliminated or changed. Table 10.1 compares the final production design to the early prototype design.

TABLE 10.1 Coldfinger Flange Production Costs

	Production Design	Early Design	Improvement (%)
Total cost ($)	37.58	112.40	66
Production time (hr)	0.61	1.45	58
Setup time (hr)	20.40	23.72	14
Major operations (#)	4.00	6.00	33

Producibility also focused on the key functional requirements (i.e. key characteristics) for the coldfinger flange. Alternative designs were considered. Innovative and less obvious designs were also considered. The function of every feature on the part was reviewed relative to its design requirements. This approach was used to develop the simplest part design with which to begin the manufacturing process analysis. Design analysis revealed that seals around the mounting holes were unnecessary if the system was sealed at the O-ring groove. This design change eliminated difficult to produce O-ring grooves.

The manufacturing process analysis phase consisted of identifying the optimal manufacturing method for the coldfinger flange. Part of the manufacturing plan developed is summarized below:

1. Lathes will turn most outside diameters and bore main inside diameter.
2. Lathes will turn remaining inside diameters and grooves.
3. Mill all hole patterns and off-axis holes.

Analysis showed that for this part, the optimum machines were precision Harding NC lathes and Bostomatic vertical milling centers.

The next phase was to tailor the coldfinger design to the specific requirements of the selected manufacturing machines and methods. An example of such a task centered on the production of a "gas pass" by drilling a 0.062-inch diameter hole with a 13° angle from the main housing into the coldfinger cavity. This was a very difficult feature to produce, even by the processes selected. Designers determined that a shorter, straight hole and a brazed tube would also perform the required function. This change more closely matched the manufacturing capabilities, resulting in higher part quality.

Power Converter Circuit Board

A second example of applying producibility techniques was the design of a power converter circuit board. A power converter circuit card assembly is a sophisticated power supply. Its function is to convert vehicle power into precision voltage levels. Power supplies tend to be very difficult to manufacture since they contain large, hard to assemble parts, such as thermal planes, heat sinks, and manual hardware mounting.

Owing to the cost and quality advantages of automation, producibility emphasis was placed on applying automated manufacturing methods throughout the production cycle. The ability to automate this circuit board production depended on three producibility factors:

1. **Simplification** to reduce the number parts
2. **Selection** of automatable components
3. **Spacing** of the components on the board layout, so they can be placed by automatic insertion equipment

The producibility effort started with design simplification, which revealed that some components were redundant; these were removed. Thermal analysis also showed that some of the heat sinks were not needed. Selecting automatable components and layout spacing resulted in an overall gain to 79% autoinsertability of the components, which is considered excellent for power supplies.

The following specific improvements were integrated into the power converter design during the production design effort:

- All axial component polarities were turned in the same directions, resulting in easier assembly and inspection.
- All components were laid out to improve flow solder quality and yields. When a circuit board passes over the flow solder machine, it is important that as few closely spaced leads as possible pass over the solder wave at one time to reduce the risk of solder bridging.
- All variable resistors were located along the perimeter of the circuit board to make testing and maintenance easier.
- Clocking tabs on all can-type devices, such as transistors, were turned in the same direction. This makes assembly and inspection easier resulting in fewer repairs.

The production design contains fewer components and is more automated than the engineering design. The results are summarized in Table 10.2. The reduction in quality problems was significant. Again, although both designs have similar performance characteristics, the production design is considerably more producible. Production designs are simpler (e.g. fewer operations), allow more efficient manufacturing methods (e.g. more automation), and easier to meet manufacturing requirements (e.g. tolerances) than engineering designs.

TABLE 10.2 Printed Wiring Board Design Comparison

	Production design	Early design	Improvement (%)
Total components	153	192	20
Automatable (%)	69	15	78
Quality problems	102	288	65

Access Plug

The third example is the design of an access plug for an electronic box. On some occasions, the maintenance worker needed to be able to take out the plug and the look into the box. The plug needed to seal the hole from the environment. The first design is shown in figure 10.1. The cost was estimated to be $ 25.00 each due to its low volume. Through a producibility analysis it was determined that the functions could be performed by selecting a common screw. The cost of the purchased screw was $0.50, a very large saving.

10.4   PRODUCIBILITY INFRASTRUCTURE

A collaborative multidiscipline team effort requires an effective infrastructure including managerial support and resources. Resources such as expertise, tools, time and money are needed for early and constant involvement of all team members, effective systematic trade-off analysis process, extensive prototyping and effective communication. A critical objective is to provide the "right information" to the design team at the "right time" to make informed decisions. This information can be provided in many ways including access to a manufacturing engineer who is on the team, design guidelines, computerized knowledge bases, CAD design rules, or any combination of these.

Many excellent producibility processes have been proposed. The Best Manufacturing Practices (BMP) program has formed two different producibility task forces, one in 1992 and the other in 1999. The first task force produced recommendations for measuring producibility. The second task force published additional recommendations for developing a producibility infrastructure and process. The recommendations were co-authored by experts from industry, government, and one academic. Dr. Priest was fortunate to be on both task forces. The recommendations can be found at www.bmpcoe.org. Both of these documents have proved valuable in helping many companies apply specific producibility measurement tools. The first producibility Task Force (PTF) focused on in-depth methods for measuring producibility. The second task force determined that "comprehensive and complex" analyses also needed a simplified and concise, common sense, system level perspective. The group found that "infrastructure" was the first key step needed to successfully implement producibility.

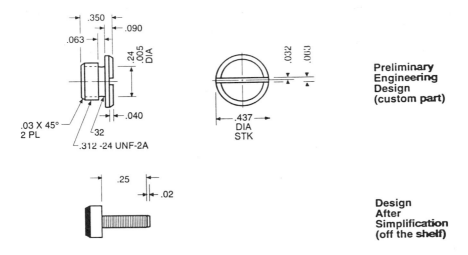

Preliminary
Engineering
Design
(custom part)

Design
After
Simplification
(off the shelf)

FIGURE 10.1 Access plug.

10.4.1 Establish A Producibility Infrastructure*

Before developing an effective infrastructure, one should understand where a company is, and where it needs to go. The 1999 PTF task force developed a matrix for an organization to "self-assess" its level of producibility development. This level or proficiency – defined as "Traditional," "Improving," and "Fully Versed in the Process" – can be measured against a multiple set of criteria to determine the company's level of producibility maturity. This matrix is shown in Table 10.3.

Responsibilities for driving the cultural changes that are necessary for an effective producibility program include:

- **Long-term commitment** to institutionalizing the application of producibility as an integral part of doing business;

- **Highly skilled people and resources** are available early in the design process;

---

* Section adapted from BMP, Producibility Systems Guidelines For Successful Companies, 1999

TABLE 10.3 Producibility Infrastructure Assessment and Maturity Matrix (adapted from BMP,1999)

Criteria	Traditional Producibility	Improving	Fully Versed In process
Organizational structure	Functional	Cross functional	Totally team integrated
CAD tool set	Simple/purchased stand-alone	Internally modified stand-alone	Integrated into CAE (toolset)
Metrics	Cost/schedule	Quality yields	Balanced scorecard of cost, quality, complexity and risk
Supplier relationships	After the design is complete	Critical suppliers phased into the process	Upfront early involvement
Process capability assessment metric	SPC	C_p & C_{pk}	Six sigma
Process information	Documented guidelines	On-line database	Knowledge-based tools
Requirement definition	Internal assessment	Voice of the customer considered	QFD/ requirement mapping

- **Education and training** employees in the producibility process to include methods and tools;
- **Empowering** teams to take action using producibility techniques;
- **Documenting** the results and lessons learned of producibility efforts as a means of validation.

The goal is to provide the development team only with what the expectations are – not how to best implement the solution. Management demonstrates its commitment to the producibility process through active engagement. It initiates producibility requirements, process, sets project goals, empowers teams, remains visible, provides managerial inputs, and commits to implement the results. Strong commitment, leadership, and resources generate success in producibility system. Program management should establish a seamless, information-rich environment in which the design team organization receives relevant and timely information to affect a producible design.

Producibility is improved by having an infrastructure that enables the generation and distribution of both expert knowledge and lessons learned, while ensuring they are applied where appropriate. Producibility design requirements and guidelines provide parameters within which the design team should operate.

10.5 PRODUCIBILITY REQUIREMENTS USED FOR OPTIMIZING DESIGN AND MANUFACTURING DECISIONS

Formal producibility requirements can ensure that the best product development decisions are made. Producibility requirements provide the informational foundation for the design team's trade-off process and product decisions. They affect the design, manufacturing, support and other disciplines. The requirements identify the level of quality, lead-time, direct cost and technical risk expected for the project. They are based on projected performance needs, new technologies, history, consensus, experience, lessons learned, statistical and non-statistical data, process measurements, and prototypes. Producibility requirements consider resource availability (parts, materials, production systems), production capacity (low versus high volume, part and volume mix), and tolerance limits. The following sections will discuss factors in developing producibility requirements. For more details reference Producibility Design-To Requirements by Hausner (1999).

10.6 TAILOR DESIGN TO THE CURRENT BUSINESS AND MANUFACTURING ENVIRONMENT

The product development team first considers the conditions of the economy, the competition and the company's "current" business and manufacturing environment. These conditions affect the design, vendor, manufacturing and producibility options that are available. Some of the key business environment factors that should be considered by the design team because they can greatly affect manufacturing and, in turn producibility, include:

- Manufacturing and business strategies
- Schedule and resource constraints
- Production volume
- Product mix
- Availability and cost of capital

- Global manufacturing and product development
- Dependence on vendors and outsourcing
- High overhead and parts cost relative to product cost
- Rate of change

The steps are to identify detailed information on each of these topics and then to determine how they affect design requirements.

Manufacturing strategies establishes the company's goals for manufacturing and determines the Producibility options available. Is the company's goal to have the highest level of performance, quality, or lowest cost? If a company has a long-term commitment to manufacturing excellence the design team can be more aggressive in identifying possible options and setting manufacturing requirements. Leading manufacturers are always improving their manufacturing capabilities and are more likely to have the infrastructure to meet difficult requirements. World-class companies generally have the environment, finances, and expertise to take on new challenges. New manufacturing processes that can significantly improve process capability, lower cost or improve quality can be considered. In contrast, a weak or non-committed manufacturing company does not have the resources for excellence nor the ability to take innovative technical risks. This causes the design team to be very conservative in determining manufacturing requirements and designing to existing low cost manufacturing methods. A successful product development team understands current manufacturing methodologies and technologies, how to design for them, where the required expertise resides, and how to identify the best vendors.

Schedule and resource constraints determine the types of producibility methods to be used. When resources such as expertise, money, time, etc. are limited, the producibility methods, manufacturing processes, and vendors that can be used are also limited. Schedule and cost constraints are facts of life. There simply never seems to be enough time or money to complete a design as one would wish. Limits on time or resources restrict the types of producibility methods that can be used. When expertise is limited within the company, training or consultants may be needed. When financial resources are limited, producibility efforts should focus on areas that can produce the greatest savings within the resource limitation. This usually means focusing only on the highest cost items. For example, low risk and low set up cost manufacturing methods should be used to minimize potential cost overruns.

When time is limited, producibility often focuses on working with vendors on purchased parts. For example, some integrated circuits, tooling, and test equipment have very long delivery lead times that must be included in the overall schedule. If specialized parts are required, its delivery lead-time may also be significant to the overall schedule. Purchasing these long lead-time items early can help to minimize the problem.

When time and money is not severely limited, companies can perform many different producibility analyses in great detail. Many companies are pursuing technology based producibility approaches such as direct analysis of CAD files, producibility software packages, expert systems and intelligent lessons learned databases.

Producibility is greatly affected by the number of products to be manufactured (i.e., its production volume or rate). Thus producibility methods and recommendations for low volume production may differ significantly from high-volume producibility. High-volume systems should be designed with automation and other high-volume production methods in mind. Producibility requirements will focus on designing to the particular automation equipment to be used. Guidelines might include tolerance requirements, top down assembly, inspectability and testability and part quality. Part quality is especially important for automated equipment so that the defective parts do not jam and shut down the equipment. Since automation equipment is usually complex and expensive and requires long delivery lead times, producibility analyses and plans for automation are made early in the design process.

If a system is projected at low production rates, its producibility should emphasize manual operations, since the volume is simply insufficient to justify expensive automation. Design efforts should concentrate on designing components and tooling to simplify manual operations. For example, mistake proofing techniques such as self-fixturing parts and tooling increases product quality by reducing the chance of human error. When the projected volume may vary from low to high, the design team should try to design for both automation and manual requirements. A summary of the differences in manufacturing approaches for low- for high volume versus low volume production is shown in Table 10.4.

To further illustrate the effects of volume, some of the process changes at one company for extremely low volumes, which is defined as 10 or less assemblies is shown below. In this example, producibility focuses on vendors performing many of the tasks to reduce technical risk.

- Only use experienced multi- layer supplier with good track record.
- Full continuity check at vendor location
- Manual planning control system, not standard MRP system
- For expensive and critical parts,
    Require full environmental tested at vendor location.
    Require vendor certification
- Assemble in prototype lab, not in production
- Manual placement and assembly
- Designer provides less formal documentation

TABLE 10.4 Differences In Low And High Volume Manufacturing

Parameters	Low Volume	High Volume
Manufacturing environment	Job shop Higher skill level Flexible process flows	Mass production Low skill level Established process flows
Level of process knowledge	Minimal knowledge of process due to low volume (i.e., lack of experience)	High volume allows statistical process control to study, optimize, and stabilize processes
Use of automation	Insufficient volume for automation	Sufficient volume to make automation cost effective. Can develop custom processes and partnerships with vendors.
Vendor relations	Little or no clout with vendors due to small volumes	Allows considerable clout with vendors. May have vendor partnerships.
Documentation	Minimal and often work off incomplete prints	Detailed documentation and procedures
Producibility objectives	Focus on saving large amounts per product by staying out of trouble (i.e., reducing technical risk) and reducing part costs	Saving small amounts of cost on every part and every process results in significant cost savings
Design for assembly	Manual Assembly	Manual and/or automated assembly

Product mix and mass customization can limit the ability of manufacturing to keep costs low. Product mix is when a number of different products are produced in one manufacturing area). Mass customization is when the customer can directly order many different configurations or options of the product. If the number of different products or options (i.e. product mix) is large, manufacturing tasks are more complex and will probably have more problems.

The key is to make different products and options as common or similar for manufacturing as possible. Each new design should be similar to or part of a family of products containing similar platforms, modules, components, vendors, manufacturing processes or assemblies. This reduces manufacturing complexity and creates a higher production volume. Significant savings can result from higher levels of quality, lower cost manufacturing methods such as group technology and automation and from the purchasing function by getting lower prices due to the larger volumes.

Producibility focuses special attention to design, part and process commonality or similarity such as

- Common product platforms can be used as the baseline for every new product design.
- Identical or very similar components may be used on several different designs, lowering production costs through standardization and economies of scale.
- Dedicated highly efficient manufacturing areas, called group technology cells, can be created to produce parts that have similar manufacturing requirements and processes. These groups of similar parts are called part families.

In short, exploiting opportunities for commonality is an important consideration during the producibility effort.

The availability and cost of capital is important since new technologies, automation and processes require large amounts of money. Capital equipment is a term usually applied to manufacturing equipment with a useful life greater than 1 or 2 years. Capital equipment costs are considered "nonrecurring" or "fixed" costs since they are incurred only once, generally prior to the start of production: In contrast, "recurring" or "variable" costs consist of most ongoing production expenses, such as material, labor, and utility costs. Recurring costs occur each time a unit is produced, and they are dependent on production volumes. When production volume is high and capital is available, the design team can consider new design and manufacturing technologies and equipment that have high potential. The expense of the new technology can be off set by the improved performance, quality, lead-time or cost. In contrast, a low production volume product will not yield enough savings in production cost and quality to pay for the initial investment in the new technology or equipment. In this case, producibility would design for manual assembly, use existing processes and limit the number of producibility analyses performed.

The emergence of the global manufacturing has created another new dimension to product competition and producibility. For manufacturing, this has often resulted in production facilities in other countries and the use of many overseas vendors. Differences in language, culture, time zones, and etc. can cause many communication and logistics problems. This makes producibility even more difficult.

The producibility methods and analyses do not change for global implementation but implementing them successfully becomes more of a challenge. The same products may be manufactured in different countries. The different manufacturing facilities can vary in terms of process capabilities, labor skills, quality, cost, logistics etc. Logistic costs due to shipping and handling vendor parts become a larger portion of the product's cost. Critical parts are often purchased from another country that is then shipped to the various facilities. Scheduling, ordering, and shipping of parts and products all over the globe also becomes much more complex. Effective communication with international vendors can be difficult due to the differences in languages, cultures, time zones, education, and government regulations. At one semiconductor company, the design was performed in the U.S.; it was manufactured in Japan, packaged in the Philippines, and then shipped to manufacturing plants all over the world.

The total cost of purchased parts for many products is a high proportion of the product's total cost and is becoming higher. It is often more than 50% of a product's total direct cost! The author worked on one project where total cost of purchased parts was 95% of the entire product's recurring cost! The project's entire work force could have been eliminated but that would only have saved 5%. When this occurs, producibility efforts best reduces part costs by focusing on working with the vendors, identifying lower cost vendors, or redesigning the product to reduce purchased parts cost.

Dependence on vendors and outsourcing to contract manufacturers is increasing. The trend is to focus manufacturing on new product introduction, complex products and high value added products and then contract out the rest of the manufacturing. Some companies are using other companies called contract manufacturers to produce their product or service. This is especially the case for foreign manufacturing or in high technology areas that require large investments to stay current. For example, most electronic companies outsource their base printed circuit boards and many outsource the circuit board assembly. This makes the outsourcing vendor a critical partner in the success or failure of a product. Selecting parts from preferred vendors with proven track records can greatly help. Choosing the best vendor and effectively communicating with the vendor are important tasks. This selection is also important for the software models that are chosen. The producibility solution is to use qualified vendors with proven technologies, quality, and ability to meet schedules.

High overhead costs are a major problem for many manufacturing companies, especially those that develop high technology products. For many companies, overhead rates can range from 100 to 300 percent. These high rates put a lot of pressure to reduce overhead by outsourcing production to lower cost manufacturers (i.e., use more vendors). In this instance, producibility efforts can focus on all areas to reduce overhead costs and working with the vendors to reduce their costs and to ensure that the product is compatible with the vendor's capabilities.

Rate of change found in many industries is changing so rapidly that significant technical risks must be taken. Changes in competition and technology are requiring companies to make major changes rapidly. For many companies constant change is normal. This causes the design team to take risks such as implementing new technologies that they might not take in a stable environment.

10.7 KNOWLEDGE AND LESSONS LEARNED DATABASES

Producibility knowledge defines a companies or suppliers manufacturing capability by delineating the capacity, limits, variability, technical risks and rules associated with each factory, vendor, technology, and manufacturing process. This information can include production volume, cost, lead-time, quality, Cp, Cpk, reliability, defect rates, risks, failure modes and tolerances. Producibility knowledge assists and coordinates all members of the development team to consider manufacturing and vendor capabilities during product development. They provide advice, information, lessons learned and instructions and can vary from suggestions to rules or requirements. The goal is to provide important manufacturing and support information to the design team in a timely matter.

As noted in earlier chapters, using lessons learned and best industry practices have been shown to be a very effective method for product development. This method identifies the "best" practices for a particular application and ensures to not repeat the same mistakes from previous projects. The Best Manufacturing Practices (BMP) program sponsors the largest and most popular repository of lessons learned for producibility. As noted in their website, www.bmpcoe.org, BMP identifies and documents best practices in industry, government, and academia. An electronic library comprised of expert systems and digital handbooks covers a variety of design topics, including ISO 9000. The automated program offers rapid access to information through an intelligent search capability. How-To cuts document search time by 95% by immediately providing critical, user-specific information.

10.8 COMPETITIVE BENCHMARKING

Design and manufacturing requirements should be competitive (i.e. as good as or better than the competition) for the product to be successful in the marketplace. Competitive benchmarking and prototyping are used to develop optimum and realistic manufacturing and support requirements. It is also used to identify "best in class" ideas for design improvements and innovative initiation. Designing products that utilize a company's existing strengths yields the greatest return. Prototyping is used as a communication tool to evaluate design ideas to help develop requirements. Prototyping allows everyone including potential customers to visualize and "touch" the design.

10.9   PROCESS CAPABILITY INFORMATION

Requirements should be based on an optimum balance between functional performance and manufacturing/vendor/support process capabilities. The development team selects design parameters that are compatible with process capabilities. For software, capability can include computer response time, memory availability, web bandwidth/speed etc. For service industries, capability includes operator/user response times, phone line availability, etc.

This requires detailed process capability information on a company's technologies and processes. Producibility's goal is to:

- **Reduce** or minimize all requirements on non-critical areas
- **Standardize** requirements within the design and with other products on non-critical areas
- **Optimized** balance of design and manufacturing requirements on critical requirements
- **Minimize** the number of defects caused by variation and
- **Compensate** for variability by reducing its effects on a product.

To enhance producibility, requirements should always be as loose and flexible as realistically possible without affecting critical performance parameters. An example of the effects of simplifying tolerance requirements on cost is shown on Figure 10.2. This demonstrates the simplification effect for machining and surface finish. For this reason, the first step is to simplify all manufacturing requirements where possible.

Producibility also considers the variability of the process when determining design requirements. Variability is often referred to as "process, vendor or part tolerance". Functional parameters include electrical parameters (e.g., voltage, timing, current), mechanical parameters, (e.g., dimension, strength, assembly fits) and software parameters (e.g., timing, user interface). Aesthetic parameters include paint quality and surface finish.

The cause of variability can come from the process such as purchased part, machine, environment, or operator, or can occur overtime such as aging and drift. Variability cannot be eliminated. The overall design of a product should compensate for and be tolerant to ever-present variations in the manufacturing processes and the parts used. When the product is placed into production, the design makes allowances for the anticipated "shifts and drifts" in the process and parts occurring over time. This can require the designer to use large design margins that reduce performance and often increase cost. The design team finds the optimum level of design requirements and manufacturing requirements. Some producibility techniques that effectively compensate for manufacturing variability include tolerance analysis, mistake proofing, Taguchi robust design, and six sigma quality methods.

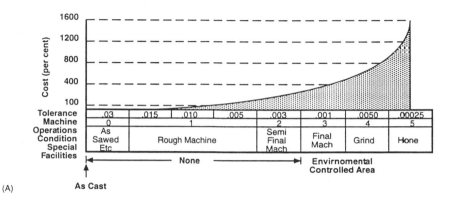

(A)

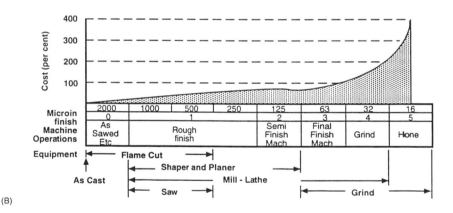

(B)

Figure 10.2 Effects of tolerance on cost: (A) machining tolerance;
(B) surface finish tolerance.

After simplification analysis has simplified the design and its requirements as much as possible, the next step is to identify information on the process capability of all manufacturing processes and vendors that might meet the design requirements. Information on a manufacturing process is determined by performing a process capability analysis. Process capability analysis statistically characterizes all process capabilities and variability in order to provide producibility information to the design team for setting requirements. Hopefully this will allow the most economical, highest quality, best understood, lowest-risk method or vendor available that meets the design's requirements to be used. Process variability is statistically described.

The metrics of process capability describe how well the process can perform overtime with terms such as limits and variability descriptors as the mean (μ or x), variance (σ), sigma (rms) and the type of statistical distribution (normal, exponential, etc.). The key descriptors are the mean and variance. Sigma is a statistical unit of measurement that describes the distribution's variance about the mean of any process or design parameter. Considering specific design requirements, a particular process or part can achieve a plus or minus sigma capability. Manufacturing typically uses the metrics of $\pm 3\sigma$ or $\pm 6\sigma$ as acceptable measures for sigma.

Cp and Cpk correlate design specifications to process capability. Knowing this information allows the design team to predict the probable occurrence of the number of defects for producibility and quality analyses. **Cp** is a ratio of the width of the acceptable values (i.e., design tolerances) to the width of the distribution (i.e., process width). **Cpk** refines Cp by the amount of process drift (i.e., the distance from the target mean to the process distribution mean). These are important and popular statistic based metrics for balancing design and manufacturing requirements. They measure the ability of a manufacturing process to meet specified design requirements. The C_p and C_{pk} measures are statistically defined as:

$$Cp = \frac{USL - LSL}{6 * sigma\,(rms)}$$

$$Cpk = Cp(1-k)$$

And

$$k = \frac{2\left|T - \overline{X}\right|}{USL - LSL}$$

Where

C_p	=	inherent process capability
USL	=	upper specification limit
LSL	=	lower specification limit

sigma (rms)	=	root-mean-square standard deviation
C_{PK}	=	non-centered process capability which includes process drift
k	=	distribution shift
dpmo	=	defects per million opportunities
T	=	target, actual center of the specification (i.e., nominal target)
$\bar{X}$	=	actual process average/mean

For a bilateral specification, one company's six sigma quality goals are:

- $Cp \geq 2.0$
- $Cpk \geq 1.5$

The design team specifies an initial acceptable level of variation (i.e., tolerance) for the requirement. This initial level of variation is based on performance and aesthetic related goals. Producibility compares the mean and the allowable variation in the design requirements to the variation found in the company's manufacturing processes, vendor parts and software. The goal is to optimize the variability for critical parameters and loosen the manufacturing requirements for all non-critical parameters.

Producibility capability studies compare the center and spread of a process with the required center and spread of a design's specifications.

10.9.1 Performing a Process Capability Study and Design of Experiments

When producibility information is not available or out of date, new process capability studies are conducted to measure, analyze, and evaluate process variables. They determine the inherent and operational ability of a process to repeatedly produce similar results. Many innovative companies consider this step to be the integral part of manufacturing high quality products. The steps in performing a process capability study are as follows:

1. Ensure that the process is properly maintained and ready for study.
2. Collect process data from the manufacturing process over different conditions.
3. Determine statistical patterns, and interpret these patterns.
4. Make design and manufacturing decisions based on the process capability information.

The process capabilities of a company's manufacturing equipment are sometimes known and already documented. When new technologies, processes, parts, materials, software, vendors, manufacturing facilities, or unique design parameters are specified, however, special process capability studies and experimentation will be required.

Design of experiments is an important method for developing process information about manufacturing and vendors. Experimentation is used when:

- Relationship between design requirements and the manufacturing process is poorly understood or unknown.
- Limited resources are available to get measures of process variation for design/process decisions.
- "Unbiased" information on variation and interactions is needed.
- Systematic statistical approach for documenting process capability is preferred.

A critical part of a process capability study is the proper collection of data. The process conditions is defined in sufficient detail to determine whether existing data is sufficient or new studies and analyses will be required. When performing a process capability study, the conditions for the test are defined. For mechanical processes, conditions to be specified might include machine feeds, speeds, coolant, fixtures, cycle time, and any other aspect that could influence the final measurement. For soldering processes, examples of controlled conditions are temperature and humidity. Process capability results are meaningless unless they can be related to a defined set of conditions and repeated in the future. When the results of a study are derived from only one machine of a particular type, the machine selected should be a representative of that type. In addition, a "machine" may actually consist of several machines (e.g., a machine with multiple spindles may show different results for supposedly identical spindles). In such cases, the study should quantify each spindle's variation. A sufficient number of tests should be run so that enough information is gained about the process. A sample of at least 50 units is desirable. The team decides how many measurements should be taken from each test and exactly how the measurements should be made and interpreted

The next step in the process is to evaluate the collected data and identify statistical patterns. A key part of this step is first to identify data variability and determine the natural and unnatural portions. The natural variation is the normal variation that occurs in a manufacturing process. Unnatural variation is the variability that is caused by some outside factor and can be eliminated by improving the manufacturing process. Several analysis tools are available to assist the team in interpreting the data. Most of these tools are relatively simple statistical methods, such as histograms and control charts. Examples of some typical histograms and control charts from a process analysis are shown in Figure 10.3. The analysis is used to identify process parameters such as distributions, patterns, trends, sudden changes, cycles, unusual single-point data, instability, drift, control limits, and percentage out of specification.

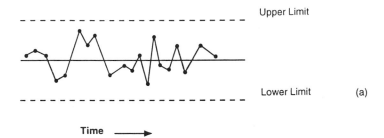

Upper Limit

Lower Limit   (a)

Time ⟶

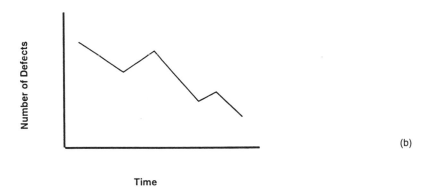

Number of Defects

Time

(b)

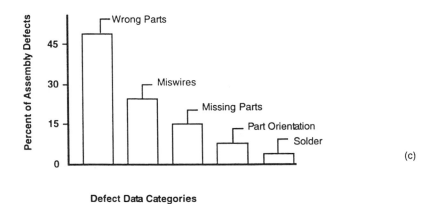

Percent of Assembly Defects

Wrong Parts

45

30   Miswires

Missing Parts

15   Part Orientation

Solder

0   (c)

Defect Data Categories

FIGURE 10.3   Histograms and control charts:
(a) Parameter monitoring (X chart);
(b) Quality monitoring;
(c) Pareto analysis of assembly defects (histogram).

Although detailed methods for identifying these parameters is beyond the scope of this text, many textbooks on data evaluation are available.

After the data have been statistically evaluated, the design team can make informed decisions. A process characterization summary is shown in Figure 10.4. If the manufacturing process can meet all design, quality, cost, and availability requirements, the process is then specified. In most cases, however, the process can only meet some of the design, quality cost, or availability requirements. The design team then makes difficult decisions between changing design requirements, identifying different manufacturing processes, or reducing the variability in the existing manufacturing process.

When the company's existing manufacturing processes are not capable of meeting the design's requirements; manufacturing must develop new processes or significantly improve an existing process. The new process development will occur concurrently with the design process

Prototyping and simulation are very useful tools in determining process capability. Verifying process capabilities and process prototyping can help reduce the risks associated with commitment to production, and investing in tooling and fixtures for an untested process or a new part design. A process prototype is created through the design of sample prototype parts that contain characteristic design features of the production part. Sample parts are then fabricated to determine process capabilities such as determining tolerances that can be held. This data is used to establish initial process capability data before production data can be gathered.

The remaining sections will briefly discuss the remaining best practices. Each of these topics will be covered in more detail in later Chapters.

10.10   MANUFACTURING FAILURE MODES

Murphy's Law is correct; what ever can go wrong will. As noted earlier, the design team considers all potential manufacturing problems and then determines ways to minimize their occurrence or at least minimize their effects. As discussed in an earlier chapter, a manufacturing or production failure mode and effects analysis (PFMEA) is a technique for evaluating the effects caused by potential failure modes. The PFMEA is a producibility analysis technique that documents the failure modes of each process and determines the effect of the failure mode on the product and manufacturing. Critical failure modes are eliminated through design improvements that can include improved processes, new vendors, new equipment, etc.

10.11   PRODUCIBILITY ANALYSES, METHODS, AND PRACTICES

A major step is to evaluate, measure and predict the level of a design's producibility. How producible is the design in terms of cost, schedule, quality, resources, risk, etc.? Most producibility evaluations and analyses include

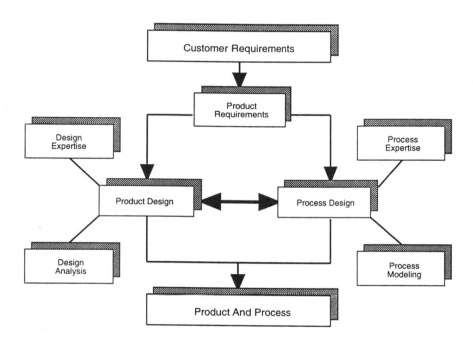

FIGURE 10.4 Process characterization

quantified measurement. Producibility metrics are an integral part of design trade off analysis and include the manufacturing parameters of cost, quality, process variability, lead-time, and technical risk. Production engineers and vendors assist the design team in the producibility analysis by identifying:

- Producibility concerns and measures such as cost, schedule and quality e.g. Cp and Cpk
- Potential manufacturing technical risks and failure modes
- Preferred manufacturing processes, materials, components, software modules, and vendors
- Manufacturing and software standards, capabilities, and limitations
- Design criteria for optimizing existing fabrication, assembly, and software production
- Production, test, inspection, packaging, and repair procedures

Since vendors play such an important role in today's manufacturing, additional questions for vendors might include:

- Can a standard item be substituted?
- Are there design or manufacturing alternatives?
- Suggestions to improve performance, quality, cost, schedule, variability, etc.

Although these prototypes are generally used by designers for evaluating form, fit and function, they can also be used in producibility analysis. Prototypes also provide an excellent method of communication between design and manufacturing. Early involvement allows sufficient time for manufacturing processes to be developed, tested, and ready when the design is ready for production. Many product designs are too complex to adequately evaluate all producibility aspects if only CAD drawings are used. Prototyping allows everyone to visualize and "touch" the design. Prototypes are also used to qualify manufacturing methods and processes. Computer simulations are used to evaluate product manufacturing. Assembly tolerances, methods and time standards, simplification analysis, process requirements, production problems, and other information can also be gathered.

Remember that designs are manufactured from documentation. Engineering requirements, drawings, procedures, and reports are the foundation for developing manufacturing plans and requirements. Communication gaps caused by poor documentation are often cited as a critical problem in producing high-quality products. Careful review of design documentation will avoid these unnecessary production problems.

The rest of the book will focus on practices and methods that can improve producibility and reliability. The following chapters are design reliability, simplification, design guidelines, producibility methods and testability. The many producibility methods to be discussed in the methods Chapter are:

- Best Manufacturing Practices "Lessons Learned" Program (BMP)
- McLeod's Producibility Assessment Worksheet (PAW)
- Boothroyd and Dewhurst Design for Assembly (DFA) and Design for Manufacturing (DFM)
- Robust Design
- Taguchi Methods
- Six Sigma Quality and Producibility
- Failure Mode Analysis (PFMEA), Isakawa Diagrams and Error Budget Analysis
- Mistake Proofing and Simplification
- Design for Quality Manufacturability (DFQM)
- Formal Vendor and Manufacturing Qualification and Certification of Processes (e.g. ISO 9000)

10.12 DESIGN RELIABILITY, QUALITY, AND TESTABILITY

High quality products require a design that has a high level of reliability and quality. **Quality for both hardware and software is a measure of how well a product satisfies a customer at a reasonable price.** Product quality is measured by sales, customer satisfaction, customer feedback, and warranty costs. Quality is achieved by knowledge, attention to detail, and continuous improvement by all of the design team (adapted from O'Conner, 1995).

Design quality is measured as how well the design meets all requirements of the customer and other groups that interact with the product. Design quality can be measured by how well the product's design performs as compared to its product requirements and to the competition.

The better the design, the better the chance manufacturing has to be successful. **Manufacturing quality is often measured as the percentage of products that meets all specified design and manufacturing requirements during a specified period of time.** This is also expressed as failures, yield or as a percentage of products with defects. Any deviation from the customer's requirements is called the "cost of quality" whether it is caused by design or manufacturing.

One purpose of test is to verify that the product or service meets all requirements at the lowest cost. The goal is to minimize testing since it is a non-value added task. Testability is the process of designing a product so that it can be tested more efficiently and effectively. It also includes self-diagnostics and self-maintenance. The process is to identify the points or items to test and making the test process as effective and efficient as possible.

Reliability is the probability or likelihood that a product will perform its intended function for a specified interval under stated conditions of use. Reliability is a projection of performance over periods of time, and is usually defined as a quantifiable design parameter such as mean time between failure (MTBF) and mean time to failure (MTTF).

Design reliability is a design discipline that uses proven design practices to improve a product's reliability. The key techniques are:

1. Multidiscipline, collaborative design process
2. Technical risk reduction
3. Commonality, simplification and standardization
4. Part, material, software, and vendor selection and qualification
5. Design analysis to improve reliability
6. Developmental testing and evaluation
7. Production reliability

Reliability, the design reliability process, and testability are discussed in later Chapters.

10.13   SUMMARY

The concepts underlying producibility are not new. Success stems from the careful consideration of manufacturing issues during the early phases of the design effort in order to anticipate and eliminate high-risk or high-cost production problems. The key to successful implementation of a producibility program is a management commitment to ensure that design concepts have been subjected to thorough analyses before production begins. Although producibility analyses may require additional "up-front" expenses, the benefits obtained will make the effort more than worthwhile. Producibility analysis techniques are only successful when the design requirements are compatible with manufacturing's capabilities and variability.

10.14   REVIEW QUESTIONS

1. What is producibility and how is it measured?
2. List the key points of the six sigma definition of quality.
3. How does the business environment affect producibility?
4. Explain how a producibility approach should be modified for very low volumes, high part cost, and high overhead situations.
5. Describe types of process capability information for different processes and products and when is design of experiments used?

10.15 SUGGESTED READINGS

1. BMP, Producibility Systems Guidelines For Successful Companies, Navso P-3687, www.bmpcoe.org, 1999 and Producibility Measurement Guidelines, NAVSO, P-3679, Department of the Navy, 1993.
2. G. Boothroyd, Assembly Automation and Product Design, Marcel Dekker, New York, 1992.
3. J.G. Bralla, Handbook of Product Design for Manufacturing, McGraw-Hill, New York, 1986.
4. M. J. Harvey and J.R. Lawson, Six Sigma Producibility Analysis and Process Characterization, Addison-Wesley, New York, 1992.
5. S. Pugh, Total Design, Addison-Wesley, New York, 1990.

10.16 REFERENCES

1. BMP, Producibility Systems Guidelines For Successful Companies, www.bmpcoe.org, 1999
2. BMP, Producibility Measurement Guidelines, NAVSO, P-3679, Department of the Navy, August, 1993, and Guidelines 1999. www.bmp.coe
3. Editor, IBM: Automated Factory - A Giant Step Forward, Modern Materials Handling, March, 1985.
6. General Electric, Manufacturing Producibility Handbook, Manufacturing Services, Schenectady, New York, 1960.
4. M.J. Harry, The Nature of Six Sigma Quality, Motorola Technical Presentations to U.S. Navy Producibility Measurement Committee, 1991.
5. W. E. Hausner, Producibility Design-To Requirements, Appendix E.2 to Producibility Systems Guidelines For Successful Companies, BMP, www.bmpcoe.org, 1999
6. E. Hegland, IBM: Focus on Quality and Cost to be Effective, Assembly Engineering, February, 1986.
7. P. O'Conner, Foreward of Electronic Component Reliability, Wiley, New York, 1995
8. D. Schraft, The Importance of Assembly-Oriented Product Design, Proceedings of the Seventh Annual Design Engineering Conference, Birmingham, England, U.K., 1984v.
9. Waterberry, IBM: Management Meeting the Automation Challenge, Assembly Engineering, February, 1986.

Chapter 11

SIMPLIFICATION: COMMONALITY AND PREFERRED METHODS

Producibility and Reliability's Major Goal

>*Simplifying a product can have a greater positive effect on cost, quality, producibility, reliability, availability, logistics and even aesthetics than any other technique. A simple or common design is easier to design, manufacture, and support than a more complex design. A simple design has fewer parts and options, is easier to build, operate and repair, and requires less non-value added processes. A common design is similar enough to a current product to allow the efficient reuse of designs, parts, software, manufacturing processes, test equipment, packaging, logistics etc. Simplification is a major design goal for all disciplines.*

Best Practices

- Simplification is a Major Goal of Product Development
- Keep It Simple – KISS and Complexity Analysis
- Limit Number of Customer Options, Features
- Product Platforms, Lines and Families
- Modularity and Scalability
- Part, Process and Vendor Reduction
- Re-engineer or Eliminate Non-Value Added Tasks
- Part Families and Group Technology
- Function Analysis and Value Engineering
- Ergonomics and Human Engineering
- Mistake Proofing and Poka Yoke
- Minimize Requirements and Effects of Variability
- Reduce Technical Risks
- Common, Standard, and Reusable Designs
- Standard or Preferred Part, Software and Vendor Lists

FIRMS SAVE MONEY USING SIMPLIFICATION!

"Businesses are making changes often subtle, sometimes seemingly silly to save millions of dollars. A single minor adjustment can mean less time on an assembly line, fewer workers to pay, a lighter load to ship, and ultimately, a better bottom line" (Schwatz, 1996).

Ford Motor Co. saved $11 billion and built better vehicles using simplification and standardization (Schwatz, 1996).

- Three types of carpeting rather than nine saved an average of $1.25 per vehicle or $8 million to $9 million a year.
- Five kinds of air filters rather than 18 saved 45 cents per vehicle or $3 million annually.
- One type of cigarette lighter instead of 14 varieties saved 16 cents per car, or $1 million per year.
- Black screws instead of color-matched painted screws on Mustang side mirrors saved $5.40 per vehicle or $740,000 per year.
- Skipping the black paint inside Explorer ashtrays saved 25 cents per vehicle or $100,000 per year.

Note that none of these changes affect the product's performance or changes the customer's thoughts on the product. Most simplification ideas lower cost and technical risk and improve quality and reliability without affecting appearance or performance.

Schwatz also reported that Breyers ice cream had a manufacturing problem with the cellophane cover sheet inside the carton's top flap. Each rectangular sheet was stamped with the words "pledge of purity" that had to be centered over the ice cream. Centering the words on the box caused many quality problems. Replacing the single pattern that had to be precisely centered with a continuously repeating pattern saved money. A repeating pattern eliminated the process of centering the cellophane sheet and eliminating the need for precision cellophane trims on the assembly line (Schwatz, 1996).

An article by Otis Port for Business Week also highlighted the importance of design simplification. A new cash register was designed with 85% fewer parts, 65% fewer suppliers, and 25% fewer components than previous products. "Putting together NCR Corp.'s new 2760 electronic cash register is a snap. One can do it in less than two minutes blindfolded. (Port, 1984)." Using design for assembly methods developed by Boothroyd and Dewhurst, the biggest gains come from eliminating all screws and other fasteners. "On a supplier's invoice, screws and bolts may run mere pennies apiece, and collectively they account for only about 5% of a typical product's bill of materials. All of the associated costs, such as the time needed to align components while screws are inserted and tightened, and the price of using those mundane parts can add up to 75% of the total assembly costs" (Port, 1984).

11.1   IMPORTANT DEFINITIONS

Design simplification is a design technique that reduces the number or complexity of manufacturing and support opportunities such as the number of tasks required or the probability of problems. The number and difficulty of the opportunities are determined by the design's requirements. Simplification tries to reduce a product's complexity. For example, reducing the number of parts will generally reduce part costs, labor costs, non-value added costs, shipping costs, failures, etc. For many products a simpler design will look better. The fewer the number of opportunities and the simpler each opportunity is to perform correctly, the more likely the product will be built and supported with higher quality at a lower cost and on schedule.

The metrics of simplification are based on the concept that a design's complexity is a function of three aspects:

1. Number of opportunities (measured by number of parts, features, lines of software code, operator options)
2. Level of difficulty for meeting each opportunity (measured by level of tolerances, tolerance fits, software timing, operator training, shipping requirements etc.)
3. Technical risk or unpredictability (measured by number of new or unproven processes, parts, vendors, software modules, or users)

Design or product commonality is a simplification technique where the number of unique or "significantly different" requirements, parts, designs, processes, vendors, and logistics methods is minimized. The level of commonality can vary from identical items to items that are only similar in one certain aspect such as a manufacturing process. There are several different types of commonality methods including product families, product platforms, part families and group technology, standardization and preferred lists. Using commonality helps to reduce vendor, inventory, and manufacturing set-up, training, shipping and support costs and reduces risks. Libraries of common parts, vendors, processes, software modules and design approaches are often available to assist the development team **The metrics of commonality are based on the number of new, different or unique items within this product and with previous products: requirements, designs, vendors, manufacturing processes, production lines, software modules, and logistics.**

The first aspect of simplification is to reduce the total number of opportunities or possibilities. The number of opportunities can be measured by the number of:

-   Product options including global permutations
-   Tasks and options necessary to manufacture or operate
-   Non-value added processes required

- Parts or different parts
- Different vendors
- Oriented surfaces/features
- Set-ups required
- Lines of software code, modules or subroutines
- Operator/user options
- Repair tasks
- Units per shipping container
- Shipping locations
- Composite summary of the metrics listed above

The second aspect is the level of difficulty in meeting each opportunity i.e. level of requirements. **It is the measure of the costs, quality, etc. that the specified level of a design requirement or task will cause.** This is how difficult the opportunity is for a particular manufacturing or support process. It can be measured by:

- Specified level of tolerance when compared to the manufacturing or service process's capability such as Cp and Cpk
- Complexity of user or repair tasks compared to historical error rates
- Number of options to perform each opportunity
- Shipping, packaging or set-up requirements compared to other products
- Number of new non-value processes required

The third and final complexity aspect is risk or uncertainty. **Technical risk is a measure of the uniqueness or newness of the design and it's required processes when compared to previously produced designs.** Risk is defined as the probability that design, manufacturing process, software, or vendor is not capable of meeting all requirements. Since new or unproven technologies, processes, vendors, and support methods increase the level of technical risks in the design, the likelihood of unexpected costs and problems also increases. As mentioned before, it can be measured by the number of new, unproven, or unique:

- Designs, parts, materials, software modules, features, technologies and options
- Vendors and logistic support methods
- Users and operators
    - Manufacturing or software programming processes
    - Design tools and testing methods used in development
    - Requirements that have historically caused problems

It should be noted that different metrics may have different relationships such as linear, exponential, non-linear, etc. For example, the complexity factor for multilayer circuit boards can increase exponentially with the number of layers.

11.2 BEST PRACTICES FOR SIMPLIFICATION AND COMMONALITY

Some people can produce innovative, simple designs, whereas, others seem to struggle. Is product simplification an art or a design technique? Are some people naturally better at product simplification than others? Unfortunately, the answer is yes! Simplification and commonality can be accomplished by using many best practices including:

- **Keeping it simple – K.I.S.S. and complexity analysis** simplifies the design and its support system.
- **Limit number of customer options** or features (i.e., features and permutations) to reduce complexity.
- **Limit number of international permutations** to reduce the number of manufacturing and service tasks.
- **Product platforms, lines and families** capitalize on manufacturing similarities to decrease manufacturing and support errors and costs.
- **Modularity** divides complex systems into separate modules having defined interfaces to simplify manufacturing, testing, repair, logistic and maintenance tasks.
- **Scalability** allows systems or products to be developed or enlarged by combining modules or duplicating designs.
- **Part reduction** includes deleting unnecessary parts, combining parts or designs, and reducing the number of different parts by standardizing common parts.
- **Process and vendor reduction** reduces complexity.
- **Re-engineering** is used to eliminate non-value added tasks.
- **Part families and group technology** also capitalize on manufacturing similarities to decrease manufacturing errors and costs.
- **Function analysis and value engineering** identify a simpler design that can perform the same functions.
- **Minimize manufacturing requirements and the effects of variability** using tolerance analysis, robust design, and six sigma quality.
- **Human engineering and error reduction techniques** reduce the total number of human error opportunities and the chance of these errors occurring.

- **Mistake proofing and Poka Yoke** is a powerful technique for avoiding simple human errors.
- **Reduce technical risks** by using proven technologies, manufacturing processes, vendors, and software products.
- **Commonality** uses proven designs, software modules, parts, materials, and vendors to reduce risks, costs, etc.
- **Standard or preferred part, material, software, and vendor lists** minimize the number of different entities.
- **Software reuse** is emphasized to save costs, reduce technical risk, and reduce lead-time.

In summary, simplification's goals are 1.) Fewer in number, 2.) Less complex, 3.) More robust, 4.) Less technical risk, and 5.) Commonality. Each of these best practices is different as far as steps, analysis, and goals; but their purpose is to accomplish one or more of these goals. Our approach is to:

1. **Identify areas** for simplification
2. **Use the identified best practices** to reduce complexity and improve the design

11.3   KEEP IT SIMPLE:
"THE K.I.S.S METHOD" AND COMPLEXITY ANALYSIS

People want simplicity in their life. No one wants complex, hard to use products. Wabi sabi is an ancient Japanese 14th century philosophy that values the simple, modest, imperfect, unique, and unconventional. "Less is more, simple is best." It is the antidote to the complexities of today's technology and our busy lives. The more basic and unpretentious a product the more beautiful it is. Wabi sabi products are also unique, unsophisticated, plain and often worn or weathered. Although true wabi sabi products are not mass-produced, the design team can incorporate the philosophy or essence into their mass-produced products.

Simplification and commonality would appear to be an easy to apply common sense approach. However, products are continuously designed with unnecessary complexity and risks. Product simplification is not a simple, one-time evaluation; rather it is a complex decision process requiring common sense, detailed analyses and trade-off studies. Complexity analysis measures the level of complexity (or simplification) of a design. Complexity measures can be the number of parts, lines of software code, number of vendors, or any of the measures listed earlier in the chapter.

The most effective method can be brainstorming meetings where the product development group tries to identify the simplest methods for a design and how to manufacture the product. The goal is to optimize the metrics described earlier. This method is often called the K.I.S.S. method (keep it simple, stupid!). The objective is to minimize the number of opportunities for cost and problems such as operator tasks, options, parts, vendors, technologies etc. The steps are:

1. Assemble a multi-disciplinary group of experienced people
2. Present detailed goals and objectives of the meeting
3. Review the overall product's requirements
4. Determine complexity metrics to be analyzed (design attributes or features, part count, process required, schedule, cost, tooling, etc.)
5. Analyze design against complexity metrics.
6. Open the meeting for discussion and encourage unique and "wild" ideas for reducing complexity
7. Extensively study all ideas for feasibility
8. Select and incorporate the best ideas

Complexity analysis is a formal method examining and measuring the complexity of a design based on historical data. For example, complex design attributes or features may require the acquisition of new machinery, processes, or personnel capabilities. Complex designs may cause schedule problems and cost overruns, as they present significant risk areas and may cause impact on the project requiring workarounds or design changes. Initial complexity analysis requires an experienced team of producibility and other disciplines to assess the various issues (and non-apparent issues) inherent to the analysis. Independent experts in all disciplines associated with the product should conduct the formal complexity analysis.

Training should focus on obtaining knowledge in understanding the interrelationships between desing, manufacturing, vendors and service. Outside instructors can be used if new designs are beyond the knowledge base of the internal organization. Software tools are available to assist in complexity analysis including Boothroyd Dewhurst, Inc. Design for Assembly (www.dfma.com). Because software tools that may be available to assess complex features of designs may not meet the individual needs of a specific project, an internal capability that matches the product line may also be developed.

11.3.1 Hardware Complexity Analysis

There are several popular methods for evaluating hardware complexity. These include number of parts, number of assembly or manufacturing processes, number of new parts/technologies, number of high-risk aspects (i.e. tight tolerances, thin walled casting, etc.) and manufacturing lead-time. For surface mount assembly, complexity is measured by the number of solder joints and the total number of components. In fact most of the opportunities listed earlier in the

Chapter can be used. One method for calculating complexity is assembly efficiency developed by Boothroyd and Dewhurst. The quantifier for the analysis is the manual assembly efficiency, E_{ma}, which is based on the theoretical minimum number of parts, N_{min}, the total manual assembly times, t_{ma}, and the basic assembly time for one part, t_a. Given by the equation:

$$E_{ma} = N_{min}t_a/t_{ma}$$

Where the variables N_{min}, t_a, and t_{ma} are calculated using values given on Boothroyd and Dewhurst's tables.

Some other successful methods are "PAW rating" by Scott McLeod, and "DPMO by Six Sigma". These are discussed in more detail in a later Chapter.

11.3.2   Software Complexity Analysis

Software complexity can be thought of as how difficult it is to maintain, change, and understand software. There are literally hundreds of complexity measures found in literature that encompass different types of structural, logical, and intra-modular complexity (Zuse, 1990). Some popular complexity measures mentioned in literature are the Measures of Halstead (which measures length, volume, difficulty, and effort), Lines of Code (LOC), and Measures of McCabe (Cyclomatic Number).

Halstead, in 1977, based most of his software complexity measures on the source code of programs. Measures of Halstead are Length (N), Volume (V), Difficulty (D), and Effort (E), where length is the number of distinct operators (n1), number of distinct operands (n2), total number of operators (N1), and total number of operands (N2) (Zuse, 1990).

Many companies measure the lines of code to determine software complexity. What determines a line of code? A line of code is any line of program text that is not a comment or blank line, regardless of the number of statements or fragments of statements on the line. They specifically include all lines containing program headers, declarations, and executable and nonexecutable statements (Zuse, 1990).

McCabe gives programmers a different way of limiting module or program size to a set number of lines. Cyclomatic number is more reasonable than limiting number of lines for controlling the programs' size (Welsby, 1984). The formal definition of McCabe's cyclomatic number as stated by Welsby:

$$MC = \text{McCabe's cyclomatic number}$$
$$= e - n + 2p$$

where

e = number of edges (or statement-to-statement arcs) in the program
n = number of nodes in the program
p = number of connected components (essentially, the number of independent modules if each has its own graph)

where

Nodes represent one or more procedural statements.
Edges (arcs) indicate potential flow from one node to another.
An edge must terminate at a node.
A predicate node is a node with two or more edges

11.4   LIMIT CUSTOMER OPTIONS OR FEATURES

Every customer wants a customized product that has his or her unique features and functionality. In contrast, simplification wants to limit the number of options and products that are available. **The goal of mass customization and design variant is to minimize the number of unique or different modules, products or parts required to still meet each customer's unique requirements.** Mass customization is very important simplification technique and critical for many products' success. Customizing the product, however, does not mean that significant manufacturing or support differences between each unique product must exist. The key techniques for mass customization are to limit the number of options for manufacturing by:

Using product platforms, lines and families, modularity, scalability, and design variants

The first method is to limit the number of options or features for manufacturing. When possible, the steps are:

1. Limit the number of product features, options, or the number of permutations of a product that can be manufactured at the factory or ordered while still satisfying the customer.
2. Change optional features to standard features.
3. Install optional features closer geographically to the customer such as in a warehouse, retail store or by the customers themselves.
4. Build flexibility into the product to easily allow options at a later time.

Most car models are made with only two or three option packages that group "options" as standard equipment. Thus, few actual options are available. Nissan announced that it was reducing the number of different steering wheels on their products from 80 to 6 and cup holder from 22 to 4. Reducing complexity understandably reduces manufacturing problems resulting in lower costs, shorter lead times, and high quality. An internal study by General Motors suggests that the bulk of the cost gap (more than $2500 per car) is directly attributable to systems complexity (Howard, 1985). Reducing the number of models and options offered lead to savings in development, labor, capital, inventory, and real estate. Quality would also improve, as there would be fewer opportunities to make mistakes.

Another way of reducing the cost of options is to transfer the location where the options are added. Computer companies eliminate special orders at the factory level by letting their stores, suppliers, and customers customize the product by adding circuit boards for extra memory and special options such as modems, sound boards, etc. Another example is T-shirt stores that stock only plain shirts. By printing the slogans on the T-shirts at the store, specialty T-shirt stores have eliminated the stocking of a potentially infinite supply of preprinted T-shirts. Now, one can go to a store and, after choosing the color and size of the shirt, have any one of hundreds of logos or slogans printed onto the shirt. This technique also reduces shipping, distribution, and logistics costs.

Limiting customer options has the connotation of being bad for the customer. However, this does not have to be the case. For example, baseball caps used to be made in eight different sizes with stretch bands. Now, they are made with adjustable plastic tabs, and one size fits all. This results in reduced manufacturing costs, the seller can display a greater selection and the caps adjustability make the user's fit more comfortable.

11.4.1   Limit Number of International Design Permutations

If not careful, designing products for different cultures and countries can result in many different product models. Onkrusit and Shaw (1993) defined a world product as a product design for the international market, whereas a standardized product is a product developed for one national market and then exported with no changes to international markets. A German subsidiary of ITT makes a world product by producing a common 'world chassis' for its TV sets. This world chassis allows assemblage of TV sets for all three color TV systems of the world (i.e., NTSC, SECAM & PAL) without changing the circuitry on the various modules. The designer's goal is to minimize the number of different product designs that are needed to sell in other countries. Important country differences include language, culture, technical standards and government regulations.

The steps are:

1. Identify requirements for different countries and cultures
2. Identify the simplest methods for meeting these requirements by minimizing the number of permutations or minimizing the cost for providing different permutations.

One very successful method for international design is to use symbols that are universally known. Automakers and electronic product companies have extensively used universal symbols for switches. When unique labels or words are needed for different countries, many companies develop add-on labels that can be added to the product at the last step of manufacturing or after shipping to the country itself. This allows one design to be used for different markets.

11.5 PRODUCT PLATFORMS, LINES AND FAMILIES

One approach for limiting options while still providing customization capability is called product platforms. Many different products can share a common platform, technology, or software module. This group of common products is often called a product family or line where each product typically addresses a single market segment but each product also uses common core capabilities.

Product platform or common architecture is a term used to designate the core technology and expertise that encompasses the key design aspect, technology, or module shared by the product family. This "core competency" or "common platform/structure" becomes the foundation for many different product designs. Many different modules are developed based on the platform's core technology or design. For example, if the platform is electric motors, the different power levels e.g. 3, 5, 10 and 12 volts would all have common processes, vendors, assembly procedures, etc These modules are also used as building blocks for new product development and upgrades. Different products are created through different combinations of the modules. For example, a screwdriver, saw, router, and drill would use the same motor. A robust product platform is the heart of successful product family, serving as the foundation for a series of closely related products (Meyer, 1993).

The objective of the platform is to create a strategically flexible product design that allows product variations without requiring changes in the overall product design every time a new product variant is needed (Ericsson and Erixon, 1999). Individual products are, therefore, the offspring of product platforms that are enhanced over time. Platforms can also be the architecture or physical structure of the product. Common computer architecture platforms are very effective method for rapidly developing application software. Lear, an automotive supplier, developed a common "core" architecture/design for the automotive instrument panel's structure that included all non-visual items such as load bearing and safety related items. Each new design can customize the driver's view of the instrument panel by selecting specific modules. Manufacturing, vendors, and design are constantly improving the platform's design and core capabilities. Core competency or capabilities tend to be of much longer duration and broader scope than single product families or individual products (Meyer, 1993). This focus on the core capabilities of the product platform allows design and manufacturing to continuously simplify the design in terms of cost, performance, and quality. Some successful companies that use product platforms are the SONY Walkman family of products, and Black and Dekker's motors and armatures used in power tools (Meyer, 1993).

11.6   MODULARITY AND SCALABILITY

Module is defined as a building block with defined interfaces. **Modularity is a simplification technique that divides a complex product into separate simple modules that are easier to design, design updates, manufacture and test independently and then assemble together**. Since many modules already exist and can be purchased directly from vendors, it allows the design team to select the best module available. The independence of each module is easier to design, production, test, repair and maintenance than a single more complex system. This is especially important for production, testing, repair and maintenance since only the affected module must be disassembled when failure occurs. To be successful, each module should have "stand-alone" test and inspection capabilities to provide integrity when the module is assembled to the other modules.

For example, new products or systems can be developed by different combinations of modules. Modularity is one of the most important techniques for software development. In the future distributed modules (i.e., objects) can exist as independently developed executables that are used and reused as self contained units anywhere on any platform. Modules are often divided based on the parts or software modules that are manufactured by other companies.

A good example of modularity is shown in the assembly of a personal computer. Rather than having one large production line and having design expertise in all areas of computers, computer assembly is typically separated into several separate modular subassemblies made on different production lines to produce each module. Other companies make many of the modules. For the personal computer, module production lines include final assembly, printed circuit boards, power supply, disk drives, and displays. A vendor almost always manufactures displays and disk drives. This allows the computer manufacturer to use the best disk drive without having to design and manufacture the disk drive. Since disk drives have common electrical interfaces, newer models can be easily integrated into the design as they are available. It also allows the notebook computer designers to use common designs for the mechanical, electrical, and software interfaces between the computer and the disk drive.

Susman (1992) states that "manufacturers can choose from extended libraries of modules/components and combine them into a system that meets the performance needs of their customers". This flexibility provides numerous producibility and reliability advantages in addition to the resulting customer satisfaction. Susman (1992) notes the following:

1. Production errors are more easily traced back to the concerned module.

2. Modules can be developed concurrently by teams and vendors that have higher levels of expertise about the module.
3. When changes are necessary, it is easy to isolate the modules in order to facilitate the changes more effectively.
4. Documentation is simplified.

"World class" vendors can be selected to ensure the best module is used in the product.

A special type of module is one that is scalable or reproducible. **Scalability is a design where modules or complete systems can be combined as needed in proportion to the level of demand or performance required.** This is important in electronic commerce, service and other software systems that have increasing levels of demand. For example, an electronic commerce system can be designed using a group of modules (servers, routers, software) for a certain traffic level, e.g. 1000 transactions per hour. A scalable design is one where modules or complete systems can be easily added later to the system if traffic levels go higher. The number of modules or complete systems added is proportional to the increase in traffic. That means if the demand doubles, an entire identical system will be added. If the demand increases by 30%, modules will be added to the system to increase capacity by 30%.

11.7 PART REDUCTION

Every part used in a design has the potential of:

1. **Failing** (i.e., reliability problem)
2. **Improperly assembled** (i.e., manufacturing problem)
3. **Delivered late** (i.e., schedule or vendor problem)
4. **Damaged in shipping** (i.e. logistics problem).
5. **Repaired improperly** (i.e. repair problem)

A significant way to increase producibility and reliability is to simply reduce the number of parts. As described by Huthwaite (1988), the ideal product has a part count of one. This is also true for software. A program with fewer lines of software code will probably be more reliable and easier to produce, and more likely to be developed on schedule.

Fewer parts reduce the number of parts to be purchased, manufactured, assembled, and inspected. Costs in inventory, material handling, purchasing and other support areas are also reduced. Daetz (1987) arrived at some of these findings:

- Assembly time and cost are roughly proportional to the number of parts assembled, given the same type of assembly environment.

- System's cost of carrying each part number in manufacturing may range from $500 to $2500 annually.
- Establishing and qualifying a new vendor for a new part may cost almost $5000.

The part reduction process has three steps that is accomplished by:

1. **Removing** unnecessary parts
2. **Combining** several single functional parts into one multi-functional part
3. **Reducing the number of different parts** by standardizing common types of parts

The first task in part reduction is to remove unnecessary parts by checking that every item within the design is essential. One should look at the design with an objective viewpoint for eliminating unnecessary features and parts. This is often accomplished by challenging the need for each part or software module. Is it really needed? As mentioned before, function analysis or value engineering is one method to identify unnecessary functions. Special attention should be given on where there are multiples of the same or similar parts. For example, when six screws are used to hold a circuit board to a base part, the design should be evaluated to see whether four or five screws could perform the same function satisfactorily. Another area is especially true where the designer added flexibility for options but changes in technology or user requirements have now made the options unnecessary.

The second task is to incorporate the functions of several parts together into one multi-functional part. Maximizing the number of functions performed by each component will reduce the number of individual parts within the product. Many mechanical and software designs are developed with every product function having a different part or a software module. For example, screws and washers are supplied separately. This means that each part has to have its own feeding and handling equipment and the number of assembly tasks is increased. An easy answer is to combine the two part operations by selecting screws with captive washers making it a one-part operator. A better answer might be to incorporate the function of one part into another part. An example might be that instead of using 6 screws to attach a circuit board to the computer case, snaps/clips are molded into the base that allow the computer board to be snap fitted into place. The design team should look at the feasibility of combining every set of mating parts (i.e. parts that touch each other). Mating parts that are made of the same material and which do not move relative to each other can often be combined. A concern of multifunctional parts is where the complexity and related costs of the new multifunctional part are more than the several parts that it replaced.

The third and last task is to standardize common parts and software modules, such as screws, connectors or common software tasks. This is to minimize the number of different parts that are processed by manufacturing or different software modules that must be designed. One effective method is to have manufacturing review the Bill of Materials List (BOM) for all electronic and mechanical subassemblies (each generated by different designers) early in the design phase. Suggestions are then made to designers to help reduce the number of different parts used in a product. This task will be discussed in more detail in a later section called standardization.

For example, one means of reducing both part count and development cost is to use programmable devices. These devices (both memory and logic) can make it easier to try out new versions of software and change electrical designs during the development phase of a project. When the design becomes ready for production they can be replaced by less costly non-reprogrammable versions or application specific integrated circuit (ASIC) devices in production. This avoids using the more expensive ASIC type device during development.

Another method to reduce component part count is to use higher order parts when one part can do the function of a number of less complex parts. Examples include using gate arrays to implement logic functions using a standard cell design approach, application specific integrated circuits (ASIC) or custom hybrid circuits.

Another option is the use plastic for the product's base. A plastic base can incorporate tabs, ledges, and posts to support and snap fits to fasten other parts. This eliminates separate brackets and fasteners. A plastic base can also eliminate fasteners by incorporating snap fits.

11.8   PROCESS/VENDOR REDUCTION AND RE-ENGINEERING

Reducing the number of processes, tasks, and vendors can significantly reduce lead-time and costs. Every manufacturing process requires set-up time and costs, process time, waiting time and material handling. Reducing just one of these processes can realize gains from each of these areas. One study showed that 85% of a part's process time is spent waiting on a machine or operator. Each process or task also has a probability for causing defects i.e. lowering quality.

Re-engineering is a process used to eliminate unnecessary (i.e. non-value added) tasks and processes including the design process itself. Any task that is eliminated will save money and time. Unnecessary documentation, packaging, and regulations are targeted. The re-engineering process documents the existing development process and then subjectively identifies items to consider for elimination or modification. Network diagrams are often used for illustration. The key steps for re-engineering are:

1.   Develop an "as is" process flow diagram of all processes, tasks, and steps including setup, waiting, and paperwork (i.e. network)

2. Identify and incorporate design changes that can simplify the process
3. Identify and eliminate non-value added processes or steps
4. Identify and replace processes with high cost, risk, lead-time, setup time or process time.

Reducing the number of vendors can reduce purchasing, contract, accounting, shipping, and inventory costs. Each new vendor requires considerable paperwork and communication. The steps for vendor reduction are:

1. Identify all vendors currently used and the number of their products that are purchased
2. Identify new or current high quality vendors that can produce a large number of different products needed
3. Reduce number of vendors by eliminating vendors are used for the same part/material

11.9   PART FAMILIES AND GROUP TECHNOLOGY

Group technology is a technique in which similar parts, assemblies or modules are identified and grouped together to be manufactured in a common production area. These similar parts are called part families. For manufacturing, part families have similar process flows and requirements. The purpose of group technology is to capitalize on manufacturing similarities in order to gain efficiency in manufacturing. A plant producing a large number of different parts is able to group the majority of these parts into a few distinct part families that share common design and manufacturing characteristics. Therefore, each member of a given group would be produced similarly to every other member of the part family. This results in higher production volumes and greater manufacturing efficiencies. Since the production line treats each member of the part family the same, many of the advantages of mass production can be implemented. The common production flow provides a basis for automation and quality improvement. Although certain design limitations may be imposed by group technology, the benefits of reduced lead times and improved quality should more than offset this inconvenience. The steps are to:

1. Evaluate all current designs and identify part families
2. Develop design guidelines and common parameters for each part family based on the manufacturing capabilities
3. Use the guideline for all new designs

The team designs to the common parameters of the company's part families in order to realize the benefits of group technology. Although each member of the part family is different, the use of similar manufacturing processes provides a higher quality, lower cost part. An example of a part family of sheet metal parts is shown in Figure 11.1. These parts may have very different design functions, but for manufacturing they are similar in process sequence and requirements.

11.10   FUNCTION ANALYSIS AND VALUE ENGINEERING

Function analysis or value engineering are unique design approaches started in the 1950's for product simplification. They can identify new ideas, approaches, or methods that provide the best functional balance between cost and performance. A multidiscipline team, following a systematic format usually conducts value-engineering studies. The goal is to identify better ideas and remove unnecessary costs in the design.

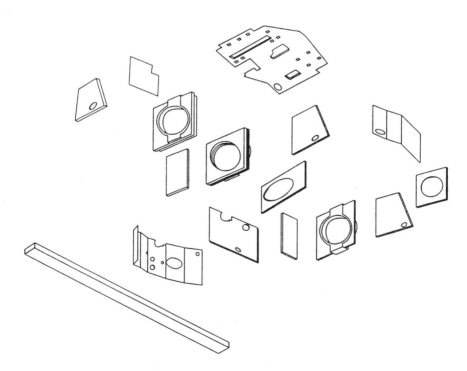

Note: Part family: sheet metal. Manufacturing limitations: (1) no single parts larger than 4 X 4 feet; (2) no machining or grinding operations; (3) no tolerances less than 0.001 inch (adapted from Priest, 1988).

FIGURE 11.1 Commonality using part families

Lawrence D. Miles developed the value engineering technique during World War II at General Electric. Miles was assigned to purchasing and was often faced with finding an alternative part, process, or material to perform the design function. Often, the use of the alternative method resulted in cost savings. From this experience the value analysis technique was developed. The first value seminars were taught in 1952, and two years later the U.S. Navy Bureau of Ships established a value program. The Navy program analyzed engineering drawings prior to execution. Therefore, the Navy called its program value engineering. As a result of the Navy's success, many companies have established programs.

A value engineering study has four basic phases. The format's objective is to restrict the length of the study and force a concise definition of each design component and its function. High-cost items in both product design and product life cycle are identified, and cost-cutting efforts are then focused on these items. The steps for value engineering are:

1. Analyze each design and part to determine its primary functions (i.e., what function does it perform?).
2. Compare these functions to the specified requirements.
3. Determine the cost of performing each function, and identify other approaches that could perform the same function for less cost.
4. Simplify the design so that it provides only the required functions at the lowest cost.

The greatest benefits from value engineering result when the study is performed at the conclusion of the conceptual design stage. Table 11.1 shows an example of a value engineering analysis for an electronics case. The analysis would at first focus on the items that are not specifically required. Should they be deleted? Since they are not required, should the lightweight and access doors be eliminated? The second focus would look at the high cost items. Is each function's cost proportional to its value? For example, the first item supporting the circuit boards appears to have a high cost. Is there a cheaper method to perform this function?

TABLE 11.1 Value Engineering Analysis

Design's primary function	Specified requirement	Cost of each function ($)
Support printed circuit boards	Yes	$24.00
Protect circuit boards in aircraft environment	Yes	37.00
Light weight for portability	No	15.00
Access doors and handles for maintenance	No	7.50
	Total cost	$83.50

11.11 ERGONOMICS AND HUMAN ENGINEERING

Ergonomics or human engineering is a design discipline that seeks to ensure compatibility between a design and its user's capabilities and limitations. The objective is to simplify the human's tasks to achieve maximum human efficiency (and hence, acceptable systems performance) in system development, fabrication, operation, and maintenance. Common measures include task time (minutes necessary to complete a task) and human error rate (number of human errors in a specified period of time). The steps for using human engineering for simplification are to:

1. Identify all of the people that will be affected or have contact with the product for all stages of use including operators, repair, manufacturing, disposal, etc.
2. Identify environmental conditions
3. Simplify the human interface design using:
   - Functional task allocation to determine which tasks should be performed by humans and which by the product
   - Task analysis for determining human task requirements to identify and eliminate potential problems and errors.
   - Design guidelines that have been proven to improve human performance and reduce errors
   - Prototypes to validate and verify all of the above tasks to optimize human performance

Most defects are invariably caused by human error whether by the production worker, designer of the work environment, designer of the production equipment, or the design itself. Some believe that for simple tasks, an error will occur at the rate of 1 per 1,000 opportunities. Since most products have thousands of error opportunities, human error cannot be reduced to zero. The objective of human error reduction is to reduce the number of opportunities for human errors (e.g. number of parts) and reduce each opportunity's chance of occurring (e.g. complexity of an individual part)

A designer's goal is then to minimize the opportunities for human error. The steps for error reduction are:

- Simplifying the worker's, programmer's, or user's tasks by specifying a simple design
- Increasing the identification of errors when they occur
- Reducing error opportunities using a systematic approach of identifying and resolving error opportunities using historical manufacturing and design information.

Each opportunity does not have the same probability of generating an error. Metrics are kept during the development process, and an analysis of the errors or defects must be done to eliminate or modify the opportunities that have the highest probability of generating faults.

11.12   MISTAKE PROOFING AND POKA YOKE

Mistake proofing is one of the most powerful techniques for avoiding simple human errors at work. A Japanese manufacturing engineer, Shigeo Shingo, used worker task simplification to improve quality and eventually eliminate quality control inspections. The methods were called "fool-proofing." Shingo later came up with the term poka-yoke, generally translated as "mistake-proofing" or "fail-safing" (Shimbun, 1987).

The goals for mistake proofing are:

- Parts cannot be manufactured or assembled wrong by the operator or machine
- Obvious when mistakes are made
- Self-tooling/locating features are built into parts
- Parts are self-securing when assembled

Mistake proofing features are designed into the product and/or devices are installed at the manufacturing operation. It is a low cost method of insuring high quality.

The two steps to implement mistake proofing are:

1. Identify all possible human errors (i.e., parts missing, misassembled, wrong parts, etc.)
2. Modify design or work area so that an operator has only one method to perform a task.

One of the best mistake-proofing feature is to incorporate guide pins of different sizes or shapes into the design to eliminate positional errors in assembly. Some examples of mistake proofing methods are shown in figures 11.2 and 11.3. Additional key strategies to reduce human error in manufacturing are comprehensive test and inspection and design margins that can compensate for when these errors occur. An article by Ayers (1988) emphasizes that manufacturing can catch 70-80% of all defects with inspection and increasing design margins and redundancies can reduce human errors. For example, 70-80% of spot weld defects do not matter in robotic spot weld applications, since the designer specifies 10% more welds than are needed

Housing

mounting slot

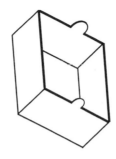

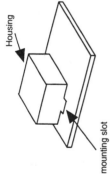

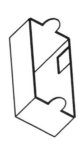

Different size tabs limits
installation to only one way

Same size tabs can be
mounted either way

FIGURE 11.2 Poka yoke for assembly.

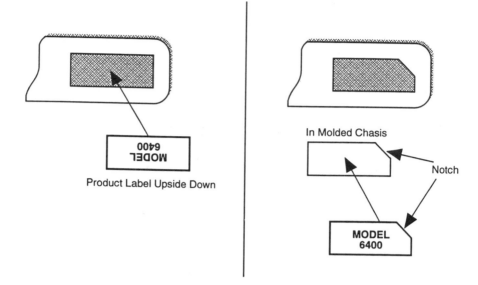

FIGURE 11.3   Poka yoke for labels.

In summary, some key mistake proofing practices used by one company for error reduction in assembly in priority order are (adapted from Freeman, 1990):

1. Design each part so that it can only be installed one way
2. The assembly is designed so that if a part is missing or incorrectly installed, the subsequent part cannot be installed
3. Make the correct installation of each part essential to the function of the product, so that a functional test will detect any assembly error
4. If a part is not verifiable by any of the above methods, access to the part is provided so that its correct installation can be verified

11.13   MINIMIZE MANUFACTURING REQUIREMENTS AND DESIGN FOR PREFERRED PROCESSES

One popular method of simplification is to simply reduce all product manufacturing requirements and tolerances where possible. This is especially true for non-critical requirements. For example, reducing a design tolerance from .002in. to .005in. can greatly reduce the price and number of defects for a part. Reducing a design's manufacturing requirements (i.e., level of precision) allows manufacturing or the vendor to use less costly and more reliable processes, flexibility to use different processes, and improve quality.

The key task is to find an optimum balance between design requirements and manufacturing capabilities. This is accomplished for many designs by performing tolerance analysis of the design's requirements to identify the minimum tolerances necessary to still meet both the designs' performance requirements and manufacturing capabilities. This assessment includes manufacturing capabilities and tolerance analyses of assemblies and piece parts to ensure proper fits and clearances.

Another objective is to tailor the design for preferred manufacturing processes and vendors. Preferred processes are those that have superior performance including lower cost, higher quality, more available, and less lead-time.

11.14   REDUCE TECHNICAL RISKS

Technical risk assessment for producibility identifies the parts, assemblies, software, vendors and processes that present the highest probability for causing problems. Risk is defined as the probability that a manufacturing process, software, or vendor is not capable of meeting all requirements. The process is manually performed by an evaluator(s) who reviews all documents and drawings. Risks are assessed for all levels, starting at the bottom or piece part level and continuing up through the final product. Some design parameters that typically lead to higher risks in manufacturing are:

- Extremely tight tolerances
- Close tolerance fits on part assemblies
- Thin-walled sections in castings
- Unique fastening methods such as adhesives, welding, tape, or brazing
- Lubricated areas
- Seals and bearings
- Brittle or fragile materials
- New manufacturing processes, vendors or technologies

11.15   COMMONALITY, STANDARDIZATION, AND REUSABILITY

One of the greatest wastes of time and talent occurs when the "not designed here" syndrome is in effect. Designing a new product, part, or software when an identical or similar product is already available is counterproductive. Whenever possible, the design team should build off of the previous knowledge gained from earlier designed parts, systems, or software that have been proven to meet the requirements. In most cases, the technical risk, cycle time, reliability, producibility, availability, and cost of previously designed products are more favorable and have been demonstrated. The necessity of designing the product, finding suitable vendors, writing part specifications and testing is removed. The steps are to:

1. Identify the types of parts, designs and software that the design team will need
2. Identify existing "best" parts, designs or vendors to be used in the design
3. Provide this information to the design team in the most effective and efficient manner including lists of preferred parts and vendors and software based design and libraries

Standard designs are one method of commonality. These designs may be published in books or available in CAD data or library files. The designer performs a library search in order to find a design or software code that could meet the design requirements. If a suitable design is found, significant savings are realized. For example, in electrical engineering a number of standard circuits are available in both military and civilian publications. In computer science, libraries of proven software code are available.

A special type of standard design is a parameterized design that is reusable by virtue of its built-in flexibility to be adapted to different requirements. An example of a parameterized design is an electronic circuit where only modifying the software instructions can change the outputs. This allows the hardware and software to be used in more than one design.

11.15.1 ASIC Software Design Libraries

Designers creating application specialized integrated circuits (ASIC), with gate complexity of over a million gates, are increasingly using software libraries of common function models to build their designs (McLeod, 1995). A library's model contains simulation and synthesis descriptions for gates, flip-flops, more complex constructs such as counters, registers, adders, as well as very complex functions such as data paths, memory, micro controllers, etc. An important part resides in the leaf cells used to describe the physical geometry of each logic function. Module generators from library vendors can automatically create a data book of standard cell logic functions for its customers. Most library vendors also have compilers that create RAM and ROM memory designs.

11.15.2 FORD'S Better Idea

In one successful example (Healey, 1995), Ford Motor engineers took just six months instead of the normal 18 to 24 months to adapt a V-8 engine for the Explorer vehicle, which was originally designed to accommodate a smaller V-6. "They did that by starting with a currently successful product, the existing V-8 engine used in Ford's F-series pickups. To increase it to a competitive 210 horsepower (hp) from 190 hp in the pickup, engineers used standardized, off-the-shelf, hot-rod parts that Ford sells to automotive racecars. This streamlined the time-consuming test process by, banking on the V-8's long record of quality and performance in the pickup, Mustang, Thunderbird, and other cars (Healey, 1995)." This resulted in a lead-time saving of 12 months!

11.16   COMMON, STANDARD OR PREFERRED PART, MATERIAL, SOFTWARE, AND VENDOR LISTS

The best choice for a specific part, material, or software invariably comes from a group of similar products that have been used on previous designs, demonstrated the ability to consistently meet specified requirements at a reasonable cost, and from vendors that delivered on schedule. Lists of approved parts, materials, software, and vendors are the most widely used technique for standardization and usually available in most companies. Electronic and mechanical assemblies use large quantities of purchased parts, including screws, rivets, resistors, and electrical components. Significant cost reductions in inventory and material handling are possible when standard parts are selected. For example, procurement and testing of a non-standard integrated circuit part can take up to a year and cost up to $45,000.

As long as the supplier's production lines stay open, standard parts are preferred and should be selected routinely before picking any other non-standard equivalent. Standard lists should be implemented early in a project, since many products are selected at the beginning of a design effort. The parts and vendor lists can be included in the CAD tools available to the designers.

In practice, the preferred parts list concept has not worked well on projects that use high technology and have a design development time of two years or less. Since a part has to be a proven technology to get on the parts list, it can already be past its peak of the life cycle when chosen for the design. If the development time of a product is longer that the mature lifetime of the parts chosen, those parts will become obsolete about the time the product gets ready for production. For example, selecting today's fastest microprocessor will not be very good if the product is still two years from production.

For many projects, the preferred vendor concept works better. If a vendor has proved that it produces quality parts/software, then the project can design the vendor's product into their product when they are in the advanced information stage from the vendor. In this way the product will be at the peak of their life cycle when the product is ready for production. Many vendors keep a common interface for their products. This allows newer versions of the parts or software to be completely compatible with the design making easier updates and revisions. This can also help in upgrading the product in the field such as when computer owners update the size of their disk drives.

World Wide Common Parts

General Motors is one of many auto companies that is working on global car programs that will accelerate the use of common parts regardless of where a car is sold. (Dallas Morning News, 1995) Using common engineering and manufacturing processes, GM builds the global car in several variations and several countries. Except for variations in chassis, engines, and transmissions, the goal is to make other parts common across the entire group of cars. GM gets the benefit of cost savings, but keeps the flexibility to preserve each vehicle's identity.

11.17   SOFTWARE REUSE

The development team traditionally designs a complete system and then produces requirements to guide software designers. So much information must be translated between the design teams that a number of errors normally occur. Producibility and reliability is increased when previously developed and proven software components are used.

The system designer has the ability to select software components that have previously been written and tested by software designers (allowing for increased producibility, less technical risk and better reliability). By selecting from the library of existing components, the translation errors between system and software designers are reduced to only those errors occurring between components or in the development of components that are not available through the library. One can simply choose software components from what is available. As the library becomes rich with varieties of components, coordination with software increasingly becomes a matter of modification to existing components for a slightly different functionality from the library component. The design team has more control over the system's development. Producibility and reliability are increased through the system's control.

11.18   SIMPLIFICATION STEPS FOR NOTEBOOK COMPUTER ASSEMBLY

The following seven simplification steps for assembly are a compilation of many different methods that have evolved over many years.

1. Use modular and common assemblies
2. Eliminate, combine, and standardize
3. Minimize fasteners and joining methods
4. Provide accessibility and process in one plane
5. Top down assembly

6.   Parts should be easy to handle and install
7.   Keep assembly processes simple

The following section will look more closely at each of these seven techniques as well as real world examples for a notebook computer. Much of the following section is from a student report by McKenna, (1996).

Modular and Common Assemblies

Modular design concepts have proven successful in the computer industry. Much of this modularity is based on different companies focusing on certain modules used in personal computers such as the display, printed circuit board, disk drive, PCMIA, and modem. Designers and consumers are able to select these modules from different manufacturers. This market of modules then allows companies to do competitive benchmarking effectively for each module.

Eliminate, Combine, and Standardize

Huthwaite (1988) correctly theorizes that "the ideal product has a part count of one". This is true not only because of the reliability improvements but in addition, using smaller numbers of parts means reduced storage costs, reduced labor costs, shorter lead times, and fewer parts to keep track of, which subsequently brings producibility benefits. The three key steps are:

1.   **Eliminate Parts**: Simply eliminate those parts that are deemed unnecessary to the functioning of the product. Figure 11.4 shows an example of evaluating the method to secure a circuit board to the base.
2.   **Combine Parts**: Multifunctional parts represent the combination of several parts with individual functions into one part. One criterion used to identify areas for combining parts is where mating parts are made from the same material and the mating parts do not move relative to each other. Plastic parts are often used to accomplish this goal. An example of this is parts for a support bracket shown in figure 11.5. Multifunctional parts have proved to be powerful tools in design improvement, but their designs often require great creative skills. The design team must be careful that the new multifunctional part is not so complex as to create new producibility problems. If so, the new part could outweigh the assembly savings. If you are not careful, complexity of the multifunctional part can lead to expensive tooling costs, new materials or processes that are too costly or have too much technical risk, or the capital equipment required for the more complex multifunctional part may be very expensive.

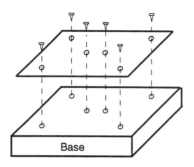   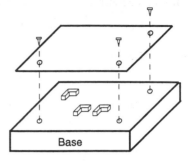

Design Approach **A**   Design Approach **B***

6 Screws   3 Screws with
                                               Supports
                                               Incorporated into
                                               ˴ Base

                                               * 50% Reduction in the
                                               Number of Parts

FIGURE 11.4   Eliminate parts.

3. **Standardize Parts:** The third step is to standardize parts where
    possible. Ettlie (1990) notes that this technique is important in
    both reducing the complexity of a design and in "controlling
    proliferation of information" throughout the manufacturing system.
    Savings are also realized in capital equipment as standardization
    reduces the need for additional equipment and allows one robot
    gripper to install multiple parts for multiple configurations.

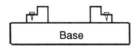   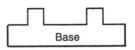

Two Support   Support Brackets
Brackets are   are Part of
Fastened to Base   Multi-Functional
(5 parts)   Base
                                               (1 Part)

FIGURE 11.5   Combine parts.

Minimize Fasteners and Joining Methods

Many producibility experts feel that fasteners are inherently a problem. Fasteners increase opportunities for poor quality and require monitoring for presence. Design teams are always shocked when they find out the cost of fasteners used in a product. A comparison of the labor cost for fasteners found in one company's line of computers ranged from 9% to 27% as shown in Table 11.3. For one model that was studied, 10% of all repair costs were attributed to fasteners! First, the design team should ensure that each fastener is absolutely necessary. Will the product perform just as well with one less screw? The second objective should be to incorporate the fastener's function into one of the other parts in the assembly such as snap fits or press fits. Standardization is the third step since limiting all of the fasteners to one size can allow manufacturing to automate the process. Although simple fasteners may cost little to buy, the process of attaching them in assembly may be costly and time consuming.

Joining recommendations from Freeman (1989) include:

- Locking snaps are preferred.
- With screws, use vertical downward screw insertion only.
- Avoid gluing, soldering, and welding; these processes rely on operator skill and therefore are of inconsistent quality.
- Design for "disassembly", assemblies will have to be removed in the field. Fasteners are desirable when repair tasks are expected in the future. Snap and press fits often break when repaired.

Provide Accessibility and Process in One Plane

Accessibility in an assembly process is needed to keep our options open in terms of the assembly and repair approach to be used. For example, if the accessibility for a part's assembly is limited, manufacturing is limited in terms of what machines that can be used. Robot manipulators and end effectors often require more access room than manual assembly. Restricted vision and manual access can create quality assurance problems as well as reduce the potential for process improvement. When processing in multiple planes, re-orientation of

TABLE 11.3. Cost of Fasteners in Computers

Computer Products	Total Assembly Cost	Total Fastener Cost	Percent of Assembly Total Cost
Low Profile	36.17	3.31	9.2
Big Box	33.87	5.28	15.6
Tower	60.07	16.10	26.8
Mid Tower	60.07	6.77	10.7
RL	63.07	6.77	10.7

parts proves to be an expensive element. To reduce these costs most authors suggest that all required assembly operations be performed on one surface before re-orienting.

Texas Instruments Inc. effectively incorporated the previous two concepts into the redesign of a reticule assembly for a thermal gun sight (Boothroyd et al, 1994). The majority of the fasteners in the original design were eliminated and through the use of self-securing parts like press fit. Reorientations were eliminated by using a cam to provide the conversion from rotational to linear movement. Assembly time was reduced by 85% (Boothroyd, et al, 1994).

Top Down Assembly

Top down or z-axis assembly is the preferred assembly method since it uses gravity to help the manufacturing process. An effective means of using top down assembly is to cradle successive components and lock-in-place with top components. A notebook computer is usually assembled in a top-down method as shown in Figure 11.6.

Northern Telecom was able to effectively integrate this approach into the assembly of their Harmony telephone design as reported by Ettlie (1990). "This product line incorporated top down assembly which they termed vertical build. In this process, each module was assembled one on top of the other in succession, without any components being inserted horizontally. Because of the simplicity of the assembly, they were able to use robots where previously it could only be done manually. By using this assembly approach, they were able to effectively reduce lead times from 20 to 3 days and product costs by 300% from the previous telephone model" (Ettlie, 1990).

Parts should be Easy to Handle and Install

When parts are difficult to handle or install, manufacturing problems occur. This problem exists for both manual and automated assemblies. Parts that are too large, small, heavy, delicate, slippery, asymmetrical, or non-rigid may provide difficulty to the assembly system. Four points are typically stressed to improve handling and installation:

1. **Provide self-aligning guide surfaces and chamfers**. Design features into parts that aid in part alignment and resolving tolerance problems. Guide surfaces and chamfers are the most popular method. Their purpose is to help the operator or the automated system line up the parts to be inserted. Self-aligning features provide a simple yet effective means of self-guided insertion. These types of features can provide corrective action in certain assembly operations, which helps to reduce the potential risks of tolerances. This is an effective method for mistake proofing.

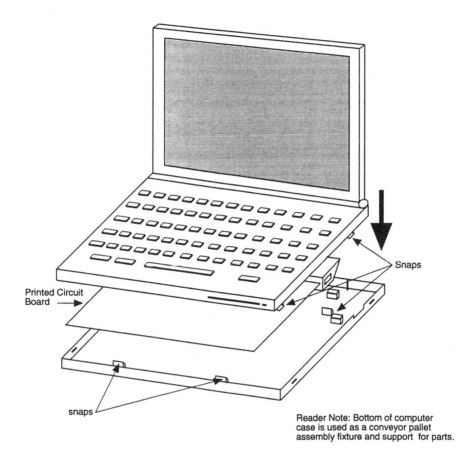

Snaps

Printed Circuit
Board ——▶

snaps

Reader Note: Bottom of computer
case is used as a conveyor pallet
assembly fixture and support for parts.

FIGURE 11.6   Top-down assembly.

2. **Design for recognizable symmetry/asymmetry.** Symmetrical parts are important because they cost far less to handle and orient automatically. This reduces the cost of part feeder tooling and part programming that asymmetrical parts require. Symmetrical parts also help to reduce improper assembly problems. When parts cannot be symmetric, the design team should design parts that are obviously asymmetric for the operator or machine. This requires each part to have easily identifiable features. Production operators can visually or by touch recognize parts without confusion, and automated assembly systems can orient parts with minimal tooling and fixturing. This was also discussed in mistake proofing.

3. **Avoid non-rigid, flexible parts.** Non-rigid parts (e.g., cables) are difficult to assemble because they are so difficult to direct (i.e., position and orient) to a specified location. Automated assembly systems especially have a problem with flexible parts since they are not able to guide parts through visual feedback and subsequent mental adjustments the way humans can. Non-rigid parts include belts, cables, and flex circuits.

4. **Design out potential part nesting/snagging.** Nesting/snagging is a drawback of using parts that can stick together such as cups and flanges, and certain non-rigid parts such as wires and springs. The design team must either reduce the use of these types of parts, redesign the parts to prevent snagging or separate their locations from one another in order to keep them from intertwining.

Keep Assembly Processes Simple

Simplify the manufacturing process itself. There are four methods.

1. Reducing the number of different processes
2. Use preferred processes
3. Avoid non-preferred processes such as gluing and welding
4. Eliminating the need for hand tools or special tooling

Sun Microsystems Computer Co. was able to greatly increase its manufacturing flexibility through the simplification of its assembly and material handling systems (Laughlin, 1995). This new method was so simple that each unit could be assembled in less than 15 minutes by a single worker. Because the process was so simplified, the necessary operations could be performed by and taught to even the most unskilled workers.

11.19 SUMMARY

A complex product is difficult to manufacture efficiently with high quality. The major focus of this chapter is getting back to the design basics of design simplification. The systematic application of these key design practices can significantly reduce the technical risk involved with product development and can result in a quality product with reduced design, production and support costs. Although these concepts are easy to understand, their effective implementation into the design process requires teamwork.

11.20 REVIEW QUESTIONS

1. What are the three functions that make up the metrics of simplification?
2. What is mistake proofing? Give two examples.
3. From your experience, what are some international parameters that must be considered and how do they affect manufacturing?
4. Explain the value engineering process.
5. List the major methods to reduce parts.
6. When is a list of preferred vendors better than using a list of preferred parts?
7. What areas of simplification and standardization are appropriate for software products? Explain.

11.21 SUGGESTED READING

1. B. Freeman, The Hewlett-Packard Deskjet: Flexible Assembly and Design for Manufacturability, p. 50-54, CIM Review, Fall, 1990.

11.22 REFERENCES

1. R.V. Ayres, Complexity, Reliability, and Design: Manufacturing Implications, Manufacturing Review, Vol 1 (1), p. 26-35, March 1988.
2. Daetz, The Effect of Product Design on Product Quality and Product Cost, Quality Progress, p. 63, June, 1987.
3. Dallas Morning News, GM Global Program Will Use Common Parts p. 7H, March 26, 1995.
4. A. Ericsson and G. Erixon, Controlling Design Variants: Modular Product Platforms, Society of Manufacturing Engineers, Dearborn, Michigan, 1999.
5. J.E. Ettlie, and H.W. Stoll, Managing the Design-Manufacturing Process, McGraw-Hill, Inc., New York, 1990.
6. B. Freeman, The Hewlett-Packard Deskjet: Flexible Assembly and Design for Manufacturability, p. 50-54, CIM Review, Fall, 1990.
7. J.R. Healey, Engineers Took Cram Course on Design Change, USA Today, p 4B, .August 22, 1995.
8. Howard, Don't Automate-Eliminate, Wall Street Journal, October 1985.
9. B. Huthwaite, Design for Competitiveness, Bart Huthwaite Workshops, Troy Engineering, Rochester, MI, 1988.
10. K.K. Laughlin, Increasing Competitiveness With a Simplified Cellular Process, Industrial Engineering Magazine, April, 1995.
11. R.T. McKenna, Design Simplification for Producibility Improvement Unpublished Student Report, Industrial and Manufacturing Systems Engineering, U.T.A., 1995.

12. J. McLeod, Deep Submicron Silicon: Changing the ASIC Design Process, Advertising Section, Integrated System Design, p 5-8, September, 1995.
13. M.H. Meyer and J.M. Utterback, The Product Family and the Dynamics of Core Capability, Sloan Management Review, 34(3): p. 29-47, 1993.
14. S. Onkvisit and J. Shaw, Industrial and Organizational Marketing, Merrill Publishing, New York, 1988.
15. O. Port, The Best Engineered Part is No Part at All, Business Week, p. 150, May 8, 1989.
16. K. Schwatz, It's the Little Things That Cost a Lot, Associated Press, Dallas Morning News, p. 2B, March 23, 1996.
17. N.K. Shimbun, ed., Poka-Yoke Improving Product Quality by Preventing Defects, Productivity Press, Cambridge, Mass, 1987.
18. G.I. Susman, Integrating Design and Manufacturing for Competitive Advantage, Oxford University Press, New York, 1992.
19. S.D. Welsby, Use of Selected Software Complexity Measures in Introductory Programming Courses, M.S. Thesis, University of Missouri-Rolla, 1984.
20. H. Zuse, Software complexity: Measures and Methods, New York: Walter de Gruyter & Co., 1991.

Chapter 12

PRODUCIBILITY GUIDELINES AND MEASUREMENT

Critical Information for Collaborative Development

Guidelines are an effective method for coordinating and informing the collaborative design team. They are the most commonly used method for implementing producibility practices and can vary from simple suggestions to absolute rules. Producibility analyses are then used to evaluate proposed designs, validate that the design meets all producibility requirements, and provide feedback. Both guidelines and analyses should provide accurate, timely, and quantified information in a format that the entire design team can easily use in design trade-off analysis. Guidelines and analyses can predict a design's level of producibility and identify potential problems and provide suggestions for improvement. Since each process, product, vendor and company has different requirements and capabilities, producibility guidelines and analyses must be tailored and specialized for each application.

Best Practices

- Producibility Guidelines and Analyses
  - Accurate
  - Timely
  - Easy–to-use and concise
  - Easy to update
- Design for Preferred Manufacturing Methods
- Producibility Analyses
- Producibility Measurement
- Software Tools

PRODUCIBILITY GUIDELINES

This example describes the development and later revisions of the guidelines used for Hewlett-Packard's DeskJet printer. Initially, the company developed a typical long list of design guidelines. Several of their guidelines, however, caused problems. The problem causing guidelines were: (adapted from Freeman, 1990).

1. One dimensional assembly operation - build from the bottom up
2. Choose efficient joining methods without using screws
3. Avoid the need for part orientation and make parts symmetrical or very asymmetrical; design so as to avoid tangling and nesting problems
4. Minimize the total number of parts and the number of different parts used in the product

The most important result from this paper was the necessary changes made to the guidelines after using them. After developing several products, the authors found several "out of the ordinary" findings that required guideline modifications (adapted from Freeman, 1990).

1. **Top down assembly caused problems with respect to rework and service.** The printer received high marks for initial serviceability because it is easy to remove the cover and mechanisms for access to the printed circuit board. Unfortunately if further disassembly is required, serviceability is difficult and time consuming.
2. **Engineers found screws to be effective and easily automatable.** A common guideline in design for assembly guidelines is to eliminate all screws in the design. When designed into a product correctly, the Vancouver division has consistently averaged low failure rates for screws.
3. **Parts tangling (e.g. parts that stick together) should be considered for all parts not just spring.** Flexible parts such as belts, cables, and flex circuits are difficult to assemble and even harder to design for assembly.
4. **Eventually, a multi-functional part becomes so complex that it becomes difficult to produce and starts failing to hold tolerances and features.** Multifunctional parts, however, can lead to increased part complexity and decreased part and fabrication process consistency.

This chapter concentrates on the effective use of producibility design guidelines, rules, analysis and measures.

12.1   IMPORTANT DEFINITIONS

There are several designations or terms used that should be clarified. **Guideline or best practice is a recommendation or suggestion that the design team "should use if appropriate".** Guidelines assist someone to move in a certain path or course of action. For collaboration, they also help to provide coordination. **Requirement, rule or standard is a recommendation that the design team "must use".** For collaboration, requirements provide control and coordination.

Producibility guidelines assist and coordinate all members of the development team to consider manufacturing and vendor capabilities when determining design requirements. Guidelines provide advice, information, lessons learned and instructions. They can vary from suggestions to rules or requirements. The goal of producibility guidelines is to provide important manufacturing and support information to the design team in a timely matter.

Producibility rules or standards are specific requirements that must be met by the development team. The requirements are usually quantified. A producibility requirement example is that all component spacing for printed circuit boards must be .025 inches or more to allow for automation.

Producibility measurement quantifies how well a design meets producibility requirements. The premise is that when you quantify a design's level of producibility, the development team will have a better idea whether the design meets all requirements, how "good" or "producible" the design is and what areas need to be improved.

12.2   BEST DESIGN PRACTICES

Guidelines, analyses, and measurement provide coordination and communication for improving a design. Best practices for implementing producibility guidelines and measures are:

- **Producibility guidelines and analyses** provide cost effective, accurate, quantified, timely, and easy to use information for the design team to measure the level of producibility, identify problems, and offer suggestions for improvement. They include specific producibility criteria for the particular application and environment. They include the lessons learned from previous projects. As with any documentation, they should also have
  - Clarity
  - Conformity to policy, convention, and standards
  - Completeness
  - Integrity
  - Brevity

- High quality
- Retrievability
- Proper format

- **Design for preferred manufacturing processes** is emphasized to minimize technical risks and improve manufacturing.
- **Producibility analyses** evaluate the design's level of producibility and provide feedback to the design team.
- **Producibility measures** quantify a design's producibility for comparing alternative design approaches, verifying that a design meets all requirements and identifying potential producibility problems and areas of technical risk in a design.
- **Producibility software** is used where possible to improve implementation and collaboration.

12.3   PRODUCIBILITY GUIDELINES AND RULES

Today's collaborative design environment requires an extraordinary amount of technical data to be distributed between the design team members. This communication is often confusing because each discipline has unique terminology and methods and technology is changing so rapidly. Guidelines try to simplify communication by summarizing information in a domain format tailored to the design team rather than the expert who developed the data.

Guidelines and rules were the first and are the most commonly used methods for producibility. No one knows their first use but they became very popular in the early 1950's. Shown below an excerpt of some early high-level guidelines.

Basic Producibility Guidelines
(Adapted from General Electric Producibility Guidelines 1953)

1.   **Eliminate** unnecessary functions, parts, part characteristics, excessive or expensive materials, excessive scrap, and unwarranted component or product size.

2.   **Simplify** the design by modifying costly tolerances, difficult to produce physical characteristics or finishes, and complicated assemblies with the objective of reducing the cost of manufacturing processes.

3.   **Combine** functions into a minimum number of parts or subassemblies, thereby, reducing material, tooling handling and process costs.

4.   **Standardize** by using common parts, processes, and methods across all models and even across product lines to permit the use of higher volume processes that normally result in lower part cost.

5.   **Facilitate** manufacturing by adding design characteristics that make it possible to employ inexpensive material, semi or fully automatic equipment, lower cost manual operations (method improvement), new low cost processes and tooling and better vendors.

Producibility design guidelines provide parameters within which the design team should operate. Design guidelines can come in many forms, from published checklists and standards to case-based reasoning incorporated in the designer's workstation. It might include knowledge-based tools, knowledge delivery and/or process gates, information refinement and processing to generate the knowledge. To incorporate these guidelines, time and resources are allocated to collect, classify, continually update and store them. The presentation of the material can be on paper or in the computer. It can be formatted different ways such as simple lists of words, mathematical formulas, or illustrated in drawings or photographs. Drawing upon experience (or the experience of others) is the best source of information for developing guidelines. Sources to consider include process capability studies, vendor history, product requirements, material properties, risk analyses, lessons learned from similar products, process variance, assembly process analysis, process/product FMEA, designs of experiment, published checklists, employee surveys, and consultants.

Most companies have developed some type of producibility information and have documented it in books, notebooks, or in a computer database. The advantages of the printed guidelines are the lower cost to write and publish the guidelines; ability to customize for the specialized nature of each company, ease of getting the information to the design team without extensive training and that most of this producibility information already exists throughout the plant.

Guidelines can also be used to incorporate changes in design parameters and tolerances that facilitate ease of production for a company's particular manufacturing process. This step assumes that the previous techniques of product simplification, standardization, and component selection have been incorporated. This phase of design for production is a systematic "customizing" effort aimed at maximizing production efficiency through product design. Specific guidelines are individually developed based on specific capabilities. Since each manufacturing process and company has different requirements and capabilities, its producibility guidelines also vary for each process.

The disadvantages of published guidelines are when they are not up dated frequently or ignored by some of the team members. Many guidelines are published, sent to the design team and then stay on a bookshelf without being used. Producibility guidelines are designed for ease of update and use. The important characteristics of successful guidelines are useful, accurate, up to date, easy to use, concise, brief and customized for the company.

The steps in developing guidelines and rules are:

1. **Identify informational needs** of the design team including the current marketplace and company's environment
2. **Identify the "best" formats** (e.g. photos, figures, formulas, graphs) and methods (e.g. books, knowledge database, decision support system, computer network access, search engines, computerized tool) for presenting the information to the development team
3. **Gather existing producibility information and guidelines** both in the company (e.g. process capabilities, availability, and lessons learned) and from outside sources (e.g. published documents, internet, benchmarking the competition and communicating with key vendors)
4. **Conduct analyses and experimentation** with or without prototypes for important information that cannot be found any other way.
5. **Develop and present custom guideline documentation** that effectively and efficiently meets all of the design team's needs and is easy to update.
6. **Verify and update guidelines** as problems are identified and as changes occur

The benefits of good design guidelines and documentation are immediate and long term. In addition to helping other team members, as discussed earlier, accurate and readable guidelines benefit the individual designer in successive iterations along the path to a finished design.

As noted earlier for any documentation, to be effective guidelines should have the following characteristics:

1. Accuracy
2. Clarity
3. Conformity to policy, convention, and standards
4. Completeness
5. Integrity
6. Brevity
7. High quality
8. Retrievability
9. Proper format

The key is to insure that the guidelines help to optimize the design without causing a loss in creativity or innovation. This section reviews several examples of design for producibility guidelines. The examples include design for preferred method guidelines, rules, and goals, producibility measurement or rating systems, and computer software.

12.4 DESIGN FOR PREFERRED MANUFACTURING METHODS

One major guideline for simplification is to design for preferred vendors and manufacturing methods. Some methods of manufacturing are much better in operational parameters such as cost, quality, lead-time, risk, etc. Preferred manufacturing methods are these "better" or "best" methods. Designing a product for a preferred method of manufacturing from among a number of different processes can be a difficult or an easy decision. This depends on the number of alternatives, their similarities, and the amount of information available for each process.

The goal is to incorporate parameters in the design that facilitate the use of a preferred production process. Since each manufacturing process and company has different requirements and capabilities, its producibility guidelines will vary for each process. This phase of design for production is often referred to as a systematic "customizing" of the design for the selected process. This customizing can vary from minor design changes to major changes, which results in changing the manufacturing process to be used.

12.4.1 Designing for Preferred Methods of Fabrication

This section evaluates some of the preferred and non-preferred methods for fabrication that can affect producibility. For more detailed information the reader is referred to Bralla, 1986.

Preferred methods	Non-preferred methods
Castings or plastic	Completely machined
Near net-shaped casting	Casting
Screw machine	Lathe turning
Milling	Jig bore
Turning	Milling
Standard materials	Nonstandard materials
Tolerances > +0.005	Tolerances < +0.001
6061 Aluminum	Steels, stainless steels
303 Stainless steel	Other types of stainless steels
Thermoplastics	Thermosets
Hardness < 41 Rc	Hardness > 41 Rc

Castings

For metal parts, casting technology is the preferred fabrication method. In fact, castings are often the only realistic manufacturing method available for many large and complex metal parts. The key advantage of specifying a casting over a completely machined part is the significant cost reduction in machining operations afforded by a casting. This is especially true for higher product volumes. Castings are metal patterns formed by pouring molten metal into a mold. Three types of castings normally used are investment, die, and sand. Investment castings are poured from plaster molds. They can produce very complex designs and hold relatively tight tolerances (± 0.020 inch). Die castings are produced by injecting metal into a metal cavity by a process similar to plastic injection molding. Die castings are relatively small, hold tighter tolerances than investment castings, and have the highest tooling costs. Many castings are produced in sand molds. They are specified for large parts for which surface finish and tolerance control are not as critical. The major disadvantage for casting is lead-time.

Plastics

Plastic parts provide a very inexpensive alternative to metal parts especially for high production volumes. Another major advantage is in assembly operations because the designers can incorporate several functional requirements into one part (e.g. a multi-functional base that incorporates snap fits and locator slots into the part). The disadvantages of plastics include higher tooling costs, longer lead time for the tooling, wider dimensional tolerances, higher coefficients of thermal expansion, and lower strengths than comparable metal parts. The tooling cost limits plastic parts to higher production volumes.

Stamping and Forming

Stamping or forming processes usually produces sheet metal parts. Although not as inexpensive as plastic, they offer a significant cost reduction compared with machined parts. All machined parts should be reviewed for possible conversion to sheet metal designs. An advantage of sheet metal parts is that they have tensile strength and other physical properties similar to those of metal machined parts. Tolerances are generally looser than those found in machining, usually in the ±0.010-inch range.

Machining

Machining is a broad term given to a wide variety of manufacturing technologies. All machining methods depend on metal removal by hard cutting tools. Common machining processes include milling, turning, lathes, screw machining, and jig boring. Machined parts are generally more expensive but they often have the shortest lead-time. Milling applies broadly to all machining methods in which the workpiece is stationary while the cutting tool is moving. Turning is the opposite and is usually used to produce round parts; milling can produce parts of almost any configuration. Milling is usually a more expensive process than turning. For this reason, round parts should be incorporated in the production design whenever possible.

Screw machines are specially tooled lathes that can produce high-quality round parts resulting in significant cost savings. All machined parts should be reviewed to determine whether they could be designed for screw machine manufacture. Special design requirements apply when designing a part to be machined on a screw machine. For example, the design should allow all cutter movements to come from the same direction. This means that round parts with holes in both ends would require two separate setups to produce. It is often possible to switch features between parts to enable them to be economically produced on screw machines.

A jig bore is a special type of high-precision milling machine. It can produce parts with very tight tolerances (± 0.001 inch). Its high precision, however, makes it a very expensive manufacturing process. During design development, jig bores are sometimes used to produce prototype parts with very high tolerances. Production design parts should never require manufacture on this type of machine. Parts that require fabrication on a jig bore should be redesigned for milling.

When machining is necessary for a metal part, a significant number of producibility parameters are involved. The book by Bralla (1986) is an excellent source for producibility recommendations for machining. An example of some typical overall design recommendations for machining are shown in Table 12.1 (adapted from Bralla).

Many other preferred methods not listed in this discussion can affect cost and quality. An example of how the specification for the corner design of a metal box can affect the manufacturing cost is shown in Figure 12.1.

12.4.2 Designing for Preferred Methods of Electronic Assembly

The assembly of electrical systems is typically composed of several consecutive manufacturing steps with relative few alternative manufacturing methods. A major design decision in electrical assembly is between automated and manual methods. A simple breakdown of some preferred and non-preferred methods for electronic assembly is as follows:

Preferred methods	Non-preferred methods
Automation	Manual assembly
Self-fixture assembly	Non-fixture assembly
Flexible circuits	Wiring and cabling
Standard hardware	Non-standard hardware
Top-down assembly	Multi-directional assembly
Modular assembly	Non-modular

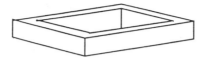

Corner Type	Finish	Proportional Cost
Free Corner	Chromate Dip	100% (Standard)
Heli - Arc	Grind smooth & chromate	200%
Spot - Weld 2 per Corner	Chromate Dip	160%
	Chromate Dip	180%
Tubular Rivet 2 per Corner	Chromate Dip	170%
Soft Solder	Tin Plate	210%
Dip Braze	Chromate Dip	210%
	Tin Plate	250%

FIGURE 12.1 Cost comparisons of different corner designs.

TABLE 12.1 Design Recommendations for Machined Parts

1. If possible, avoid machining operations

2. Specify the most liberal surface finish and dimensional tolerances possible

3. Design the part for easy fixturing and secure holding during machining operations

4. Avoid sharp corners and sharp points in cutting tools

5. Use in-stock materials whenever possible

6. Avoid interrupted cuts

7. Design the part to be rigid enough to withstand the forces of clamping and machining without distortion

8. Avoid tapers and contours

9. Avoid the use of hardened or difficult-to-machine materials

10. Place machined surfaces in the same plane or, if they are cylindrical, with the same diameter to reduce the number of operations required

11. Provide tooling access room

12. Design so that standard cutters can be used

13. Avoid having parting lines or draft surfaces serve as clamping or locating surfaces

14. Expect burrs, provide relief space for them

Source: Adapted from Bralla, 1986.

Design for Automated Printed Circuit Board Assembly

A printed circuit board is a good example of a complex assembly process that benefits from a design for producibility effort. Automatic component insertion methods are always preferred to manual insertion, regardless of production volume. The low cost and high quality of automatic insertion makes it the single largest goal in producibility design efforts. Automatic component insertion equipment has been available for many years and is commonly applied throughout the electronics industry. It is basically an automated system that places individual electronic components on a raw printed circuit board. Equipment is available to insert most types of axial, radial, dual in-line package, and surface-mounted components. The key design steps are to:

1. Select only those components that are compatible with the automated equipment.

2. Select only from vendors that can provide components in automatable packaging

3. Layout and design the raw circuit board that is compatible with the automated equipment

An example of some design requirements for a raw circuit board is explained in Table 12.2. Component spacing is standardized since nonstandard spacing requires special costly fabrication and special inspection tooling.

The key to producibility, as mentioned earlier, is to have all components autoinserted. Autoinsertion, however, requires special design considerations when locating the parts on the board. The design team determines the process capability of the autoinsertion machines and other related processes. Specific layout design requirements that allow the automatic insertion of three types of components are shown in Figure 12.2. These would be updated frequently and customized for each application.

If possible, miscellaneous parts, such as terminals and clips, should not be installed on the board prior to automatic insertion. Terminals or clips that are installed prior to automatic insertion should be oriented and placed according to automatic insertion equipment tooling requirements.

Design for Manual Printed Circuit Board Assembly

Although automated assembly technologies are receiving widespread attention, much of the low volume assembly in the electronics industry is still done during

TABLE 12.2 Design Requirements for Printed Circuit Board Assembly

All components must be auto-insertable and available from vendors in auto-insertable packaging

All component holes to be on a 0.100-inch grid

No components to be placed near a masked area (0.100 inch from component body edge to masked area), including both ends of jack tips, and all edges of heat sinks and connectors

All components oriented in well-ordered rows, with the rows parallel to the connector

Similar components grouped together whenever possible

Polarity standardized for like components

When common mounting centers are not used for axial components, organize rows so that the shortest centers are on one end and the longest centers are on the opposite end

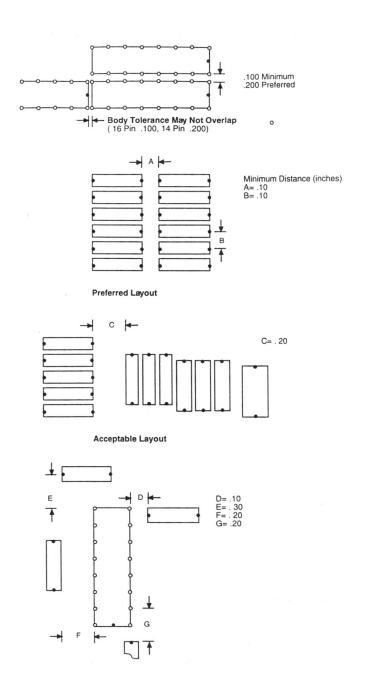

FIGURE 12.2 Spacing between components for automation.

the design process. For example, the sequence of assembly steps is manually. Ample consideration to the human element should also be given closely analyzed to ensure that all operations can be easily accomplished manually. Another step for manual assembly is to identify the potential producibility and quality problems (i.e., failure modes) associated with design requirements and manufacturing processes. This process is similar to that of a design's failure modes and effect analysis. That is, its purpose is to identify all potential failure modes caused by manufacturing and their effects on quality. This can include all failures induced by parts, vendors, equipment, procedures, personnel, and materials. In electronic assembly when using manual insertion methods, component selection can have significant impact on production quality. Some components that historically cause failure mode problems in manual component insertion are listed in Table 12.3.

12.4.3   Design for Preferred Methods of Manual Assembly

The most important step is to simplify the design in order to minimize the number of assembly operations to be performed or automated. Whether or not the assembly process will be automated, the design team should design every product for robotic and automated assembly since later design changes will be too expensive.

When designing for manual assembly, the design team should focus on

TABLE 12.3 Production Failure Modes Caused by Component Selection

Components that Historically Cause Assembly Problems	Failure Mode
	Quality defect
Can component	Tilt, float, short to another component
Small lay-down axial components (body diameter smaller than 0.080 inch)	Solder in the lead bend
Polarity (axial or radial components)	Stuff error
	Labor Problem
Heat-sensitive component	Hand solder after flow solder or vapor phase
Heavy hardware-mounted component	Install and hand solder after flow solder
Large components	Uses excessive board space
Component lead spacing not on 0.1-inch grid	Requires extra preparation

the following critical areas of producibility:

1. Adequate physical and visual access (i.e., accessibility)
2. Foolproof assembly (i.e., parts cannot be assembled incorrectly, self-orientation, guide pins for critical part fits, self-securing, obvious when parts are missing)
3. Minimal assembly steps and adjustments eliminated
4. One-directional, top-down assembly to a base part
5. Ease of repair and inspection
6. Safety factors
7. Nonstandard assembly practices and techniques eliminated
8. Tolerance accumulations and dimensional tolerances for ease in mating parts

Special attention is also given to the standardization and ease of assembly of mechanical hardware and electrical components throughout the system.

Some specific areas of investigation are as follows:

1. Screws
2. Nut plates
3. Spacers
4. Card guides
5. Resistors, diodes, etc.
6. Connectors
7. Wire
8. Test points

Design for Accessibility in Manual Assembly

As noted earlier, accessibility is a main consideration in any assembly process. Anyone can build a model of a ship faster and easier on the kitchen table than inside a bottle. Poor accessibility for manual assembly causes problems, including the following:

1. Worker fatigue
2. Decreased quality
3. Decreased productivity
4. Increased manufacturing times and cost
5. Increased safety problems
6. Increased training time

Efforts made during the product design can increase accessibility and reduce these problems. An approach to increasing accessibility is to divide complex assemblies into modules (i.e., subassemblies). A good example of this principle would be the assembly of the overwing fairing on the B-1 bomber. The fairing was originally designed to be built without subassemblies and then sent for sealant and paint. The sealant operation was complicated due to a fillet seal on the inside seam of the fairing. The workspace was so cramped that the sealer had to use a special sealant tip and a mirror. The work was done by holding a small mirror deep inside the pan and, with the other hand, moving the sealant gun in a tight, enclosed area. The worker had to work "backwards" owing to the reversed image in the mirror. This portion of the sealant operation typically lasted over an hour. A simple design change was made to wait to install the skin panel that enclosed the location of the fillet seal until after the fillet had been performed. This ended all the accessibility problems and reduced the sealing operation to under 5-min.

To illustrate the effects of design on accessibility, a list of typical factors used to determine the increase in assembly time caused by poor accessibility are given in Table 12.4. The factors are broken down by work area, work, and worker position. The values selected under the classifications of area, work, and position are to be added to the accessible value of 1.00 to arrive at the total accessibility factor for the work under consideration. For example, the labor cost penalty can be as much as 200%!

12.4.4   Design for Automated and Robotic Assembly

The problems with automating assembly are considerable. Rossi in 1985 highlighted this problem when he stated: "Often designs are made in such a fashion that you just can't access a certain area with a robot. Humans can get around obstacles and operate within those designs easily, but robots can't because they're not quite as flexible as human beings. Today, what happens is that users try to apply a robot to something that's been designed without robotic assembly in mind. Either the robot cannot handle it at all or the users find that they have got to put a lot of additional engineering design into a particular workcell, or perhaps into an end effector, in order to get around the problem. All this does is add to the price tag, and cost is very much in consideration when one is trying to sell these systems. Key design tasks for robotic and automated assembly are to:

1. Ensure high-quality parts
2. Minimize the use of fasteners and cables
3. Provide accessibility to install parts
4. Select parts and vendors that can be used in part feeders, pallets or some robot friendly method for material handling

Assembly operations usually require part feeders, end-of-arm tooling fixtures, and a material-handling system. Except in the case of robots with vision

TABLE 12.4 Accessibility Factors Affecting Assembly and Repair Time[a]

Factor[b]	Accessibility description

I. **Accessibility modifier for work area**

0.0 Generally accessible with little or no difficulty in obtaining access to work area

0.1 Area sufficiently restricted on two sides so that operator must exercise care in moving body members

0.1 Operator must work with arms extended or head tilted back to see work

0.4 Area usually restricted on three sides with cramped space in which to work surrounded by small shell

1.0 Operator entirely enclosed or arms and shoulders entirely enclosed

II. **Accessibility modifier for work task**

0.0 Little or no restriction on use of hands or tools

0.2 Moderate restriction on use of hands or tools

0.6 Considerable restriction to use of hands or tools; work under other installations

1.5 Very difficult to reach or use tools; most of work in "blind" location

III. **Accessibility modifier for worker position**

0.0 Stand on floor

0.1 Stand on ladder

0.2 Stooping from standing position

0.2 Stretch over obstacle

0.2 Lie on back

0.5 Lie on side

[a]Time to assemble = (standard time) × (1.0 + accessibility modifiers).
[b]Add the factors to an accessible value of 1.00 for the total accessibility factor.

or special sensors, parts with which the robot will interact must be precisely located, which requires additional tooling. Table 12.5 shows one author's recommendations for design rules.

12.5   PRODUCIBILITY MEASUREMENT

There are many different producibility analyses that the design team can use to predict a design's level of producibility before it is turned over to manufacturing. Measures are based on the product's producibility requirements and on other special measures that design or manufacturing have found to predict manufacturing performance. These analyses numerically quantify the design's level of producibility to be used in trade-off analysis. Lord Raleigh is often

TABLE 12.5 Qualitative Goals for Robotic Assembly

The product should have a base part on ,which to build assemblies

Base shape should facilitate orientation and be stable enough to allow parts to be assembled on or within it, without becoming dislocated

Where possible, incorporate parts into subassemblies that can then be put together to form a final product

Parts should be capable of being inserted from above; if possible, parts should be able to be added to an assembly or subassembly in layers

If possible, use guide pins to simplify layering of parts

Parts should accommodate insert with straight-line motion

Design so that any tool required for insertion or tightening can easily reach the required location and perform the required task

If possible, all parts for the same assembly should accommodate handling by a single gripper; if not, explore the possibility of multiple-device tooling

Design parts to prevent dislocation, once installed

Avoid use of bolt-and-nut assembly

Where possible, parts should be able to be pushed or snapped together; if screws are required, they should all be the same size and have conical or oval points

Parts should be compliant and self-aligning using chamfers and tapered parts

Source: Adapted From Stauffer, 1984.

quoted as "If you cannot express something in numbers, then you do not have sufficient knowledge of the subject".

Most product level producibility requirements are too high a level for many lower level design decisions. Because of this many "specialized measurement systems have been developed to assist the design team. In addition several methods have been proposed using decision theoretic and maximization of expected utility techniques (Priest and Sanchez, 1990, Burnell and Priest, 1991, and Hausner, 1999). Developing a producibility measurement system obviously requires a higher level of knowledge about the topic than simple guidelines. The steps are to:

1. Identify key design and manufacturing requirements and parameters that affect a design's producibility.
2. Quantify the*se* relationships.

3. Develop an effective, easy to use measurement system for the design team to evaluate and predict a design's level of producibility to be used in trade-off analyses.

The goal is to develop a quantified measurement system that can accurately and easily measure and rate the producibility of a design. This allows the comparison of two different designs. The most well-known rating systems are the Assembleability Evaluation Method developed by Hitachi and modified by General Electric (1953) and the Design for Assembly method developed by Boothroyd and Dewhurst (1999). Many companies use these systems.

12.5.1 Producibility Measurement Systems for Assembly

A simple example of a rating system for the producibility of a printed wiring board is illustrated in Table 12.6. The purpose is to show how a "specialized lower level" measurement system might work. The higher the score, the higher the cost, the lower the score, the more producible the design. Using this type of system, a design can be "rated" and then compared to other designs and standards. The score can highlight potential problem areas.

TABLE 12.6 A Producibility Rating System

Specific objective	Meets objective	Meets over 90% of objective	Meets under 90% of objective
1. There are no wires/cables	0	5	10
2. All similar components are oriented in the same direction	0	1	3
3. All component can be auto-inserted	0	5	10
4. Polarized components are readily identifiable	0	5	10
5. All components are on an X or Y-axis	0	5	10
7. Component insertion area is 18 × 18 inches or less	0	10	20

[a]Operation: Insert components on printed wiring board.
Source: Adapted From Maczka, 1984

12.6   PRODUCIBILITY SOFTWARE TOOLS

There are many advantages of using software in producibility analysis. Systems that can directly access the CAD database can result in better communication between the design team. They also provide consistent implementation of producibility methods between projects, better documentation for collaborative work, knowledge bases, faster responses and lessons learned of the company. The BMP Task Force developed a list of software categories, which are discussed on the next page (adapted from BMP, 1999).

1.   **Design for producibility and manufacturability** - Producibility software can improve communication between the members of the design team because it provides direct access to the CAD database. The software allows users to simulate and model process parameters and assembly issues including tolerance, form, fit, function, costing and manufacturing. It then analyzes and evaluates each component and makes recommendations for redesign based on optimizing performance. The software provides cost data, design guidance and producibility analysis to identify and reduce costs at every stage of a product's life cycle. It is also capable of working with part geometry input from feature-based CAD models. Since the software provides preliminary cost estimates early in the design life cycle, it can identify key cost drivers that impact manufacturing costs, quality and cycle time, and greatly improve the design process. The most popular example is the suite of software products by BDI (www.dfma.com)

2.   **Statistical process control and statistical quality control** - A statistical process control/statistical quality control system collects critical production data from manufacturing and vendors in a database. The design team can compare quality results against preset control limits, such as C_p and C_{pk}, then present this information in statistical process control and statistical quality control charts, graphic screens, and reports in order to facilitate corrective action. A fully integrated system can merge design, manufacturing and production management systems such as Enterprise Resource Planning systems.

3.   **Simulation and virtual reality** - Simulation and virtual reality software provides interactive 2D or 3D graphic simulation software to model conceptual and detail designs at every stage of development from design through manufacturing planning to production. It allows the user to change the model, to perform "what if" analysis, and run cost tradeoff studies to evaluate different design and manufacturing alternatives before building prototypes or modifying existing design so that critical producibility parameters such as capacity, throughput, cycle time, production yields, costs, and quality are optimized. The software is capable of simulating work cells using libraries of manufacturing resource components such as human operators; material parts and components; robots, machine tools, work benches, gantries, weld guns, etc.

4.   **Tolerance analysis** - Tolerance analysis software performs tolerance analysis and tolerance allocation to help identify all contributors to

both geometric and dimensional tolerance that impact manufacturing processes and cost. It is capable of evaluating tolerance specifications of design to reduce the chances of assembly interference between adjacent mating components or potential stack-up tolerance between mating parts in complex assembly. The software calculates the percent contribution and sensitivity of critical dimensions in assembly to changes in constraints, then compares them to current design databases and evaluates them for their impact on form, fit and function as tolerances are updated or changed.

     5.   **Miscellaneous Producibility Software** - Other miscellaneous producibility software includes Quality Function Deployment, Design for Assembly, Design for Manufacturing, Failure Modes and Effects Analysis, Risk Assessment, Risk Analysis and Risk Management, Design Tradeoff Analysis, and Complexity Analysis.

12.6.1   Producibility Knowledge Database System

     In complex and company, product, and process applications, a specific producibility domain model is created. The domain model includes the tasks and the ontology needed to understand the requirements and to build the desired database. Ontology describes the structure, semantics and syntax of the domain being studied. It includes the entities, their activities, their objectives, and their relationships which, used together, define the domain. Task and domain analysis methods are used to build the ontology of the domain model. These methods include interviews with experts, observation of applications in use, and written and video material. The domain model provides a shared ontology between users (e.g. design team, manufacturing) and software designers that allow the user to perform continuous metric-based evaluations of the developed products.

     One method of developing domain models and to capture requirements is to use scenarios. A scenario is a sequence of interactions and activities that occur between the to-be developed system and the user. A scenario captures externally visible behavior of a system. Typically, one creates two collections of scenarios, each called a view. The "as-is" view assists software developers in understanding the current domain and can be used to justify further software development. The "to-be" view is a description of what a group of users would like to have, i.e., user requirements and desires. Actions the system performs in the "to-be" view are extracted from the scenario descriptions. These actions are called tasks. An example producibility task is to "calculate C_p".

     Tasks are parameterized and added to a database of reference requirements and later linked to software components that implement the tasks. The next time similar tasks are required to be developed, reference requirements and their associated components can be selected thus shortening the development schedule. A software designer can begin requirements elicitation with a series of questions based on the reference requirements, such as "Do you need to predict producibility?" If the answer is yes, more detailed questions are asked about producibility tasks, using the contents of the reference requirements database and the domain model as guides. A simplified scenario, patterned after (McGraw and Harbison, 1997), for predicting producibility by the quality discipline is illustrated in Table 12.7.

TABLE 12.7 Scenario for the Quality Discipline

Scenario No. 1:	Updated version of current power supply model PS367
Objective:	Determine the level of quality of the new power supply's design for producibility trade-off analyses
Task Responsibility:	Manufacturing or Quality Engineer
Major Task Reference (Task 3, Level 1):	Evaluate and measure producibility for quality
Goal:	Predict level of quality (C_p and C_{pk}) and identify problems/concerns
Subtask descriptions (Level 3):	1. Determine design requirements and select manufacturing processes from CAD database 2. Identify manufacturing process capabilities for each design requirement. These are statistical distributions for the required processes, parts, and supplies 3. Calculate C_p, C_{pk}. 4. List largest defect contributors in predicted defect rate 5. Distribute results to team members
Stimulus/cue:	All CAD and purchased part data is available
Schedule temporal aspects:	1 week
Personnel resources:	1 week
Equipment/Software:	Statistical Software and CAD database
Information requirements:	1. Manufacturing quality distribution database for selected processes, parts, and supplies 2. CAD and supplier information including process plan.
Output recipients:	All design and support team members

12.6.2   Expert System For Producibility Example

Another computerized method for producibility is to develop an expert system for the designer. The expert system to be discussed is the Printed Control Board (PCB) Expert System for Producibility (ESP) developed initially by Texas Instruments, Inc. Its main objectives are to evaluate and rate board designs on manufacturability and to provide design recommendations that optimize fabrication and assembly.

First, a completely manual rating system that could be used to evaluate a board's design was created. With this manual rating system, the design team could identify manufacturing problems before the circuit board design was finalized by the drafting center. Next, the expert system's knowledge base was created, with input from a panel of experts because the domain of the problem was so large that no one person knew all the details of the system. The database included factors such as component dimensions, manufacturing tolerances, quality, and labor factors. An extensive set of design guidelines and a component
analysis system could then be developed to contain design-related electrical information. The functions and features of the database are:

- **Density Analysis** – a simple comparison of the available circuit board space compared to the part sizes in the parts list to determine which space is available on the board.
- **Height Analysis** – a look at the individual mounted part heights and comparison of these values to the space that is available as determined by the mechanical engineers who have done the board stack up analysis.
- **Automation Analysis** – an analysis of the ability to automatically insert parts by part type, placement on the board, and by availability of placement equipment.
- **Quality/labor Analysis** – a calculation of the number of parts having special manufacturing requirements that could add cost to the product or reduce the reliability of the product.
- **Standard Parts Usage Analysis** – a comparison of the parts selected on a design to a standard parts database, which could be the company's standard parts database, or a customized version.
- **Parts Analysis Ratings** – a determination of the overall board design rating using the results of the previous analyses.
- **Cost Analysis** – an estimate of the labor cost to manufacture a board.
- **What-If Analysis** – an analysis that reruns the entire series to evaluate different parts, layout or manufacture process changes. Each what-if analysis result could be saved to different files for later comparison.

With the Texas Instruments Personal Consultant computer, the model initially consisted of 216 combined rules, 147 parameters, and 1720 lines-of user-written IQLISP code which provided additional features to the expert system, such as iteration, printed reports, multiple recommendations, and rules. A feasibility model was tested against prior, manually reviewed designs and was found to successfully identify producibility problems. When the system is completely finished, the knowledge base is expected to grow to approximately 1500 rules. This expert system parallels the design flow, enabling real-time iterative design analysis by showing how a change in the board design could affect the manufacturability of the board. This system is now available through the BDI Company (www.DFMA.com).

12.7   SUMMARY

Producibility guidelines and analysis is an immensely broad subject that varies depending on the product type, technologies, process capabilities, marketplace and the company environment. The intent of this chapter is not to provide cookbook guidelines to train the design team in all areas of producibility. The goal is to make the design team aware of the necessity of producibility guidelines, rules and measures and its key principles and major considerations. The product development team, once sensitized to the capabilities of manufacturing, can significantly improve the producibility of their designs.

12.8   REVIEW QUESTIONS

1.   Discuss some of the advantages and disadvantages of documented producibility guidelines.
2.   What characteristics do design guidelines need to have in order to be effective?
3.   List the steps used in developing good guidelines and rules.
4.   Discuss briefly the following preferred methods of fabrication: castings, plastics, stamping and forming, machining.
5.   What are some of the preferred methods of electronic assembly?
6.   Discuss some of the considerations to keep in mind when designing manual assembly processes.
7.   Discuss some of the considerations to keep in mind when designing automated assembly processes.
8.   How can a producibility rating system be used in manufacturing? In support systems? In the service industry?

12.9   SUGGESTED READINGS

1.   MIL-HDBK-727, Military Handbook: Design Guidance for Producibility, Department of Defense, Washington, D.C., April 1984.

2. BMP, Producibility Systems Guidelines For Successful Companies, www.bmpcoe.org, 1999
3. BMP, Producibility Measurement Guidelines, NAVSO, P-3679, Department of the Navy, August, 1993, and Guidelines 1999. www.bmpcoe.org

12.10   REFERENCES

1. BMP, Producibility Measurement Guidelines, NAVSO, P-3679, Department of the Navy, August, 1993, and Guidelines 1999. www.bmpcoe.org
2. Boothroyd and Dewhurst, Design for Assembly, BDI, www.dfma.com, 1999.
3. G. Bralla, Handbook of Product Design for Manufacturing, McGraw-Hill, New York, 1986.
4. L. J. Burnell, J. Priest and K. Briggs, An Intelligent Decision Theoretic Approach to Producibility, Journal of Intelligent Manufacturing, Vol 2, p. 189-196, 1991.
5. B. Freeman, The Hewlett-Packard Deskjct: Flexible Assembly and Design for Manufacturability, p. 50-54, CIM Review, Fall, 1990.
6. W. R. Hausner, Producibility Design-To Requirements, Appendix E.2 to BMP, Producibility Systems Guidelines For Successful Companies, www.bmpcoe.org, 1999
7. J. Maczka, GE has designs on assembly. Assembly Engineering, June 1984.
8. J. Maczka, Designing For Robot Assembly Links Products To Process. Appliance Manufacturer, February 1986.
9. K. McGraw, and K. Harbison, User-centered Requirements: The Scenario-based Engineering Process, Mahwah, NJ: Lawrence Erlbaum & Associates, 1997.
10. J. W. Priest and J. Sanchez, An Empirical Methodology for Measuring Producibility Early In Product development, International Journal of Computer Integrated Manufacturing, 1990.
11. Rossi, Dialogues, Manufacturing Engineering, October: 41 1985.
12. N. Stauffer, Robotic assembly, Robotics Today, October 1984.

Chapter 13

SUCCESSFUL PRODUCIBILITY METHODS USED IN INDUSTRY

Comprehensive Methods for Producibility

As noted earlier, we believe that the best producibility approach includes the use of many different analyses and techniques. Several groups have developed comprehensive methods that are well documented and ready for use. This chapter looks at examples of some of the most popular and successfully used methods in industry. Although the examples are brief they give the reader a sense of each method's objectives and process. Even though each of the methods has a different focus, approach and measurement method; their goal is still the same, to improve producibility!

Best Practices in Industry

- Best Manufacturing Practices (BMP) Program
- Producibility Assessment Worksheet (PAW)
- Boothroyd and Dewhurst Design for Assembly (DFA) and Design for Manufacturing (DFM)
- Robust Design
- Taguchi Methods
- Six Sigma Quality and Producibility
- Production Failure Mode Analysis (PFMEA), Root Cause, Isakawa Diagrams and Error Budget Analysis
- Mistake Proofing and Simplification
- Design for Quality (DFQM)
- Formal Methods for Vendor and Manufacturing Qualification and Certification of Processes (e.g. ISO 9000)

PIONEERING WORK OF BOOTHROYD AND DEWHURST

Design for Assembly is the best-known commercially available producibility method. The development of the original design for assembly (DFA) method stemmed from earlier work in the 1960s on the automatic handling of small parts. As reported by Dr. Boothroyd, a group technology classification system was used to catalogue solutions for the automatic handling of small parts. Then, in the mid-seventies, the US National Science Foundation (NSF) awarded a grant to extend this approach to design for manufacture (DFM) and design for assembly (DFA). This allowed the effects of product design features that affect assembly times and manufacturing costs to be classified and quantified. The DFA time standards for small mechanical products resulting from the NSF supported research were first published in handbook form in the late seventies (Boothroyd, 1999). A company, Boothroyd Dewhurst Inc. (BDI), was formed as the result of the development of the DFA software in 1982. Using the DFA methodology, the design team can begin with a standard set of guidelines, quantify the design for assembly difficulty and cost and then re-design for improvements.

According to Boothroyd (1999), a major breakthrough in DFA implementation was made in 1988 when Ford Motor Company reported that the DFA software had helped them save billions of dollars on their Taurus line of automobiles. Eleven major companies have reported annual savings totaling over 1.4 billion dollars. A study of 117 case studies showed the following top 4 producibility improvements: (Boothroyd, 1999).

Category	No. of Cases	Average % Reduction
Part count	100	54
Assembly time	65	60
Product cost	31	50
Assembly cost	20	45

In recognition of their contribution to American manufacturing competitiveness, President Bush awarded the National Medal of Technology to Drs. Boothroyd and Dewhurst.

Each of the methods described in this chapter has unique advantages and disadvantages. The best method for a particular project depends on many factors including the company, product, resources, schedule etc.

13.1   BEST PRACTICES

The key comprehensive methods that are discussed in this chapter are as follows:

- **Best Manufacturing Practices (BMP) Program** provides a large "lessons learned" computer database for producibility and manufacturing where best methods can be identified, improved upon, and implemented.
- **McLeod's Producibility Assessment Worksheets** are used to identify technical risk in the early evaluation of the design's producibility.
- **Boothroyd and Dewhurst DFMA** uses producibility measurement such as Design for Assembly's efficiency rating to reduce manufacturing cost and improve quality.
- **Robust design such as Taguchi methods** uses cost of quality and design of experiments to optimize producibility and manufacturing related decisions.
- **Motorola's Six Sigma quality and producibility** emphasizes quality by using statistical distributions of process variability to predict the number of defects for a design.
- **Production failure mode analysis (PFMEA), root cause, Isakawa diagrams and error budget analysis** emphasizes quality by identifying potential manufacturing induced failures, how these failures could affect performance, and evaluates methods to decrease their occurrence.
- **Mistake Proofing and Simplification** uses design features and process changes to eliminate the chance of manufacturing errors.
- **Design for quality manufacturing (DFQA)** uses common error catalysts to evaluate a design from a quality perspective.
- **Manufacturing qualification and certification** reduces technical risk through a formal process such as ISO 9000 for approving manufacturing and vendor processes that includes the measurement of their process variability.

Although each of these methods is different, all of them are proven to improve producibility. In this chapter we will continue to use our notebook computer example to show how all of the various methods might be used in a product development project.

13.2   BEST MANUFACTURING PRACTICES PROGRAM

As noted in earlier chapters, using lessons learned and best industry practices are one of the best techniques for product development. This method identifies the "best" practices for a particular application and ensures to not repeat the same mistakes from previous projects. The Best Manufacturing Practices (BMP) program sponsors the largest and most popular repository of lessons learned for producibility. As noted in their website, www.bmpcoe.org, the program has changed American Industry by sharing information with other companies, including competitors. Their unique, innovative, technology transfer program is committed to strengthening the U.S. industrial base.

The BMP program, sponsored by the Office of Naval Research, began in 1985 by identifying, researching, and promoting exceptional manufacturing practices, methods, and procedures in design, test, production, facilities, logistics, and management with a focus on the technical risks highlighted in the Department of Defense's 4245-7.m Transition from Development to Production Manual. The primary steps are to identify best practices, document them, and then encourage industry, government, and academia to share information about them. By fostering the sharing of information across industry lines, BMP has become a national resource in helping companies identify their weak areas and examine how other companies have improved similar situations. This sharing of best practices allows companies to learn from others' attempts and to avoid costly and time-consuming duplication.

In-depth, on-site, voluntary surveys of manufacturing and design operations represent the heart of the BMP program. Through its surveys, BMP identifies and documents best practices in industry, government, and academia; encourages the sharing of information through technology transfer efforts; and helps strengthen the competitiveness of America's Industrial base. BMP publishes its findings in survey reports, and distributes the information electronically and in hard copy throughout the U.S. and Canada. The information is also provided through several interactive services including CD-ROMs, BMPnet, and the BMP Website (www.bmpcoe.org). Exchange of additional data is between companies at their discretion.

The best practices are summarized into How–To Books. These are located in the Know How software tool, an electronic library comprised of expert systems and digital handbooks covering a variety of product development topics, including ISO 9000. The automated program offers rapid access to information through an intelligent search capability. How-To cuts document search time by 95% by immediately providing critical, user-specific information.

13.2.1   Best Practices Example

Designing the printed circuit board (i.e. printed wiring board) is a major design and producibility task for the notebook computer example. A first step would be to identify the "best practices" for design and producibility for performing these tasks using the database. We would then compare our design practices with the best practices to identify needed changes and improvements to our product development and manufacturing processes.

Using producibility and circuit board as key words, we found an excerpt from the BMP database at www.bmpcoe.org as shown below.

Texas Instruments, DS&EG (now Raytheon Systems) -Dallas, TX
Original Date: 11/01/1991 Revision Date: 06/24/1998
Information: Printed Wiring Board Producibility

TI DSEG uses an Electronic Design Automation (EDA) operation to focus on procedures and supplementary tools to enhance PWB producibility. EDA at TI DSEG provides electronic drafting services for PWB placement and routing and generation of assembly documentation. The basic tools used by EDA are PC Cards for interactive layout and routing, ASI Prance for autorouting, and AutoCAD for mechanical documentation. All tools run on PC compatibles interconnected by a LAN to a SUN Microsystems File Server through PC Network File System software.

The emphasis on producibility begins with PWB Design Process orientation for personnel on new projects. This orientation attempts to explain the reasons for process requirements and guidelines and relates to their impact on producibility and ultimately cycle time and cost. During orientation, the design team is provided with a well documented process description which includes roles and responsibilities, data requirements, and deliverables, most particularly a form used by the design team to provide layout personnel with data on the type of design and various board parameters.

Several tools are available to the design team to evaluate the producibility of the intended design. Manufacturing and assembly producibility engineers are co-located with the design team to assist in these capabilities to perform pre-placement and thermal analysis.

Producibility efforts are reinforced since a PWB design must also meet a set of minimum design criteria before it can be released to layout. These criteria are evaluated by the PWB Expert System for Producibility (PESP), and the MIL-STD-2000 Checker. The minimum design criteria include board density, board dimensions, compliance to MIL-STD-2000, and usage of components from the DSEG Master Engineering Parts Library, and percentage of autoinsertable components. These criteria are also used as metrics to gauge the design process.

PESP is a computerized, knowledge-based analysis tool utilized prior to board layout. The software uses the component list generated from the schematic database. The designer can then interactively determine if there is sufficient board space to reasonably place all the components (board density). Height and ease-of-assembly analysis can also be evaluated. The system has provisions to build the input component list if one does not exist.

Numbers and types of components, board size, board thickness, and number of layers can be changed and evaluated in real time. Since the expert system is essentially a standalone system, required schematic modifications must be made on the designer's host workstation. Future plans are being discussed to integrate the expert system with the workstation.

After layout is complete, the PWB design is translated into TI Common Format. Common Format is a neutral database format used to drive subsequent manufacturing processes. It allows manufacturing to work with product designs originated on different design and layout systems. Common Format also drives several TI-developed producibility tools such as Thermal Analysis; AutoVer, for automated verification of the artwork data; and Automatic Electrical Test User Interface, a final check of layout to netlist continuity. Improvements in the layout process and implementation of the LAN and file server have helped the TI DSEG EDA to reduce its average PWB design cycle from 40 to 25 days with a five-day quick-turn service for prototype designs.

For our example we would review and compare our current process with the above process. The comparison might include hardware, software, tasks, training, etc. Of particular note for evaluation is the training orientation for new personnel, co-locating all personnel working on the development team, buying the PWB Expert System for Producibility, and incorporating the TI or some other common format database.

13.3   PRODUCIBILITY ASSESSMENT WORKSHEET *

During the late 80's, Scott McLeod recognized a need for a simple and easy to use producibility measurement system for trade-off analysis to be used early in the design process. The emphasis was on identifying technical risk early in the process. With this goal, he developed single page worksheets called Producibility Assessment Work Sheets (PAW) for quickly evaluating a design for different manufacturing processes. These sheets are copyrighted. To keep it simple, each worksheet is a single page. When used early in product development, these worksheets provide a quick and easy method for comparing design alternatives and a communication link between disciplines.

13.3.1   PAW Example

For our notebook computer, we will need a power supply assembly case. In the conceptual design phase we would trade-off various design approaches to determine the best manufacturing method. The following case study is adapted from the U.S. Navy's Producibility Measurement Committee and later published in their guidelines.

Step 1. A meeting is held with various functional disciplines in new product development to identify the requirements. The team develops preliminary sketches of possible frame case structures for assessment.

---

* Section adapted from NAVSO P-3679, 1993.

Step 2. Upon reviewing the design and obtaining schedule, design-to-cost (DTC) goals, and quantities, the team selects three possible designs/manufacturing processes for the case.

 1. Thin sheet metal with fastening operations
 2. Sand casting with secondary machining and fastening operations
 3. Plastic with snap fits for fastening operations (i.e. no fasteners).

Step 3. The team enters these selections on the evaluation form called a PAW (Figure 13.1) under the above design alternatives, lines 1, 2, and 3. It is important to remember that the processes selected are not assessed against each other. The processes are assessed against the requirements of the design candidates.

Step 4. The evaluator assesses method 1 against the criteria in step 1, examines the design, and based on experience in this area, selects one of the five criteria. The numeric value assigned to the selection is placed in method column 1. This process is continued until all five categories have been completed. A summary of values is then calculated. Research may be needed for certain categories in the assessment. This may require coordination with vendors and other designers.

Step 5. After the assessment of method 1, the team proceeds to assess method 2. This process is repeated until all of the selected design methods are assessed and numerical values are summarized.

Step 6. The numeric values for each method are compared and evaluated.

Step 7. The quantitative numbers are then communicated to all of the design team.

Method 3 shows to have the greatest probability of success. The quantified result may or may not be the final answer. The selection of plastic assumes that the production volume is sufficient to pay for the injection molding tooling set up costs. What if the production volume changes from the original 12,000 per year to only 1000 per year? The design team can return to the PAW worksheet and see that production method 1 is just a few points behind 3. Because of the change in production volume, the design to cost requirements would be changed. This might raise method 1 in "C2" from .6 to 9. This would also affect the assessment of "C4" from .7 to .9. This is now satisfactory.

Part name: _____ Power Supply Assembly Case _____

Supplier: _____ In-House _____

| Program Phase |
| _____ Concept Exploration |
| _____ Demonstration Validation |
| _._ Full Development |
| _____ Production |

DTC Goal _____ $5.00 _____

Quantity _____ 6000 per year _____

	C1	C2	C3	C4	C5	C6	C7	C8	Producibility / risk assesment value (PAV)
M1	9	6	9	7	7	9	5	7	5.9
M2	9	7	7	7	6	7	7	7	5.7
M3	9	9	7	9	5	7	9	7	6.2
M4									

Producibility / Risk assesement matrix

Procedure or process selection

M1 Sheet metal with fastening operation

M2 Sand castings with secondary machining operations

M3 Plastic with snap fits for fastening _____

M4 _____

MECHANICAL

C1 technical
.9 .... Minimal or no consequence
.7 .... Small reduction in technical performance
.5 .... Some reduction in technical performance
.3 .... Significant degradation in tech. Performance
.1 .... Technical goals achievement unlikely

C5 design
.9 .... Existing / simple / manufacturing engineer's involved
.7 .... Minor redesign for assembly required
.5 .... Moderate redesign / possible assembly problems
.3 .... Complex design / specialized assembly equip. Required
.1 .... State of the art / needs R&D /MFG. Eng.'s not involved

C2 desing to cost (dtc)
.9 .... Budget no exceeded
.7 .... Exceeds 1 - 5% in dtc
.5 .... Exceeds 6 - 15% in dtc
.3 .... Exceeds 16 - 30% in dtc
.1 .... DTC goals cannot be achieved (>31%)

C6 process
.9 .... Proven mature in-house process
.7 .... Minor experience with process in house
.5 .... Experience available locally
.3 .... Experience available, but not proven yet
.1 .... No experience, needs R&D

C3 schedule
.9 .... Negligible impact on program
.7 .... Minor slip (<1 mo.)
.5 .... Moderate slip (<3 mo.)
.3 .... Significant slip (3-6 mos.)
.1 .... Stretch out of program >6 months likely

C7 Inspection
.9 .... Minimal / use of statistical process control (SPC)
.7 .... Minor testing or gauging / floor inspector available
.5 .... Check fixtures accurate and required available
.3 .... Requires extensive testing at every workstation
.1 .... 100 Percent inspection required / no SPC

C4 tooling
.9 .... Dedicated fixturing / flexible manufacturing centers
.7 .... Significant fixturing / cnc and standard tools
.5 .... Moderate fixturing / manual machines
.3 .... Minor fixturing / manual machines / pins and clamps
.1 .... Simple fixturing / m anual clamping

C8 materials
.9 .... Readily available / off-shelf components
.7 .... 1-3 Month order some components
.5 .... 3-9 Month order some components
.3 .... 9-12 Month order / special ord. componts
.1 .... 12-18 Month order

FIGURE 13.1 Mechanical PAW

13.4 BOOTHROYD AND DEWHURST DESIGN FOR ASSEMBLY

Boothroyd and Dewhurst's Design for Assembly (DFA and DFMA are registered trademarks of Boothroyd Dewhurst, Inc.) is a producibility analysis tool to help choose simple and cost effective ways to assemble a product. DFA is the most known and widely used software supported system that can be purchased. Their company, BDI, also has developed software called Design for Manufacturing (DFM) that can analyze the producibility of electronic circuit boards and other processes. (See www.dfma.com)

This discussion will focus on the assembly process. The methods use a system of tables to quantify alternate assembly designs in order to optimize the product design for ease of assembly and cost effectiveness.

Our example and this discussion will consider only manual assembly method. It should be noted that the robotics and automated transfer analysis is similar to the manual analysis in that it uses various tables with decision variables quantified at each step. Since most of this work is copyrighted, this discussion is an overview. More details are available in Boothroyd (1992) or through their web site. The basic phases of this method are:

1. Choose an assembly method based on production volume
2. Incorporate general design guidelines to simplify assembly
3. Quantify producibility by using DFA analysis procedure

For the notebook computer we might use DFA to evaluate the final assembly tasks. This includes installing the printed circuit board, power supply assembly, disk drives, and battery into the computer's case. Note that vendors make these modules so their participation will be required.

Choose an Assembly Method

During design it is helpful to know which assembly method will be used. The assembly methods evaluated include special-purpose indexing, special purpose free transfer, single station one robot arm, single station two robot arms, multi-station with robots, and manual bench assembly. Using the variables of annual production volume, number of parts in the assembly, and total number of parts, "Assembly Method Selection Charts" are used to guide to an assembly method. For example, low volume and small number of parts indicates using the manual assembly method.

For our example with a production volume that may vary from 12,000 to only 1000, the final assembly method will be manual as determined by management.

Incorporate General Design Guidelines

General assembly guidelines are then chosen for the notebook computer design team depending on the assembly method e.g. manual assembly for our example. These guidelines are similar to those discussed in the Chapter on simplification and mistake proofing. Manual assembly is classified by 1.) Part handling, 2.) Insertion and fastening, and 3.) General

For part handling:
- Design parts that are symmetrical
- If parts can not be symmetrical, make them obviously asymmetrical
- For parts stored in bulk, make sure parts will not tangle or jam
- Avoid slippery, delicate, flexible, sharp, splintering, very small or very large parts

For insertion and fastening:
- Design for little or no resistance to insertion
- Standardize parts
- Use pyramid assembly
- Avoid parts that have to be held down for insertion
- Design for self location
- Use common fasteners, preferable snap fittings, plastic bending, riveting, and screwing, in that order
- Avoid the need to reposition assemblies

General guidelines include:
- Avoid connections
- Design for unrestricted access
- Avoid adjustments
- Use kinematic design principals

DFA Analysis Procedure

After the general guidelines have been incorporated into the design and conceptual design has begun, the design team can start quantifying the ease of assembly. The quantifier for the analysis is a measure of simplicity called the manual assembly efficiency, E_{ma}. It is based on the theoretical minimum number of parts, N_{min}, the total manual assembly times, t_{ma}, and the basic assembly time for one part, t_a. Given by the equation:

$$E_{ma} = N_{min} \, t_a \, / \, t_{ma}$$

The variables N_{min}, t_a, and t_{ma} are calculated using values given on tables. (See Boothroyd, 1992)

The DFA analysis procedure is as follows:

Step 1. Gather all the information about the product or assembly, such as engineering drawings and prototypes

Step 2. Take a prototype assembly apart and number each part. Begin filling out worksheet for the analysis.

Step 3. Reassemble the product, part-by-part, and complete information on the worksheet for each part. The columns include

- Part number
- Number of times
- Manual handling time
- Manual insertion time

- Operation time
- Operation costs
- Estimated minimum number of parts

Manual assembly times are based on the following variables:

- Size
- Thickness
- Weight
- Nesting
- Tangling
- Fragility
- Flexibility
- Slipperiness

- Stickiness
- Use of two hands
- Grasping tools required
- Magnification
- Mechanical assistance

Manual insertion times are based on:

- Accessibility of assembly
- Ease of assembly tool operation
- Visibility of assembly

- Positioning of assembly
- Depth of insertion

For example, the design team evaluates each part. A part that is "easy to grasp", thickness > 2mm, size > 15mm, by one hand would have a material handling time of 1.13 from the table. In contrast, a part that requires tweezers to grasp, requires optical magnification, thickness > .25 would have a material handling time of 6.35, e.g. 6x factor. Other times such as insertion would also vary depending on the part's characteristics. Designs with producibility problems such as hard to grasp or assemble will be identified and hopefully corrected.

Step 4. Summary estimated assembly times to get t_{ma}. Add theoretical minimum number of parts to get N_{min}. Use the average of 3 seconds for t_a.

Step 5. Calculate manual assembly efficiency using values obtained from step 4 and the equation

$$E_{ma} = (N_{min})(t_a)(t_{ma})$$

Step 6. Use the worksheet to identify high cost assembly items. Identify operations or parts can be eliminated, combined or reduced. For our example this would require vendor participation since they design and make the modules. Re-evaluate new design and compare efficiencies. Continue until design is optimized (or until time constraints allow.)

13.5   ROBUST DESIGN

Robust design determines levels of design and manufacturing parameters that minimizes the design's sensitivity to change or variation. Can the product operate satisfactorily when reasonable levels of variation occur? It is the ability to tolerate variation even for hard to control variables. Producibility's goal is to make a design more insensitive to variations in processes, materials, parts, vendors, production operators, and users' environments. Robustness is measured by the design or manufacturing ability to maintain a high level of consistency and uniformity when variation occurs. To be successful, robustness should be designed into the product. A simple version of robust design can be seen in the selection of an electronic component. Does the component and its interfaces have a wide operating range so that when variation occurs does it affect the product's performance? Manufacturing robustness is the flexibility to meet changes in product design, production volume, or vendors without affecting performance. There are several methods for robust design. The best known is the Taguchi method, which is now discussed.

13.6   TAGUCHI METHODS

Taguchi methods are usually used for critical product/manufacturing characteristics or for major problems. Quality Loss Function (QLF) is a starting point of the Taguchi method. The Taguchi method's goal is to design the product and manufacture process such that the quality loss is minimized. The product's quality loss is measured in terms of the distance between its actual performance value and its target performance value. It is used for the most critical parameters. Let y represent a critical performance measure of the product, m represent the target value of the product's performance, L (y) then represents the quality loss of the product's performance. As shown in Figure 13.2, Taguchi's loss function is expressed in a quadratic function form (Roy, 1990):

$$L(y) = k(y-m)^2$$

Where:

$L(y)$	=	Quality loss function
k	=	cost per unit of distance
y	=	target
m	=	actual performance

The focus is on minimizing the QLC rather than the traditional whether the product is "within specification tolerances". The quality loss of a product should be thought of as a continuous function measurement (Pradke, 1989).

This method conflicts with traditional design and quality control where any product that meets specification tolerances is good. There is a well-publicized example to illustrate this measurement (Pradke, 1989). Two factories at Sony (Sony-USA and Sony-Japan) used the same product design and components. Color density was recognized as a very important parameter for customer satisfaction. No TV was shipped by Sony-USA that was outside the tolerance limits for color density although the distribution was very uniform between the tolerance limits. In contrast, about 0.3 percent TVs shipped by Sony-Japan were out of the tolerance limits. More consumers, however, seemed to prefer the sets made by Sony-Japan because most of their sets were statistically closer to the optimum value for color density. Taguchi and Wu, (1980) explain this as the higher quality loss of the U.S. sets caused by their poor color density performance. TVs made by Sony-USA had a larger quality loss function than the quality loss of the sets made by Sony-Japan.

Another Taguchi's method is design robustness. One part of this is called parameter design. Parameter design is to distinguish between control factors and noise factors. Control factors are any factor/parameter that can be set at a particular level. Noise factors are uncontrolled since they cannot be set at specific levels because it is too difficult or expensive. Noise variation can be due to external causes (e.g., temperature or humidity in the factory), or internal causes (e.g., material properties). The objective is to identify the optimal settings of the control factors so that the robustness of the product or process will be significantly improved. One strategy is to use robustness to identify lower cost parts or modules that can still meet all design requirements.

Another focus of Genichi Taguchi (1990) is to use design of experiments to improve quality. Experiments are used as an inexpensive way to find the optimal settings of the control factors. Taguchi methods deal with averages and variability. Although most of the literature has focused on manufacturing issues, his techniques

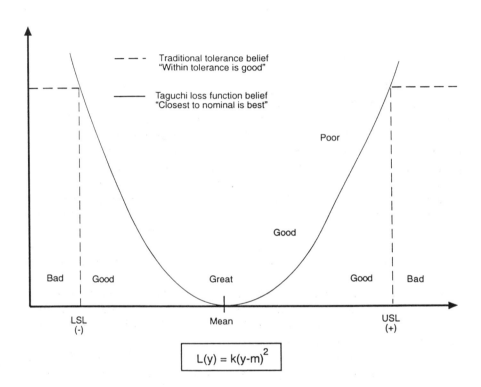

$$L(y) = k(y-m)^2$$

FIGURE 13.2   Taguchi quality loss function [L(y)]
1) K is constant
2) y - m is a deviation from the nominal
3) Loss is proportional to the square of the
   deviation from the nominal value.

are very effective for producibility. This discussion will focus on producibility issues.

The Taguchi experiment and analysis method consists of five steps:

1. Brainstorming session
2. Designing the experiment
3. Conducting the experiments
4. Analyzing the results
5. Performing confirmation tests

The following example developed by Chang (1991) and elaborated by McKenna (1995) shows how Taguchi techniques can increase robustness and improve the process of inserting integrated circuits (IC) into printed circuit boards. Using some of Genichi Taguchi's techniques, an analysis was made on a RAM expansion board.

13.6.1   Brainstorming Session

The first step is to identify all factors that might potentially improve the design's producibility. Using suggestions by Roy (1990) on Taguchi's brainstorming sessions, the focus is on the following elements.

A. **Object of the Study**. The objective of this case study was to optimize a design's producibility. Historical records for similarly assembled printed circuit boards shows that producibility is directly linked to problems with electronic part insertion reliability. It was observed that when insertion reliability falls below 99.5%, manufacturing costs rise dramatically. As a result, the reliability of component insertion is used to measure the design's producibility with the goal set at an insertion reliability of 99.995%.

B. **Design Factors/Variables and Their Levels.** All possible factors that may affect insertion reliability (and thus producibility) and each of their possible levels are identified. For the circuit board, the design factors and their levels that most significantly affect producibility are listed in Table 13.1.

C. **Noise Factors.** As defined by Taguchi, noise factors influence the response of a process, but cannot be economically controlled. These may be attributed to variations external to the process (like temperature or variation in raw material), unit-to-unit variation, or process drift. This study used two noise factors that are divided into two levels. These levels are not precise. The noise factors are then used comparatively with the design's signal to noise ratio. The goal is a "robust design" meaning it has a high Signal to Noise Ratio (S/N Ratio).

TABLE 13.1. Producibility Factors and Levels (Chang, 1991)

| | | Levels of Difficulty | |
Critical Producibility Factors	Level 1	Level 2	Level 3
A. Component Orientation	0 degree	0 or 90	--
B. Robot/Machine Repeatability	+0.0001" Company A	+0.002" Company B	+0.003" Company
C. Tolerance on Hole diameter (Tools)	+0.001" drill bit 1	+0.002" drill bit 2	+0.004" drill bit 3
D. Holding Fixture Positioning(Conveyor)	+0.002" Method 1	+0.008" Method 2	+0.010" Method 3
E. Positional Tolerance on Drilled Holes	+0.001" CNC drill	+0.002" CNC Punch	+0.010" Manually drill
F. Tolerance on Tooling Hole (hole diameter of .125")	+0.002" -0.0"	+0.004" -0.0	+0.010" -0.0"
G. Accuracy of Part Feeder	0.003" Bowl feeder	0.005" Slide Magazine	0.02" Pallet
H. Ratio of Hole-to-lead Dia.	2.5	2.0	1.6

The four possible noise conditions are provided in Table 13.2. Considering each of the four possible cases of natural variation, and working to minimize them and their influence, the designer can help assure that a design will behave in a reliable manner. If noise considerations are not taken into account in the design stages, quality and reliability problems will occur in production.

13.6.2   Designing the Experiment

It is not economically feasible to perform a separate experimental test for every possible experimental condition (there are 686 possible combinations of levels and noise factors). A method commonly used by Taguchi is called the factorial design of experiments that is used to minimize the number of experimental test combinations. Using this method, 18 different experimental designs, each under the four stated noise conditions, combine for a total of 72

TABLE 13.2. Noise Factors

Noise Factors	Level 1	Level 2
X. Variations and Deviations on Lead Diameter	Low +0.002"	High +0.01"
Y. Variation of Play Between Tooling Hole and Location Pin.	Small +0.002"	Large +0.01"

experiments. The signal to noise (S/N) ratios and design reliabilities are then analyzed to determine how the various levels affect their results. Taguchi experiments chart the trials and organize the results using an orthogonal array.

13.6.3 Conducting the Experiments

The results of each experiment give insertion reliability values and a Signal/Noise ratio. For the 18 experiments chosen, the signal to noise ratio (S/N) output is found in Table 13.3. For example, experiment number 16 has a S/N ratio of 25.765. Looking at the resulting reliabilities, the design team can compare the benefits of different designs, but cannot yet determine the optimum conditions.

TABLE 13.3. Reliabilities and S/N Ratios

Experiment Number	Reliability (%)				Signal to Noise Ratio
	X1Y1	X1Y2	X2Y1	X2Y2	
1	99.9	99.9	99.9	99.9	25.888
2	99.8	98.2	99.8	98.8	25.814
3	41.3	39.1	41.3	39.1	21.643
4	68.7	63.2	68.7	63.2	23.847
5	99.8	98.4	99.8	98.4	25.823
6	98.8	94.1	98.8	94.1	25.629
7	98.6	90.6	98.6	90.6	25.465
8	89.7	87.4	89.7	87.4	25.294
9	99.4	94.1	99.4	94.1	25.629
10	99.5	97.8	99.5	97.8	25.796
11	55.5	55.0	55.5	55.0	23.196
12	98.1	87.1	98.1	87.1	25.294
13	76.0	68.8	76.0	68.8	24.233
14	98.4	98.9	98.4	88.9	25.383
15	99.9	98.8	99.9	99.8	25.883
16	99.8	97.1	99.8	97.1	25.765
17	95.5	84.5	95.5	84.5	25.149
18	87.1	82.3	87.1	82.3	25.030

13.6.4 Analyzing the Results

Taguchi methods use the term "robust design" when referring to a design that has an optimal S/N ratio. A robust design is accomplished by making the design immune to noise factors (i.e. those that cannot be controlled). If the design has to live with the natural variation caused by noise factors, the design team simply designs around them.

In order to maximize the S/N value for the design, the design team determines how sensitive each level is to the natural variation caused by noise. Table 13.4 provides these values from the 72 experiments.

Looking at these results, the setting for each factor is based on the level with the highest S/N. From these results, the "optimal" process combination is A2, B3, C1, D2, E2, F2, G1, and H1. Using these levels, a verification experiment is run. This results in an insertion reliability of 99.9% that meets the company's requirements. The design team is not yet through, however.

Returning to Table 13.3, both Experiments 1 and 15 also met the 99.5% insertion reliability requirements, although they do not have as high of reliability as the optimum signal to noise ratio combination. So which design is best: 1, 15, or the design with the best S/N? This is where Taguchi's definition of quality is used.

The levels chosen for the optimal design may result in a product that can be produced in a very reliable manner with little influence from noise factors, but it may create producibility costs or design problems in other areas that were not considered in this example. Perhaps the optimal design measure uses a very expensive process, or is very difficult to implement, or requires technical skills that the organization does not currently have, or forces other production processes to halt.

13.6.5   Confirmation Test

The overall insertion reliability of the experimental process design has been shown. Three designs have been able to meet our reliability requirements. Now one should take producibility cost analysis one-step further.

As developed by Taguchi, each level of a design imparts its own producibility cost, or loss, to the organization. For this study, these costs are far less than the producibility cost of using a design with reliability below 99.5%, but they can be used to choose among our three remaining designs.

Using cost information at the company:

- Experiment Design 1 has a producibility cost of $640
- Experiment Design 14 has one of $360
- New optimal design's cost is $330.

Table 13.5 suggests using level three for each factor would minimize producibility loss.

TABLE 13.4 Signal to Noise Response Table

Level	A	B	C	D	E	F	G	H
1	25.003	24.605	**25.166**	24.934	25.098	25.249	**25.630**	**25.619**
2	**25.082**	25.133	25.111	**25.276**	**25.316**	**25.250**	25.624	25.044
3	--	**25.390**	24.851	24.918	24.714	24.629	23.874	24.465

TABLE 13.5. Producibility Costs for Insertion Factors

Critical Producibility Factors	Preferred Method	Level 1	Level 2	Level 3
A. Component Orientation	0 degrees	$0	$10	-
B. Robot/Machine Repeatability	Semi-auto	$200	$100	$50
C. Tolerance on Hole Diameter	Drill Bit 4	$50	$30	$20
D. Holding Fixture Positioning	Method 4	$120	$80	$50
E. Positional Tolerance on Drilled Hole	Manually	$80	$30	$10
G. Accuracy of Part Feeder	Bulk	$90	$70	$30

13.7   SIX SIGMA QUALITY AND PRODUCIBILITY

Motorola Inc. developed Six Sigma Quality and Producibility to achieve world-class quality. Its goal was to consistently produce high quality products with only 3.4 defects per million opportunities for error. This effort resulted in winning the Malcolm Baldridge National Quality Award and radically changing how many companies view and measure quality. Although it was an aggressive goal, Six Sigma allowed Motorola to become a leader in quality.

The key emphasis is to develop integrated designs and processes that meet 6 sigma quality. This is accomplished by using statistical measures of variability in design, manufacturing, and service decisions. The quality measures for this method are defects in terms of "defects per million opportunities" (dpmo), "parts per million" (ppm) and with defects per unit (dpu). Opportunities are all things that must be correct to manufacture a product or service and are assumed to be independent. In mechanical assembly the number of opportunities has been found to proportional to the number of parts given the levels of process control and the design margins are equivalent.

Unique aspects of the method were the focus on variability and inclusion of a variable (k) that modifies the process capability distribution for process drift and non-centered shifts in the distributions. They believed that drift occurs in almost every process and its shift can be approximated as 1.5 sigma. Drifts are caused by changes in the machine, operators, materials, or use of more than one machine.

For producibility, the Six Sigma measure is used to evaluate a design and predict the number of defects that should be expected in manufacturing. The likelihood of a product with no defects (i.e., success) can be called yield or rolled throughput yield. Yield is the throughput-per-opportunity, where an opportunity is arbitrarily defined as a process step or a part.

Design tolerances and process capability are evaluated to predict the Cp and Cpk for the particular design. A design with a ± 6 sigma variation can be expected to have no more than 3.4 ppm defective for each characteristic, even if the process mean shifts by as much as ± 1.5 sigma. Although the calculation of six sigma producibility requires a considerable amount of information and resources, it is one of the best measures of a design's producibility.

Six Sigma producibility is the ability to define and characterize various product and process elements that expert undue influence on the key product response parameters and then optimize those parameters in such a manner that the critical product quality, reliability, and performance characteristics display:

- **Robustness to random and systematic variations** as measured by the central tendency (μ) and variance (σ^2) of their physical elements
- **Maximize tolerances** related to the "trivial many" elements and optimize tolerances for the "vital few" (i.e., critical parameters)
- **Minimize complexity** in terms of product and process element count
- **Optimum processing and assembly characteristics** as measured by such indices as cost, lead time, etc." (Harry, 1991 and Harry and Lawson, 1992)

The key design considerations of this definition are:

- Concentrate on key design parameters
- Develop parameters that are "robust" to manufacturing variation
- Loosen tolerances on "trivial many" non-critical parameters and develop optimum tolerances for the "vital few" critical parameters. Critical parameters are those that directly relate to what the customer perceives as important.
- Minimize design complexity; especially the number of parts and processes
- Optimize design parameters for process and assembly capabilities
- This definition highlights the role of manufacturing variability in developing design requirements.

Since producibility is used to improve a design before it is manufactured, Six Sigma uses the number of defect opportunities as its predictive measure. Opportunities are any chance for nonconformance to occur. Most companies have found that the best predictors are number of parts for mechanical and electronic assembly, number of solder joints for electronic assembly, and number of critical signals for electronic performance and software.

The Steps to Six Sigma Producibility are: (Harry, 1991 and Harry and Lawson, 1992)

Step 1   Identify the critical product characteristics or key product requirements for customer satisfaction.

Step 2   Identify the critical product and process characteristics of every part, assemblies, and software for achieving the critical product characteristics identified in step 1. Methods can include fault tree, failure mode, cause and effect, simulation, and Quality Function Deployment.

Step 3   For each critical characteristic, determine whether it is controlled by the vendor, design, process, or a combination.

Step 4   Determine the maximum allowable range (e.g. design tolerance) of that characteristic that can be tolerated by the design and still operate satisfactorily.

Step 5   Determine the variation that can be expected in that characteristic based on the known capability of the process, part, or vendor (e.g. process capability). This includes knowing the process's nominal value, variability and type of distribution. Process capability studies and experiment supplies this data. Experiments, simulation, Taguchi, and SPC may be required.

Step 6   Determine Cp and Cpk for each key characteristic and process:

$$Cp = \frac{USL - LSL}{6\sigma} \quad \text{or} \quad \text{Design Tolerance / Process Capability}$$

Where:

Cp = Inherent process capability, a high value indicates that the process is inherently capable

USL = Upper specification level

LSL = Lower specification level

And for compensating for drift or differences between the design target and the process nominal capability:

$$Cpk = Cp(1-k)$$

where:

Cpk = Non-centered process capability which includes process drift, and indicates how far μ is from the nominal condition or design target.

K = Target Mean (Nominal) − Actual Process Mean /

½ (USL-LSL)

That is k is equal to the process shift divided by one half the design tolerance.

In summary, Cp and Cpk correlate design specifications to process capability. Knowing this information allows the design team to predict the probable occurrence of the number of defects for producibility and quality analyses. After the values are calculated for each process, they are compared to requirements such as recommended by Motorola, (Harry, 1991)

If Cp ≥ 2, and Cpk ≥ 1.5, then
- Design and process will meet 6 sigma quality.

If Cp < 2, and process is benchmarked as Best in Class, then:
- Develop an alternative product design that eliminates the need or increases the allowable variation for that characteristic. Best in class values are determined by benchmarking within the company, competitors, and known leaders in other industries.

If Cp < 2, and process is not Best in Class, then:
- Improve the current process or develop an alternative process design that will result in acceptable variation within the allowable range.

If Cp ≥ 2, and Cp$_k$ < 1.5, then:
- Develop an alternative process design that will result in proper centering of the characteristic.

Design and verification tests are performed to ensure that the processes and parts are meeting the required goals. Other companies use different Cp and Cpk values but the general methodology is the same.

The previous discussion was for one process. When more than one process or opportunity is involved, the Cp and Cpk for each process is established. To predict the quality for all of the processes, the term rolled throughput yield (Y_{RT}) is used:

$$y_{RT} = \prod_{i=1}^{m} y_{FT_i}$$

Where:

y_{RT} = rolled throughput yield of a product characteristic

m = number of processes or actions

y_{FT} = yield of an individual process

If a product required six processes with identical yields (Y_{FT}). of .9526 then $Y_{RT} = .9526^6 = 74.67\%$. This relationship shows that as more processes are involved, the higher level of quality that is required for each process. For more details on Six Sigma Producibility, the reader is directed to Producibility Measurement/BMP Guidelines, NAVSO P-3679 (BMP, 1993) and Six Sigma Producibility Analysis and Process Characterization (Harry and Lawson, 1992).

For the notebook computer, 6 sigma producibility would especially important for 1.) Vendor and part selection for circuit board electronic performance, 2.) Circuit board assembly (i.e. solder joints), and 3.) Overall electronic performance. Predicted levels of Cp and Cpk would be calculated using the median and variability of each part and process. Part parameters could include several important electronic parameters such as timing, resistance, and signal strength. All the parts and processes would be combined to calculate Yrt. Vendor modules, software, and service processes could also be analyzed.

13.8   PRODUCTION FAILURE MODE ANALYSIS, ROOT CAUSE, ISAKAWA FISH DIAGRAMS AND ERROR BUDGET ANALYSIS

As noted in several Chapters, another successful producibility method is to simply identify and control all potential manufacturing and quality problems (i.e., failure modes) associated with processes and vendors. These methods focus on eliminating "every potential problem". This goes beyond traditional problem solving methods such as Pareto Analysis that focuses on the most important problems. Five failure mode focused methods that can be used are:

1. Production failure mode analysis (PFMEA)
2. Root cause analysis
3. Isakawa fish diagrams
4. Error budget analysis
5. Mistake proofing and simplification
6. Design for quality manufacturability

The first four will be discussed in this section with the remaining 2 to be discussed in later sections. The PFMEA procedure is similar to that of a failure modes and effect analysis used to evaluate a design. The steps are to

1) Identify all potential failure modes (i.e. problems) and their sources (i.e. root causes) caused by design, manufacturing and vendors
2) Identify and implement cost effective changes that eliminate or reduce their effects on quality.

The method identifies sources of problems that the design team can hopefully "design out". This includes all failure modes induced by design

TABLE 13.6 Assembly Processes and Associated Defects

Part Assembly
        Missing part
        Fastening operation incorrect
        Wrong part
        Wrong orientation
        Jammed
Soldering
        Excessive heat
        Insufficient heat
        Excessive solder
        Insufficient solder

Source: Adapted from Reliability Analysis Center, 1975

requirements, equipment, procedures, personnel, vendors and materials. It provides a means of striving for manufacturing perfection (i.e. muda). Unfortunately, identifying all potential problems, their causes and then identifying solutions can be a very time consuming effort. For our notebook computer example, a list is developed of the most common problems associated with different methods for manual electronic assembly Table 13.6. Special emphasis is placed on those induced failures that could result in serious or catastrophic failures in operational use.

Other methods that are similar to failure mode analysis are called root cause analysis and cause and effect diagrams, which are often, called Ishikawa diagrams. This technique consists of defining an occurrence and identifying all the possible contributing factors. The key is to get to the root cause of the problem. An example of an Ishakawa is shown in Figure 13.3

Error budgeting is a systems analysis tool used to reduce the cumulative manufacturing error in a design and manufacturing process by initially identifying all contributors to error, determining allowable error budgets, and then ensuring that the budget is not exceeded. This technique has shown to be especially effective in the process design of high precision machine tools. Although it is not often used for producibility, it has been included in this book because of its significant potential in future producibility systems that reduce errors in design and manufacturing. Error budget analysis is a methodology for partitioning a complex manufacturing process into smaller factors that affect a design parameter.

Step 1 is to identify all possible errors that may occur for each factor. The tasks are to analyze the design and its manufacturing process in great detail to identify error possibilities, determine error budgets and then evaluate the various processes to see if the design requirements can be met. Whereas process capability analysis

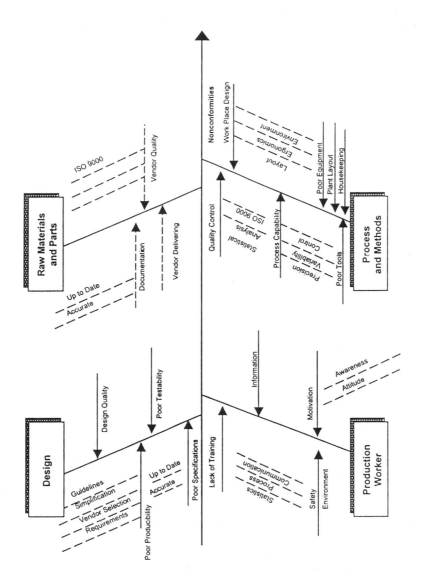

FIGURE 13.3 Ishikawa diagram.

statistically determines a process's variability, error budget analysis focuses on what caused the variability and errors (i.e., the root causes).

Step 2 is to redesign the product and the process for greater producibility with less chance of error. This may require re-tolerancing, selecting new vendors, redesigning the product to compensate for errors that cannot be avoided, and redesigning the process to reduce the source of error. This entire methodology reduces the possibility that error, which is outside of the product's tolerances, will occur.

In the error budget approach, the following steps include:

1. The initial critical design requirements are identified.
2. Identify all possible errors that may occur in the manufacturing of the product.
3. Design requirements such as tolerances are linked to each of the individual components of the manufacturing process that may affect them.
4. Modify the design or process so that the components of the error do not exceed the optimum requirements.

The goal is to remain within the established design requirements while reducing the cumulative error involved in production. For example, there are several general causes of error in the machining process (adapted from Harry, 1991):

- Geometric - sources of error effect the position of the part, the design's tolerances, how the tolerances are referenced, machine selected, and tool quality.
- Dynamic - vibration of the machine effecting accuracy
- Workpiece effects - deflection during cutting and inertial effects of motion.
- Thermal - In high precision applications the thermal effects generate the most error. Minimize the total effects of thermal conditions on the machine tool.

Once an evaluation has been performed, changes in the product's design and manufacturing processes are made to reflect the conclusions of the error budget analysis. A simple example is shown in Table 13.7.

13.9   MISTAKE PROOFING AND SIMPLIFICATION

As noted in the Chapter on simplification, mistake proofing is one of the most powerful techniques for avoiding simple human errors at work. Many

TABLE 13.7 Error Budget Analysis

Error	Source of Error	Error Budget	Recommended Course of Action
placement of component on circuit board	1. machine vibration	± 0.002 in. of true	Reduce vibration with use of dampening devices
	2. tool positioning	± 0.002 in. of true	Re-evaluate error checking on machine tool
	3. Part tolerance	± 0.003	Identify vendors with high tolerances
	4. Part feeders	± 0.002	Reduce vibration and improve location device

companies have developed major quality programs centered on mistake proofing both the design and the entire manufacturing process including vendors. Mistake proofing features are either designed into the product and/or devices are installed at the manufacturing/service operation. It is a low cost method of insuring the highest levels quality i.e. zero defects. As previously noted, the initial goals for mistake proofing are:

- Parts cannot be manufactured, assembled, packaged, shipped, serviced, etc. wrong by the operator, machine, or process
- Obvious when mistakes are made
- Self-tooling and self-locating features are built into parts
- Parts are self-securing when assembled

Its effectiveness and relative low cost makes mistake proofing a popular method in industry. Just like most methods, mistake proofing requires a team effort. Just like failure mode analysis (PFMEA), every possible thing that can go wrong must be identified and then eliminated or controlled. The designer's role is to conduct analyses, incorporate mistake proof features into the design and accommodate fail-safe process control into the design. Manufacturing and test develop fail-safe production and inspect/test methods to identify and control the process when errors are made. Vendors adhere to high levels of quality with no process changes without customer notification.

One approach is to identify items/parts by their characteristics such as:

- Weight
- Dimension
- Shape
- Color
- Location/position

- Orientation
- Electrical output characteristic

The next step is to identify methods that detect any deviations from the fixed values of these characteristics such as:

- **Weight**   Automated precision scale
- **Dimension**   100% Go/No go gage, computer vision
- **Shape**   Part outline, locating features, computer vision
- **Color**   Color coded, high contrast
- **Location/position**   Self-locating,   champers,   self-locking, nesting, high contrast, computer vision
- **Orientation**   Same as location
- **Electrical output characteristic**   Electrical test
- **Assembly**   All fasteners identical, reduce number of parts, all self-locating and fixture, cannot mix parts due to mechanical differences, ease of identification, tolerances allow easy assembly, all parts symmetrical or asymmetrical

Eliminating every possible error and the 100% measurement of important characteristics and allows the highest levels of quality. For our example we would consider mistake proofing every step from vendors to shipping. For the packaging process, we would identify all possible mistakes such as wrong or improperly installed box, packaging material, manuals, or mailing label. Design incorporates unique features for each type of computer using different color packaging/manuals, unique, self-locating features, and machine-readable labels. Manufacturing would install automated scales to detect missing items (i.e. deviations from norms), and finally computer vision to check labels, product type and proper installation. Vendors such as the printer of the manuals would also implement mechanism to ensure correct documentation that is correctly shipped to the factory.

13.10   DESIGN FOR QUALITY MANUFACTURING

A comprehensive effort to develop a methodology to focus on evaluating a design from a quality perspective was developed by Sanchoy Das and others (Das et. al., 2000). They identified and then classified the occurrence of the inherent defect process found in manufacturing and then developed a producibility measure called the quality manufacturability index (QM-Index). The method is currently limited to assembly operations.

The key elements are classes of defects, specific defects influencing factors, factor variables, and error catalyst. As noted by these authors and others, quality problems can be aggregated into representations of several defect classes of that are commonly seen. The likelihood of an occurrence of a defect is influenced by several features/factors/events that are inherent in the product's design. Although these lists of defect classes and factors may be large, a smaller number of most common occurring factors were identified. The method then quantified each identified factor and their related error function. The error functions are used to measure the overall QM-Index of the design. Although the method is too involved for an in depth discussion, there are several interesting conclusions of the method that should be reviewed.

Some overall criteria, findings and recommendations are (Das et.al., 2000):

1. Keep total number of parts to under 40 and preferably under 25

2. Assemble entire product in one location

3. Procured subassemblies are considered as one part

4. 90% of common quality defects can be traced to a few physical defect classes

The standard classes of assembly quality defects are:

- Misplaced or missing part

- Part alignment

- Part interference

- Fastener related problems

- Assembly nonconformity

- Damaged parts

To calculate the index, four types of information are required: general design data, mating relationship matrix, part factor variables, and mating factor variables. General design data includes number of parts and product envelope. The part factor variables include 29 data inputs such as assembly method, part dimensions, material type, geometry, symmetry type, flexibility, and minimum clearance. Mating factors have 20 data inputs including positional relationship, direction, intensity, fasteners, accessibility, etc. This data is evaluated with the error catalyst database information. Error catalysts are design-assembly situations that promote the occurrence of quality defects. The strength, probability or criticality of each catalyst is of course related to the design-assembly data. They use a 0 to 1 scale to rate the likelihood of the defect occurring.

For our notebook computer this method could be used for all assembly operations. It would be especially helpful if formal quantified producibility measures were required in the trade-off process.

13.11   VENDOR AND MANUFACTURING QUALIFICATION

Qualification is a formal process that verifies the capabilities of the selected processes and vendors prior to the start of production. This is accomplished by

1. Formal certification of the process and vendor such as using international standards (e.g. ISO 9000 or approved company certification).
2. Actual testing of the process or vendor using prototype parts,
3. Examination of data from current production of similar parts or processes

One important method is formal certification. The two kinds are pre-approved company wide certifications and certification for the particular part or process. Company wide certifications can include ISO standards 9000-4, which is determined by the published ISO requirements and an extensive audit process. ISO 9000 focus quality management to the actual process in order to support continuous improvement and provide customer satisfaction. All levels and processes in an organization are measured and evaluated. Other special certifications for preferred vendor lists are made by the purchasing company and usually require vendor visits and documentation requirements. This type is based on meeting quality-related requirements such as quality documentation, defect control, quality assurance procedure, and statistical quality control.

Using prototypes to evaluate a manufacturing process or vendor part under actual manufacturing conditions is one of the best methods to reduce technical risk. Although "paper" analyses may show that the selected process or vendor can adequately produce the parts, actual conditions may expose conditions that were not visible during the earlier analyses. An actual checkout of the process capability and availability using prototypes provides the design team with the knowledge that all possibilities were examined prior to production. These tests should be performed on the manufacturing process under similar conditions to assure the design's producibility. Some companies have established a manufacturing-type laboratory equipped and staffed to perform this type of manufacturing research and development. The laboratory provides the manufacturing methods, machinery, and equipment needed to support engineering design and verify process capabilities. In addition to identifying potential manufacturing problems, manufacturing laboratories perform research on improving manufacturing methods, techniques, processes, machinery, and equipment through testing and experimentation.

The last and least preferred method is when to evaluate data from similar processes or parts. Evaluating similar manufacturing data is only used when actual prototypes or processes are not available. The key is to focus on the important parameters and their statistical distributions.

For our notebook computer example we would require all parts to come from ISO 9000 qualified vendors.

13.12 SUMMARY

Many producibility methods have been proven to be successful in industry. This chapter has briefly discussed several methods that we believe are exceptional in improving quality and reducing cost. The reader should remember that to ensure a "producibility" design we need infrastructure, communication and many of these methods to be used. Producibility is not a simple quick analysis but rather a long evolving process that continues throughout product development.

13.13 REVIEW QUESTIONS

1. List the key reasons for using each of the methods in the chapter.
2. Explain Taguchi's methods use of experimental design to determine quality loss and improve producibility.
3. Define the major steps in performing a Six Sigma analysis.
4. Describe how error budget analysis could be used for producibility?

13.14 SUGGESTED READINGS

1. Producibility Measurement Guidelines, NAVSO P-3679, www.bmpcoe, Department of the U.S. Navy, 1993.
2. Producibility System Guidelines for Successful Companies, NAVSO P-3687, 1999.
3. M. J. Harry and J.R Lawson, Six Sigma Producibility Analysis and Process Characterization, Addison-Wesley, Reading, Mass. 1992.
4. M.J. Harry, New book on 6 sigma to be released in 2000.
5. Best Manufacturing Practices Program (BMP), www.bmpcoe.org
6. BDI Company, www.dfma.com

13.15 REFERENCES

1. Best Manufacturing Practices Program (BMP), Producibility Measurement Guidelines, NAVSO P-3679, Department of the U.S. Navy, and www.bmpcoe.org, 1993.
2. G. Boothroyd, Development of DFMA and Its Impact on US Industry, Proceedings of Bath, 1999.
3. G. Boothroyd, Assemble Automation and Product Design, Marcel Dekker, New York, 1992.
4. G. A. Chang, Optimization Techniques for Producibility Using Taguchi Methods and Simulation, The University of Texas at Arlington, 1991.

5. S.K. Das, V. Datla and S. Gami, DFQM – an approach fr improving the quality of assembled products, International Journal of Production Research, Vol 38(2), p. 457-477, 2000.
6. M. J. Harry and J.R Lawson, Six Sigma Producibility Analysis and Process Characterization, Addison-Wesley, Reading, Mass. 1992.
7. M.J. Harry, The Nature of Six Sigma Quality, Motorola Technical Presentations to U.S. Navy Producibility Measurement Committee, 1991.
8. R.T. McKenna, A Case Study on Taguchi Producibility Improvement Topics Class Report, Department of Industrial and Manufacturing Systems Engineering, U.T.A., April, 1995.
9. S. McLeod, Producibility Measurement Committee, Assessment Worksheet Guidelines, NAVSO P-3679, and www.bmpcoe.org, 1993.
10. M.S. Phadke, Quality Engineering Using Robust Design, Prentice Hall, Englewood Cliffs, New Jersey.
11. R. Roy, A Primer on the Taguchi Method, Van Nostrand Reinhold, 1990.
12. G. Taguchi and D. Clausing, Robust Quality, Harvard Business Review, January-February, p. 65-72, 1990.
13. S. Taguchi, Taguchi Methods: Quality Engineering, Dearborn, MI, American Supplier Institute Press, 1988.
14. G. Taguchi, and Wu Yuin, Introduction to Off-Line Quality Control, Central Japan Quality Control Association.

Chapter 14

RELIABILITY: STRATEGIES AND PRACTICES

Design in Reliability

Reliability is a major part of a customer's perception of quality. Failures result in customer dissatisfaction and costly redesign, manufacturing and repair efforts. Failures increase warranty costs, reduce sales, and lower the corporation's image. Reliability is a design parameter associated with the ability or inability of a product to perform as expected over a period of time. It greatly affects the success or failure to meet other design requirements such as life cycle cost, producibility, warranty costs, and schedules. Problems in reliability can usually be traced to inadequate requirements, design, analysis, testing, and support. A successful reliability approach focuses on implementing proven design practices that improve reliability.

Best Practices

- Accurate Reliability Models For Trade Off Analysis
- Reliability Design Practices
- Multidiscipline Collaborative Design Process
- Technical Risk Reduction
- Simplification, Commonality, and Standardization
- Part, Material, Software, and Vendor Selection and Qualification
- Design Analysis To Improve Reliability
- Developmental Testing and Evaluation
- Production Reliability and Producibility

FAMILY OF OSCILLOSCOPES

Producers of oscilloscopes are in a highly competitive market that does not tolerate high purchase or maintenance costs. One company decided to develop a family of oscilloscopes with increased capability and reliability three times greater than existing models (Wheeler, 1986). To accomplish this goal, reliability and quality engineers were involved at the start of the design decision-making efforts. The initial task of the development group was to perform analyses of past equipment failures. The result of this effort provided information that identified that many failures were caused by temperature stress. The design process obviously focused on reducing stress To realize their company goal of improving reliability by 300%, it was determined that the new design would have to limit the oscilloscope's internal temperature increase to 10° C or less. As reported by Wheeler (1981), the stress analysis revealed the following as design goals:

1. Hold stress levels to 50% of a component's rated stress to decrease failure rates. Examples: A bipolar integrated circuit (IC) at 70% has twice the probability of failure than one at 50%. A dielectric capacitor operating at 70% of its rated voltage fails five times more frequently than one at 50%.

2. Reduce temperature increase. Examples: The same IC fails twice as frequently at 85°C as it does at 70°C. The same capacitor fails twice as frequently at 75°C as it does at 60°C.

A 300% improvement in reliability was accomplished by the different disciplines combining to meet the desired reliability objective (Wheeler, 1981).

Too many people think of reliability as a numerical predictive measure developed by specialists using complex mathematical methods. A successful reliability approach focuses on design methodologies and actions that improve reliability, rather than focusing on numerical measures. For many large companies, reliability and quality are too often considered as a support function. Reliability and quality must be an integral part of design's decision-making process to be successful.

The objective of this chapter is to provide a product development team with an understanding of the basic design fundamentals of reliability and quality for both hardware and software oriented products.

14.1   IMPORTANT DEFINITIONS

Reliability is a design parameter associated with the ability or inability of a product to perform as expected over a period of time. High reliability means that a product continues to meet the customer's quality expectations over its intended life. Unlike many design parameters, reliability has used a quantified measure or metric for many years. Reliability is a projection of performance over periods of time, and is usually defined as a quantifiable design parameter such as mean time between failure (MTBF) and mean time to failure (MTTF).

Software reliability is a design parameter associated with the ability to perform all functions as expected over a period of time and continues to perform when foreseeable input or usage violations occur. A software failure is when the service expected by the customer is not met. Software reliability uses the same measurement parameters as hardware, such as mean time between failure (MTBF) and mean time to failure (MTTF).

Reliability can be formally defined as the probability or likelihood that a product will perform its intended function for a specified interval under stated conditions of use (DOD, 1983). The key elements of this definition are as follows:

- **Probability.** Statistical prediction, rather than absolute knowledge, about the future performance of an item is based on a statistical distribution of a population of similar items. Probability is used due to the variation in materials, suppliers, quality, manufacturing, customer use, etc.
- **Performance of its intended function.** The design can meet all functional requirements when it is properly manufactured and no failures have occurred.
- **Specified interval.** Since almost everything eventually fails, the period of use is constrained by defining the interval in such terms as time, number of uses, operating time, or storage life.
- **Conditions of use.** The importance of properly using the item as intended is obvious. For example, a television set cannot be expected to operate outdoors for a very long period of time or software that was written for one version of an operating system cannot be expected to necessarily work for another version.

This definition of reliability assumes that the product was properly functioning at the start of the time period. Another assumption is that the product is still in the useful period of its life cycle.

A complete design specification for reliability would be as follows: "A notebook computer shall have a reliability of at least 0.999 for operating without any failures during 1000, thirty minute uses in an office environment, after it is properly charged and passes a built-in test prior to each use." This specification illustrates all the characteristics previously discussed:

1. Probability: 0.999
2. Performance: no failures
3. Specified interval: 1000, thirty minute uses
4. Conditions of Use:
    a. Office environment
    b. Properly charged and passes a built in test prior to each use

14.2   BEST PRACTICES FOR DESIGN RELIABILITY

To be successful, the development team must understand the fundamentals of reliability and its relationship to the design process. The key practices discussed in this chapter are as follows:

- **Accurate and detailed reliability and availability models** are used in reliability and trade-off analysis to evaluate and improve a design's reliability.
- **Proven design reliability practices** are an integral part of the design process such as:

    - Multidiscipline collaborative design
    - Technical risk reduction
    - Design simplification and standardization
    - Part, material, software and vendor selection and qualification
    - Design analysis that improve reliability and reduce the effects of stress
    - Developmental testing and evaluation
    - Production reliability

14.3   ACCURATE RELIABILITY MODELS ARE USED IN TRADE-OFF ANALYSIS

There are significant differences in how different reliability measures and models are calculated and used in design. Since this book is oriented toward design practices rather than the statistical theory and modeling used for predicting reliability, only a brief overview of reliability modeling and mathematics is presented. Additional information on reliability models and mathematics can be obtained from the references at the end of the chapter. This section will review the major concepts of reliability modeling.

Measures of Reliability

Reliable life is a measure of how long an item can be expected to perform satisfactorily, and is often expressed in units of time, years, or such operating parameters as the number of cycles. The variation around the stated value can be defined as a minimum, mean, or median value.

Mean time to failure (MTTF) is another measure of how long an item will perform satisfactorily and is a commonly used reliability parameter for items that are not repairable, have a limited life, or arc subject to mechanical wear out. This parameter describes the expected average life of an item and is very important in establishing product warranties. A manufacturer of air conditioners would not offer a 5-year warranty on a unit if the compressor has a mean time to failure of 4 years. Items that are operated in a cyclic mode, such as a light switch, can be described in mean cycles to failure.

Mean time between failures (MTBF) is a measure of how long between repairs that an item will perform satisfactorily and is the most popular used reliability measure for repairable items. It is used for items that are expected to fail and be repaired prior to wearing out. An automobile may have a mean time between failures of 8000 miles, but its useful life after repair and adjustment may be 150,000 miles. Software products often use MTBF for reliability prediction. For example, software may occasionally lock up (i.e., fail), but when reinstalled or minor configuration changes are made, the software will operate successfully for a long period of time.

Failure rate is the average number of failures from a group of items (i.e. a population) expected in a given period of time. Failure rate is usually used to define the reliability of individual parts or software modules rather than products consisting of several parts. For example, the failure rate for an integrated circuit may be one failure per million hours of operation, a large computer system consisting of 10,000 integrated circuits could then be expected to fail, on the average, every 100 hr.

The design specification for the previously mentioned notebook computer can be stated in various terms and using various types of nomenclature. Assuming that the system is repairable, its reliability may be stated as follows:

- Probability of success: R = 0.999
- Reliable life > 5 years
- Mean time to failure (expected to be the battery): MTTF = 2 years
- Mean time between failures: MTBF = 999 uses or 499 hours
- Failure rate (λ) = 0.001 average number of failures in one use.

Unfortunately, reliability cannot be used to predict discrete events. Only averages or probabilities of failure can be developed for a period of time. Reliability can estimate only whether a device will operate for a specified period of time. For example, in theory a computer's hard disk is extremely reliable with a mean time between failures of 300,000 hours. If a computer is run for eight hours a day, every day, the disk will crash in over 102 years. The reality, however, is that it could crash next week, wiping out your records and the software needed to boot up the computer.

Design reliability is a design discipline that uses proven design practices and analyses to improve a product's reliability. Unfortunately, there can be confusion between quality and reliability since they are sometimes used interchangeably. They are different, and it is important that this difference be understood. Quality control is a time-zero measurement of the quality of a product, whereas reliability is a time-dependent measurement of quality. Reliability can be considered as the product's quality over its useful life. Metrics of a successful design reliability program are customer satisfaction, warranty costs, failure rates, and other traditional reliability measures

A non-conformance, defect, or fault is an imperfection in either the design or manufactured structure of the product. The imperfection can be a characteristic that is too far from its nominal value target. Imperfections may be measured in continuous physical units (e.g., volts, response time) or judged in terms of meeting or not meeting a requirement (e.g., missing part, good paint job). They can also include both functional parameters that may result in a functional failure and aesthetic parameters, which may cause customer dissatisfaction but not a failure. An example of an aesthetic parameter is the quality of the paint color on an automobile. If the paint color does not meet company standards, the customer may be dissatisfied but it would not result in a functional failure. Hardware commonly uses the term "defect" and "nonconformance". Software typically uses the terms "nonconformance" and "fault".

A failure is when a product cannot perform its intended function until a corrective action (i.e., repair, design change) has taken place. The defects that caused the failure are defined as the failure modes. A description of the parameters that caused the failure is called the failure source. Defects may or may not cause a functional failure. Some defects may only result in a failure under unusual or extreme conditions. One example is a crack in a solder joint that will only cause a failure over a long period of time when vibration, shock, and temperature cause the crack to finally result in a complete fracture. Another is software when only certain conditions would cause a failure. Reliability measures the number of failures not the number of defects.

Although a defect may not result immediately in a failure, it is still thought of as a nonconformance to specifications and must be corrected. Since testing identifies failures, many tests are designed to stimulate the product at a high enough level to cause the defect to result in a functional failure. Inspection is often limited to identifying nonconformance and defects that may or may not result in failures now or at a later date.

Software Reliability[*]

Software can fail in many ways and with varying degrees of impact to the program and its users. For example:

- The program completed execution, but did not produce the correct or desired output. This erroneous output could vary from slightly wrong to complete garbage.
- The program did not complete execution, such as abnormal program termination. The operating system detects the violation of some constraint intended to protect the operating system, the hardware or other executing programs.

A failure that results in abnormal program termination can occur for a number of reasons (e.g., faults or defects). At a high level of abstraction, abnormal program termination occurs because of external events, user termination, and internal errors.

External events that cause programs to abnormally terminate include disk crashes, power outages and operating system errors. These are events that are external to the software of interest; modification of the program may not be able to prevent this cause of failure.

User selected termination is where the user executing the program of interest intentionally or unintentionally chooses to prematurely stop execution of a program. This can be considered either as an external or internal event. User selected termination only affects the program of interest. If no modification of the program could have prevented this termination, it is considered an external event.

Internal errors are the cause of failure that can be considered an error in the software of interest. A modification to the software could prevent that failure. Some common internal problems found in software development are shown in Table 14.1. This includes programs that complete execution but produce an incorrect output. Because internal errors are contained within the program code, examining the code with other information (e.g., program execution traces) is necessary to find the cause of the failure. This can be further decomposed to three classifications of failure.

- Omissions - missing program code
- Mistakes -wrong program code
- Changing domain or environment - assumptions no longer valid

---

[*] Written by Lisa Burnell, University of Texas at Arlington, 2000.

TABLE 14.1 Common Failure Causes in Software Development

Requirements phase
    Incorrect or incomplete-requirements
    Untestable requirements
    Inconsistent or incompatible requirements
Design phase
    Deficient design representation
    Lack of structure
    Incomplete reflection of requirements
    Inaccurate approximations
Coding phase
    Missing or incorrect logic
    Misinterpretation of language constructions
    Erroneous or unjustified assumptions
    Data structure defects
    Typing errors

Source: Adapted from Lipow and Shooman, 1986

Reliability Mathematics

The reliability of a design can be predicted based on reliability models and detailed information about the design. Probabilistic parameters such as random variables, density functions, and distribution functions are often used in the development of reliability models. These models are concerned with both discrete and continuous random variables. Discrete reliability is the number of failures in a given interval of time. Continuous random reliability is the time from part installation to failure and the time between successive equipment failures. Before discussing reliability models and predictions, however, the design team should understand some basic reliability fundamentals.

The probability of failure (P_f) of an item in a group of items may be described in simple statistical terms as:

$$P_f = \frac{n_f}{n}$$

Where n_f is equal to the number of failures that have occurred after a given time and n is equal to the total number of items in the group. Reliability (R) or probability of success at that point can then be expressed as:

$$R = \frac{n - n_f}{n} = \frac{n_s}{n}$$

Where n_s is equal to the number of surviving units after a given time. Since the sum of failure and success probabilities is equal to unity, reliability is:

$$R = 1 - P_f$$

The design team needs to analyze and estimate the time-dependent effects of failures that occur during the product's life. It is usually assumed in reliability predictions that early-life failures have been eliminated and the product is still in its useful life. Only the useful life portion of the product's life cycle, therefore, is used for design analysis. This period is usually characterized by a constant failure rate related to time.

Failure rate (λ) (often called hazard rate h(t)) is a measure of the average number of failures expected from a group of items for a given period of time, t_1 to t_2. It is the "instantaneous" failure rate of the number of units that have survived so far to a specified period of time. The period of time is generally standardized such as 10E9 hours for the hazard rate for electronic components. Given a period of time (dt), the conditional probability of failure in that period of time is:

$$\lambda(t) = f(t)/R(t)$$

From this measure of failure rate (λ), the reliability of a design having an "exponential" failure distribution can be expressed as where t is equal to the time interval over which reliability is measured

$$R(t) = e^{-\lambda t}$$

A probability distribution often used for strength testing and life testing of mechanical parts, is the Weibull distribution which can be expressed as

$$R(t) = e[-t/\mu)^{\Theta}]$$

Where μ is the scale parameter and Θ is the shape parameter.

From this expression, reliability can be expressed as a fraction between 0 and 1.0 for a given failure rate and time interval. Ideally, the objective is to obtain a probability as close to 1.0 as is practical. The only term that is under the designer's direct control is the failure rate. The task is to ensure that the particular failure rate associated with the completed system is as low as possible, and to make it economically consistent with other design constraints. When the failure rate is constant, the mean time between failures for an "exponential" distribution, by definition, is the reciprocal of failure rate:

$$MTBF = \frac{1}{\lambda}$$

Reliability Models

Reliability models are used during all phases of product development for design trade-off analysis. The models also indicate to management whether the product's reliability goals can or have been met, or whether corrective action is required. In early phases of product development, where actual products are not yet available for evaluation and test, reliability models are developed with the limited design information that is available. Accurate predictions of reliability are critical for a successful design effort.

Series Reliability Model

Most products can be modeled as a serial process. Series models recognize that for the system to function properly, all subsystems or blocks must operate without failure. Blocks can represent hardware or software. The probability of two blocks functioning without failure for a given period of time is:

$$R = R_1 \times R_2$$

Using failure rates this can be simplified to:

$$\lambda = \lambda_1 + \lambda_2$$

Based on these equations, the general reliability model for a series of m blocks is shown in Figure 14.1.

The reliability of the notebook computer previously discussed is shown below. Failure of any of these major functions is considered a product failure.

Subsystems	Failure rate (Failures per 10×10^{-6})	MTBF	Reliability
Circuit Board	1400	714	0.9986
Software	650	1538	0.9993
Processor	550	1818	0.9994
Display	900	1111	0.9991
Total System	3500	286	0.9965

MECHANICAL

Producibility Assessment Worksheet

Assessment Candidate   <u>Power Supply Housing</u>

Production Method (PM)

1. <u>Sheet Metal</u>
2. <u>Casting</u>
3. <u>Investment Casting</u>
4.

	Method	PM # 1	PM # 2	PM # 3
A1 Desing				
.9 .... Existing / simple design		✓		
.7 .... Minor redesign / increase in complexity			✓	✓
.5 .... Major redesign / moderate increase complexity				
.3 .... Tech. avail. complex design / significant increase				
.1 .... State of the art research req. / highly complex				
A2 Process / method				
.9 .... Process is proven and technology exists		✓	✓	✓
.7 .... Previous experience with process				
.5 .... Process experience available				
.3 .... Process is available, but not proven yet				
.1 .... No experience with process, needs R&D				
A3 Materials (availability / machinability)				
.9 .... Readily available / aluminum alloys		✓		
.7 .... 1-3 Month order / ferrous alloys			✓	✓
.5 .... 3-9 Month order / stainless steels				
.3 .... 9-18 Month order / non-metallic (SMC, etc.)				
.1 .... 18-36 Month order / new R&D material				
A4 Tooling				
.9 .... Simple fixture		✓		
.7 .... Minor fixturing				
.5 .... Moderate fixturing			✓	✓
.3 .... Significant fixturing				
.1 .... Dedicated fixturing				
A5 Desing to cost (DTC)				
.9 .... Budget no exceeded				✓
.7 .... Exceeds 1 - 5% in DTC			✓	
.5 .... Exceeds 5 - 20% in DTC		✓		
.3 .... Exceeds 20 - 50% in DTC				
.1 .... Cost DTC goals cannot be achieved >50%				

Producibility Assessment Ratings   PM # 1 <u> 8.2 </u> PM # 2 <u> 5.6 </u> PM # 3 <u> 7.4 </u>

$$\text{For each Method} \quad \frac{(A1 + A2 + A3 + A4 + A5)}{5} = \text{Producibility Assessment Rating for that Method}$$

FIGURE 14.1   Mechanical PAW

Parallel Reliability Models

One method of increasing a product's reliability is to design redundancy into the system. Redundant or parallel systems can continue to successfully operate when only one of the two available systems is working. The reliability for two redundant elements A and B can be expressed as:

$$R = R_A + R_B - R_A R_B$$

If the elements are identical:

$$R = 2R_A - R_A^2$$

If each component of the notebook computer in this example were made redundant, the reliability of the system would be increased to 0.9988. Although this increase in reliability is good, many considerations usually make fully redundant design impracticable; among these are cost, producibility, software complexity, power, and weight.

There are, however, situations in which a sizable reliability improvement can be achieved through a combination of series and parallel designs. This is particularly true when one part of the system is a large contributor to its overall unreliability. Looking at the previous example, the circuit board has the largest impact on system reliability. If the circuit board alone were made redundant, a significant improvement in reliability could be obtained with an acceptable increase in cost. The circuit board reliability would then be increased to 0.9998 and the overall system reliability would be increased to 0.9979 from 0.9965. The designer must decide if the cost and weight of the redundant circuit board is worth the reliability improvement.

Most examples of parallel systems are in complex software products such as electronic commerce where the failure of one server or computer could cause unacceptable problems, such as in large financial losses (e.g. consumer commerce, banking, etc.) or loss of life (e.g. air traffic control).

Availability and Mean Time to Repair

Reliability does not address the period of time during which the product is out of use after a failure does occur. The concept of availability was developed to resolve this limitation by quantifying the amount of time that a product is in operational use. Availability is affected by reliability (how many times it fails), manufacturing/vendor quality (number of defects), repairability (how long does it take to repair), and logistics (how long to respond). This time period is best illustrated in an example: three different notebook computers A, B, and C, are purchased. All three systems operate failure free for 1 year, but fail at the start of the second year. System A is repaired in 1 day, system B requires 90 days to repair because of its complexity, and system C requires 90 days of downtime because the replacement parts must be ordered. The time lines are shown.

System Start	Year 1 Operation	Failure Occurrence	Time to Repair
A	year 1, no failure	day 366	1 day to repair
B	year 1, no failure	day 366	90 days to repair all repair time
C	year 1, no failure	day 366	90 days of downtime due to 89 days waiting for parts

Although every customer would rather have system A, a simple calculation of reliability shows an interesting result. Since each system has only one failure occurrence over the 2 years, the reliability of each of the three products is equal! Reliability measures the number of failures, not the amount of time that the product is available for use. Operational availability is defined as the probability that an item is properly operating at a given point in time.

Operational availability can be defined as:

$$A_o = \frac{uptime}{uptime + downtime}$$

or

$$A_o = \frac{MTBF}{MTBF + MDT}$$

Where MDT is defined as mean downtime. Mean downtime is the total time necessary to return an item to a serviceable condition. The operational availability for the three different systems is then:

	Product A	Product B	Product C
A_o	0.9986	0.8902	0.8902

A brief analysis of the results, however, shows no explanation for the different lengths of downtime between systems B and C.

This difference is identified in the calculation of a product's inherent availability, which can be defined as:

$$A_I = \frac{MTBF}{MTBF + MTTR}$$

Where MTTR is the mean time to repair the product and restore it to service. Mean time to repair is the amount of time necessary to repair the system if all of the parts, software, and equipment are available. Its relation to the total mean downtime is:

$$MDT = MTTR + \text{miscellaneous delay}$$

Miscellaneous delay, for example, can be attributed to waiting for parts on order or training time.

A final summary of the three systems is shown in Table 14.2.

Highly Reliable Software Programs

As computers assume more vital roles, the need to minimize computer failures becomes crucial. Some key design practices for high reliability is robustness, fault avoidance, tolerance, detection, and recovery. Fault tolerant systems are designed to continue to operate after a failure has occurred. The 100% availability promised by fault-tolerant systems can be vital for some companies, such as telephone companies and continuous process manufacturers. One fault-tolerant approach is the parallel duplication of key components. This can be a cost-effective way to maximize the likelihood of continuing operation. In applications with high penalties for shutdown or loss of control, the enhanced reliability is worth the price.

The terms of redundancy and fault tolerance are sometimes used interchangeably. Some distinctions can be made:

- Redundant systems have individually specified duplicate components and means for detecting failures and switching to backup devices.
- Fault-tolerant modules have internally redundant parallel components and integral logic for identifying and bypassing faults without affecting the output.

TABLE 14.2 Comparison of Reliability and Availability

	System A	System B	System C
MTBF (days)	730	730	730
MDT (days)	1	90	90
MTTR (days)	1	90	1
Operational availability A_O	99.8%	89.0%	89.0%
Inherent availability A_I	99.8%	89.0%	99.8%

Designing in fault tolerance is done in many different ways depending upon cost constraints, time schedules, and the abilities of the designer. Fault tolerance is implemented through combinations of hardware and software. The use and type of fault-tolerant systems must be defined early in the specification process in order to properly design the different software programs for the overall system.

14.4   RELIABILITY LIFE CHARACTERISTICS

A product's failure rate and reliability change over time. Reliability life characteristics or mortality curves describe the manner in which failures occur throughout their lives. The product's life characteristics vary depending on the type of product being evaluated. Differences between hardware and software life characteristics are especially great and are discussed in the following sections.

Hardware Reliability Life Characteristics

A number of different failure distributions are used for hardware reliability predictions and calculations, such as the Weibull, normal, gamma, and exponential. Exponential distribution is used almost exclusively for electronic equipment and is used in this discussion.

Historically, the failure of hardware oriented products over their total lifetimes can be classified into three major types of failures:

- **Quality failures** occur early in the life cycle and are due to quality defects caused by manufacturing, vendors or design.
- **Stress related failures** occur at random over the total system lifetime and are caused by the application of stresses that exceed the design's strength.
- **Wear out failure** occurs when the product reaches the end of its effective life and begins to degenerate (i.e., wear out).

When the three failure mechanisms are summed over the lifetime of the product, the famous bathtub curve for reliability is generated. A plot of a typical population's failure rate over time is shown in Figure 14.2. The three distinctively different failure rate types are shown in the bathtub curve. These correspond to three different periods in the life of the population. Although many individual parts or software modules do not have a bathtub-shaped mortality curve, most have reliability life characteristic curves that are represented by at least one of the periods. Each of these periods is discussed below.

When a product or part is first developed, the population exhibits a high failure rate during the "infant mortality period." This may be due to a number of causes: poor design, inadequate test and evaluation, faulty manufacture, or transportation damage. This failure rate decreases rapidly during this early life and then stabilizes at an approximately constant value. A prime example is a new

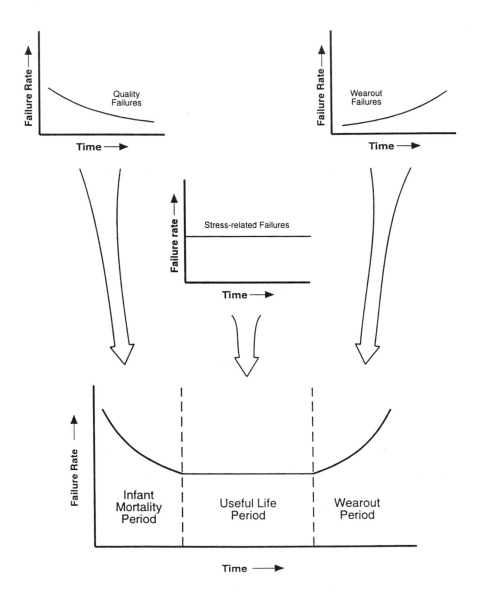

FIGURE 14.2 Product reliability life characteristic.

automobile in the first few months, when the purchaser finds hidden defects that escaped the inspectors at the factory. To alleviate this problem, many manufacturers test their products prior to sale in order to minimize the number of initial failures. In electronics, this type of testing is called "burn-in".

Common reasons for early failures are: (adapted from Smith, 1976)

- Poorly defined and inadequate requirements for design, manufacturing, vendors or test
- Misapplication of part (picked wrong component for this application)
- Designs that are difficult to manufacture
- Dirt or contamination on surfaces or in materials
- Chemical and oxide impurities in metal or insulation
- Voids, cracks, or thin spots in insulation or coatings
- Incorrect positioning of parts
- Improper packaging and shipping
- High stress levels such as electrical overstress (EOS) and electrostatic discharge (ESD)

Early failures usually reflect the "producibility" of the product, vendor selection and the quality control of the manufacturing process. Early failures can be minimized by designing for producibility, using qualified manufacturing processes and vendors, and performing quality tests prior to shipment.

After the infant mortality period, the population reaches its lowest failure rate level. This normally constant failure rate period is called the useful life period. Most failures during this period are characterized as stress related failures. The length of this period varies among products, but its failure distribution is usually modeled as an exponential function. This is the most significant period for design and reliability prediction activities.

During the useful life period, failures usually result when the stress applied to a part is greater than the strength of the part at that given time. Any type of environmental condition or user activity can cause this stress. The strengths of individual items in a population may vary because of slight differences in materials or manufacturing processes. This is true of almost all items, from bolts to sophisticated integrated circuits. The strength characteristic of a population of parts is usually normally distributed. The stress to which a specific part is subjected varies with its application. For example, two identical transistors, when used in the same circuit application, are subjected to slightly different stresses unless their supply voltages, bias, and operating temperatures are exactly the same. The effects of temperature and voltage stresses on a part's failure rate are shown in Figure 14.3. Note that any increase in the level of stress or combination of stresses can reduce reliability. Temperature, humidity, vibration, shock, and altitude contribute to the failure rate if inadequate design

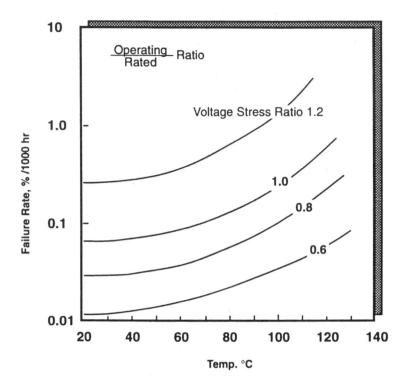

FIGURE 14.3   Effects of stress on failure rate.

margins are specified. This is one of the key design parameters for developing a reliable product.The last period, the wear out period, occurs when the population begins to deteriorate due to the effects of aging. During this period the failure rate starts to increase. When the failure rate becomes unacceptably high, replacement or overhaul of the item is required. Short-life parts should be identified during the design process so that replacement plans and procedures can be implemented. These failures are caused by deterioration of the design strength. As listed by Smith (1976), deterioration results from a number of well-documented chemical and physical phenomena:

- Corrosion or oxidation
- Insulation breakdown or leakage
- Ionic migration of metals
- Frictional wear or fatigue
- Shrinkage and cracking in plastics

- Drying out or out gassing of lubricants and epoxies

Many of these factors are the reasons that an automobile will eventually wear out. There is, however, no way to predict how long an individual automobile can be effectively repaired before it wears out. The wear out point varies among cars even if they all receive the same maintenance.

Software Reliability Life Characteristics

Software reliability life characteristics are different from hardware. Some software models only have the first two phases of the traditional bathtub curve: infant mortality and useful life. Software does not physically deteriorate or wear out in a "classical" way such as hardware. This type of software reliability life model is shown in Figure 14.4. Most software programs require several revisions, modifications, and re-installations throughout the product's life. These tasks may be required for software improvements or changes that occurred in the hardware applications or software interfaces. These changes will affect the software's reliability. Because

of these common updates, many authors revise the software life model for revisions as shown in Figure 14.5.

Some common reasons for early failures in the software infant mortality phase are:

- Poorly defined software and user interface requirements
- Design errors
- Coding errors
- Inadequate test and evaluation

Just as in hardware, the design uses proven software design practices and testing to minimize the affects of these errors. During the software's useful life, failures can be due to human variability or errors, traffic volume, data corruption, electrical interference, defects not caught in test, or changes in the hardware system and other software products. Design margins, error recovery designs, and strict configuration control are the best methods to minimize these types of problems.

14.5   RELIABILITY PREDICTION

Reliability prediction is a process for estimating the reliability of a design prior to its actual operation. This prediction of reliability provides a measure whereby designers can assess technical progress and offer a quantitative basis by which design alternatives can be evaluated. Reliability predictions provide a basis for evaluating reliability growth in developmental testing and for planning in technical risk assessment, maintenance, logistics, and warranties.

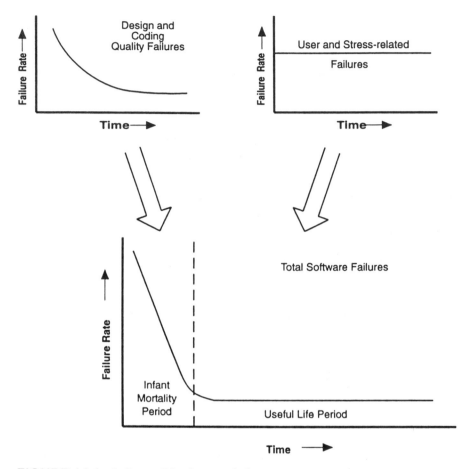

FIGURE 14.4 Software life characteristic.

Although reliability prediction is a key parameter in product development, the development team must remember that it is only a prediction and only as good as the estimates and assumptions that were used in its development.

A prediction of reliability is usually determined by combining the reliability of individual items (e.g., parts, software modules) in the product. The quality of this prediction depends on how accurately the model reflects the overall system and the availability of individual item reliability information. Many theoretical and statistical procedures are used for reliability predictions.

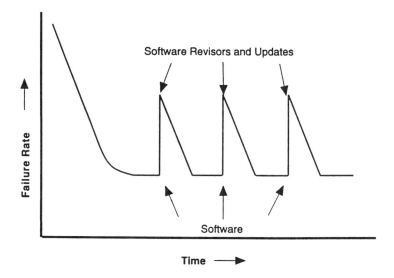

FIGURE 14.5 Revised software reliability life characteristic.

Note: Every software revision usually
causes more failures.

14.5.1   Hardware Reliability Prediction

The four most commonly used hardware reliability prediction techniques are:

1.   Parts count method
2.   Stress analysis method
3.   Design similarity method
4.   Physics of failure or reliability physics

The parts count method is a simple, easy-to-use method for hardware based on an estimate of the number of parts by each part type. The total number of parts in each part type is summarized in a table. These numbers are then multiplied by the typical failure rate for that part type. This method is usually used during proposals and early in the design process since it requires only general design and failure rate information. An example of the parts count method is:

$$\lambda = \sum_{i=1}^{n} N_i (\lambda_G \pi_Q)$$

for a given equipment environment where:

λ = total equipment failure rate (failures/10 hr.)
λ_G = generic failure rate for the i th generic part (failures/10^6 hr.)
π_Q = quality factor for the i th generic part
N_i = quantity of i th generic part
n = number of different generic part categories

The second method is called the stress analysis method. This method takes into account the effects of environmental and other stresses on a part's failure rate for a particular type of application. It requires more detailed design information such as environmental stress and part reliability information based on the levels of stress and other factors. Each part's failure rate is based on an analysis of the particular design application. An example of one stress analysis method with its factors is:

$$\lambda = (\lambda_B)(\pi_E)(\pi_A)(\pi_Q)...(\pi_N)$$

Where

λ = predicted failure rate
λ_B = base failure rate
π_E = environmental adjustment factor
π_A = application adjustment factor
π_N = additional adjustment factors

Both the parts count method and the stress analysis method relies on published part failure rates based on historical data. When new technologies, applications, or unique parts are selected for which little supporting data are available, failure rates must be estimated. Two commonly used methods are design similarity and physics of failure. The design similarity method uses comparative evaluations to develop failure rate information by extrapolating failure data from similar or comparable items. The quality of this method depends on the level of similarity of the two items. An example would be the transition from a 32 MB RAM device to a 64 MB RAM. If only minor changes in design, manufacturing processes, and technologies are in the new 64 MB RAM, reliability information of the older 32 MB RAM device can still be used as a baseline.

If the failure mechanisms are different from the ones previously known due to configuration changes, material changes, or introduction of new technologies, this method will not be accurate. Therefore, predicting reliability according to similar design may be unreliable.

The last method is called the physics of failure or reliability physics method. Reliability predictions can be derived from or based on a study of the physics or mechanics of real failure mechanisms. This model gives insight into or control over actual failure mechanisms. This is especially important when designing and qualifying a new product or a product with new technologies. From the viewpoint of physics of failure, the reliability characteristics of an electronic system should be defined by the failure mode, failure mechanism, and failure site (Hu, 1994). This method requires extensive testing to determine the failure modes and nature of the stress mechanisms. Special testing, however, may not be feasible because of the extra cost, lack of available items to test, and the extended time period constraints required obtaining statistically valid data. Physics of failure is based on acceleration transform models that correlate the potential failure mechanism in an operational environment to the test time or cycles in accelerated qualification tests.

14.5.2 Software Reliability Prediction

For software, the three most common techniques are based on error history and measured by:

1. Counting the lines of code
2. Measuring the complexity of the code
3. Reliability test and evaluation

The easiest method is to simply counting the lines of code. Using historical data on the number of faults found per thousand lines of code, an analyst predicts a new software program's reliability by counting the lines of code. In general, the number of faults is an increasing function of the total number of lines of code. The larger the program, the more entities the programmer has to deal with and the greater the chance of error. The reliability of a program can be estimated by using known failure rates. For example, a 10,000 line program with an estimated failure rate of 0.001 failures per line of code would be expected to contain 10 faults. This number is then reduced by debugging and testing. Although it is easy to implement, the method is not very accurate.

The second method is to measure the complexity of the software program by counting parameters that have historically had an effect on a product's reliability. The parameters can include the number of variables, nested levels, branches, subroutines, etc. Complexity measures can also include the number of user inputs, user outputs, subroutines, files, and external interfaces.

The last type of method for software prediction uses testing results to estimate the number of faults that remain in the product. Its compares the number of faults found with the amount of testing that was performed. Before testing starts, the number of faults is estimated. As testing progresses all failures are recorded and compared to the estimated number of errors. One method is called the cumulative failure profile. This graphical profile uses historical test results to predict the number of defects per unit of test time that should be found. The new software is then tested where these results are compared to previous test results. The failure rate at any time can be calculated from this graph and then plotted over a time frame projected in the future.

Reliability growth models are used to predict a product's reliability and when a particular level may be attained. The model provides a way of assessing how fast the software reliability is improving with time. Similar to hardware, the software is tested using a statistical model approach and the reliability is measured. The reliability measurements are compared with the growth model. Reliability predictions are then computed. By measuring the number of faults during test, software reliability improvement can be used as a benchmark against additional testing and future products. An example of some software reliability prediction models for testing as recommended by Lawrence and Persons (1995) are:

- **Error count models**. When using test intervals and the number of failures in the test intervals as input data. The Schneidewind model is generally preferred.
- **Time domain models**. When using time between failures as input data. There are three subcategories:
  - Exponential time domain models. The Jelinski-Moranda model is recommended.
  - Non-exponential time domain models. The Musa-Okumoto Logarithmic Poisson model is recommended.
  - Baysian time domain models. The Littlewood-Verrall model is recommended.

14.5.3   Predicting Reliability During the Design Phases

It must be remembered that a reliability prediction is just that: a prediction of the reliability potential of a product based upon available design information. This does not mean that the product will exhibit this level of reliability in the field. However, a reliability prediction that does not exceed its reliability goals is a good indicator that the system design has problems and will not meet its reliability requirements later in production.

In the conceptual design phase, models are based on part counts, published part failure rates, and design similarities from previous product usage to evaluate feasibility of reliability approach and trade-off analyses to analyze conceptual design approaches. Software uses predicted lines of code, number of modules or functions, and similarity with previous experience. In the detailed design phase, models are based on stress level models or physics of failure used for design trade-off, manufacturing and logistics analysis planning and to identify reliability problems. During test and evaluation hardware and software models are based on current failure data including number of failures and failure modes test to identify problem areas for design improvements and support methods. For operational use and warranty, models use number of failures in field and results from their analyses to identify failure modes for future products, identify if changes are needed in support procedures or product recalls.

14.6   DESIGN FOR RELIABILITY

Design reliability is a design discipline that uses proven design practices to improve a product's reliability. The next few sections review some of the key practices for ensuring high levels of reliability. The key techniques are:

1. Multidiscipline, collaborative design process
2. Technical risk reduction
3. Commonality, simplification and standardization
4. Part, material, software, and vendor selection and qualification
5. Design analysis to improve reliability
6. Developmental testing and evaluation
7. Production reliability

14.7   MULTIDISCIPLINE COLLABORATIVE DESIGN

As mentioned throughout this book, reliable products require a multidiscipline design team that focuses on design actions. A successful approach includes early and constant involvement, effective and open communication, creative environment, and a systematic trade-off analysis procedure. These actions include performing tradeoff analyses of the different parameters to identify the best design.

14.8   TECHNICAL RISK REDUCTION

As also discussed in earlier chapters, successful products generally push the limits of current processes and technologies. Poor reliability is often caused by unexpected problems in the design, vendors, or manufacturing. Identification, assessment and resolution of all technical risks are essential for resolving problems or to at least minimize their effect when a problem occurs. Early detection is critical because it is easier and cheaper to make changes early in a design. This requires a systematic methodology of technical risk assessment and management.

Technical risk management identifies and controls uncertainty found in product development. Identification and assessment of technical risks are essential for identifying and resolving potential problems to ensure that the proposed system will work as intended and be reliable when it reaches the user. The steps are to:

1. Systematically identify areas of potential technical risk for failures
2. Determine the level of risk and probability for each area
3. Identify and incorporate solutions that reduce the risk
4. Continue to monitor progress on minimizing technical risk

As mentioned in Chapter 2, technical risk management can document technical progress, identify technical problems early, assess their impact, and provide sufficient information for managerial planning and control. Technical risks and methods for reducing risk can be found at Department of Defense (DOD). Transition from Development to Production. DOD 4245.7-M, September, 1985. www.bmpcoe.org The document provides a "template" that describes what risk exists, how to identify the level of risk and outlines what can be done to reduce it. The term "template" is used to define the proper tools and techniques required to assess and balance the technical adequacy of a product transitioning from development to production. Table 14.3 lists the reliability checklist questions for identifying the level of technical risk for a computer power supply.

The next step is to evaluate and determine the "level" of risk for each area based on the answers to the questions. One method is to use checklists that compare current design information to items that have historically caused problems. Subjective judgment of management and the design team is often used to determine the severity of the risk.

The last step is to identify actions that for reducing areas of high technical risk. This can include design or vendor changes, closer management scrutiny, etc.

14.9   SIMPLIFICATION AND COMMONALITY

The increasing complexity of today's products increases the statistical probability of failure. Therefore, the first step in achieving reliability is to simplify the product as much as possible without sacrificing performance. This is a key design trade-off, since it enhances not only reliability, but also cost and producibility. The important design goal is to reduce the number of items and improve the reliability of each item. For hardware, we reduce the number of parts and select more reliable parts. For service companies, we focus on the number, complexity, and reliability of transactions. Finally for software we

TABLE 14.3 Template Recommendations for Power Supply Reliability

---

A.3 RELIABILITY CHECKLIST

COMPONENT LEVEL

(1) Is the item an off-the-shelf device or especially developed for a particular function?

(2) Have requirements for component quality level been observed Cp and Cpk?

(3) Have component derating guidelines been observed?

(4) What is the failure history of this item?

(5) Is the item critical; i.e., would its failure result in system failure?

(6) What are the possible modes of failure?

(7) What steps have been taken in the item application or system design to eliminate or minimize the effects of these modes of failure?

(8) Is it possible to introduce the concepts of redundancy and/or use the item at derated performance levels?

(9) What is known about its storage life, operating time or cycles; i.e., how much time or cycles, operating and non-operating, may be accumulated without significantly degrading its reliability?

(10) If the item is a newly developed design, what are its critical weaknesses, and what provision has been made in the design so that modifications can be made at the earliest possible time if these or other weaknesses show up in testing?

(11) Is it physically and functionally compatible with its neighboring components; i.e., will the physical location affect its performance or reliability?

(12) Are any unusual quality control or vendor problems expected?

Source: adapted from (www.bmpcoe.org)

reduce the lines of code. Fewer parts or lines of code usually translate into higher reliability. Commonality and standardization are also important. The team has the ability to select software components that have previously been written and tested by software designers (allowing for increased producibility, less technical risk and better reliability). By selecting from the library of existing components, the translation errors between system and software designers are reduced to only those errors occurring in the development of components that are not available through the library.

14.10   PART, MATERIAL, SOFTWARE, AND VENDOR SELECTION AND QUALIFICATION

A product is not very reliable if some parts or materials selected are not capable of surviving the stresses of the environment or application or if they are known to have a high failure rate due to poor design or manufacturing. Software that has not been proven to be reliable can degrade reliability of the system. Since most products contain purchased materials, parts, and software, the proper selection and qualification of high quality vendors that provide highly reliable products cannot be overemphasized. An important consideration is the level of reliability required. Use of higher reliability parts and software (i.e., lower failure rate) usually causes a product's cost to be higher. The key task is to select parts, materials, software, and vendors that a history of high reliability and are appropriate for the product's other requirements.

14.11   DESIGN ANALYSIS TO IMPROVE RELIABILITY

Design reliability is the application of analyses to identify potential problems and implement design actions that minimizes or eliminates the problem. In a study by Rockwell International (Willoughby, Jr., 1988), reliability design analyses identified 333 reliability improvements for the design of a radio. A breakdown of the analyses is shown below.

Note that for electronic products, thermal stress is often the most important area for analysis. This would also be expected for our notebook computer example.

Design Analysis	Number of Reliability Improvements
Thermal Stress	233
Worst Case	44
Electrical Stress	31
Sneak Circuit	8
Dynamic	8
Failure Mode and Fault Tree	5
Thermal	4

Perhaps the largest positive impact the design team can make on a product's reliability is to reduce the effects of stress. Stress can be caused by temperature, current, voltage, environment, user, traffic volume, bandwidth, etc. This stress reduction can be accomplished either by reducing stresses or by selecting another component that can withstand higher levels of stress. A combination is generally used. For example, U.S. Navy reliability studies of systems currently in the fleet indicate that a failure rate improvement factor of 12 can be achieved by reducing temperature related stresses (e.g., transistor junction temperature) by only 30°C (Willoughby, Jr., 1985). Although the advantages of reducing stress for all parts may not be as spectacular as those indicated for transistors, the reliability improvement potential is significant. As noted earlier, derating is one managerial method to reduce stress and can be described as selecting parts to be operated at a stress level less severe than for which it is rated.

As a basic design philosophy, a design's reliability or failure rate can be improved in anyone of the following ways

- **Increase average strength** of the product by increasing the design's capability for resisting stress.
- **Decrease the level of stress** placed on the design through system modifications, such as packaging, fans, or heatsinks.
- **Decrease extreme variations of stress and strength** by limiting conditions of use or improving manufacturing methods.

Important design analyses for improving reliability that have been discussed earlier include:

- Derating criteria and design margins
- Thermal analyses
- Failure mode and effects analysis
- Fault tree analysis

14.12   DEVELOPMENTAL TESTING AND EVALUATION

Predictions may indicate a high level of reliability, but it takes test and evaluation to actually verify that a design meets its reliability, performance, and other design requirements. The objective of developmental testing is to identify problems so that they can be corrected. Testing used to improve reliability is called by different names including reliability growth, reliability assessment, reliability development, or test analyze and fix (TAAF) testing. The essence of reliability growth testing is failure corrective action, not reliability measurement. The reliability of a product improves through testing only when failures are corrected through changes to the actual design or manufacturing process. This is called reliability growth and was discussed in the Chapter on test and evaluation. Test and evaluation are also important in software and service oriented products.

14.13   PRODUCTION RELIABILITY AND PRODUCIBILITY

Historically, reliability degrades when a design progresses from developmental testing to production and installation. Since the developmental prototypes were hand built by highly trained technicians and programmers, many problems are encountered when the transition to manufacturing is made. This degradation is caused by a combination of problems, including those introduced by users, manufacturing processes, defects attributable to purchased parts, manufacturing errors, inefficiency of quality control, lack of configuration control, changes in the design, and unique conditions not encountered in the test program. Significant levels of management planning and support are required in order to maintain product reliability during the transition to the production phase.

To minimize the number of reliability problems in production, the steps in the design process should include

- Active reliability, producibility, manufacturing, and quality design efforts throughout development
- Extensive technical risk assessment and lessons learned program
- Development of a design that is easy to repair, inspect, and test
- Selecting best vendors
- Accurate design and manufacturing documentation
- Allocation of necessary manufacturing and test resources
- Quality control of vendors, parts, and software
- Thorough quality control with feedback to manufacturing and design
- Analysis and corrective action for all manufacturing and reliability problems

Only if these tasks are accomplished will the transition to production be accomplished without degradation in the inherent reliability.

14.14   DESIGN FOR RELIABILITY AT TEXAS INSTRUMENTS

A successful design for reliability effort by Texas Instruments was to develop a family of forward-looking infrared radars (FLIR). These are very complex electronic and mechanical systems used for night vision in an aircraft environment. The key design practices listed earlier where part of this success.

1. Multidiscipline design and technical risk reduction
2. Design simplification and standardization

3. Part, material, software, and vendor selection and qualification
4. Design analysis to reduce the effects of stress
5. Developmental testing and evaluation
6. Production reliability

Multidiscipline Design And Technical Risk Reduction

Reliability was a major decision parameter for the multidiscipline team from the very first of the program. One of the first steps to reduce technical risk and improve the design's reliability and producibility was to modularize the system by designing common modules where possible. Some of the more significant results of the trade-off studies between reliability, producibility, cost, and performance are highlighted in the following discussion.

Design Simplification

Design simplification techniques were used for design improvement. A video amplifier was simplified by replacing the discrete preamplifier, postamplifier, and gain control consisting of 20 components by just one integrated circuit. For a scan module, an oscillating scan mode was selected to replace previous rotary techniques. This change led to the replacement of hard to manufacture lubricated ball bearings with unlubricated flexible pivots that were more producible and compatible with system motion.

Part, Material, Software, And Vendor Selection and Qualification

Part selection was limited to qualified standard parts or approved nonstandard parts. Strict adherence to selecting only qualified parts resulted in only 39 unique part types that did not meet these requirements. These parts were approved only after testing the parts to typical military requirements and performing a reliability study on the capacitors, which showed that the capacitors could meet all reliability requirements. This standardization resulted in high levels of production quality and field reliability.

Derating and Design Analysis To Reduce The Effects Of Stress

Stress analysis was a major part of the reliability design effort. Program reliability objectives required that part stresses should be held to as low a level as was consistent with derating guidelines and circuit requirements. Over 87% of all electronic parts were operated at less than 10% of their rated capacity at 60°C ambient temperature.

Developmental Testing And Evaluation

A number of tests were run on the system and individual modules. Examples include:

- Early critical module evaluation test to promote growth
- Over 8000 hours of common module system burn-in
- Over 18,000 hours of formal reliability demonstrations in common module system
- Nine formal common module system environmental qualifications
- Failure analysis and corrective action program
- Over 30,000 hours accumulated on early common module systems in field application
- Failures were analyzed and corrective action taken where applicable.

In summary, the following factors were considered critical to the success of this program (Grimes, 1983):

1. Early and thorough reliability involvement in all phases of design
2. Management and customer emphasis on reliability
3. Early reliability growth and development tests
4. Comprehensive failure analysis and corrective action programs

14.15   SUMMARY

This chapter has presented a brief overview of the fundamentals of reliability, with particular emphasis on design. The material presented here gives the product development team an understanding of reliability and its importance to customer satisfaction. Two points must be emphasized: reliability and producibility are mutually supportive and must be designed into the product from the start.

14.16   REVIEW QUESTIONS

1. Define the concept of reliability.
2. Why is it important to consider reliability early in the conceptual design stage?
3. List and describe at least three factors that have a definite impact on product reliability.
4. Contrast the concepts of quality and reliability.
5. Define the concepts of "MTTF" and "MTBF". Provide appropriate units of measure for these two concepts for different types of products.

6. What is the main purpose in using reliability models during the product development life cycle?
7. List and describe the major types of hardware and software failures that a product may have over its total lifetime.
8. List and describe the methods more commonly used for reliability prediction of hardware and software.
9. Can you provide an example where you recommend applying each one of the methods listed in question 8?
10. Provide at least five practical recommendations to reduce or eliminate production reliability problems.

14.17   SUGGESTED READINGS

1. Department of Defense (DOD). Transition from Development to Production. DOD 4245.7-M, September, 1985. www.bmpcoe.org
2. F. Jenson, Electronic Component Reliability, Wiley, New York, 1995.

14.18   REFERENCES

1. Department of Defense (DOD), Reliability Prediction of Electronic Equipment, MIL-HDBK-217E, Washington, D.C., June 1983.
2. Grimes, High Quality and Reliability: Essential for Military, Commercial and Consumer Product, Equipment Group Engineering Journal, Texas Instruments Inc., 6(4), July-August: 35-47, 1983.
3. J.M Hu, Physics-of-Failure-Based Reliability Qualification of Automotive Electronics, Communications in RMS, Vol. 1(2) 21-33, July, 1994.
4. J.D. Lawrence and W.L. Persons, Survey of Industry Methods for Producing Highly Reliable Software, Lawrence Livermore National Laboratory; NUREG/CR-6278 or VCRL-ID-11724, 1995.
5. Lipow and M. L. Shooman, Software Reliability. Consolidated Lecture Notes, Tutorial Sessions, 1986 Annual Reliability and Maintainability Symposium. 1996.
6. P. O'Conner, Foreward of Electronic Component Reliability, Wiley, New York, 1995.
7. Reliability Analysis Center (RAC), Reliability Design Handbook, No. RDH3712, IIT Research Institute, Chicago, Illinois, 1975.
8. C.O. Smith, Introduction to Reliability in Design, McGraw-Hill, New York, 1971.
9. E. Wheeler, Quality Starts with Design, Quality Magazine, May: 22-23, 1986.
10. J. Willoughby, Jr., Certification of the Technical Disciplines in the Design and Manufacturing Process, Presentation to the Defense Science Board, Washington, D.C., 1985.

Chapter 15

TESTABILITY: DESIGN FOR TEST AND INSPECTION

Insure Efficient and Complete Verification

Test and inspection are the last critical step for producing a high quality, low cost product. They represent the last chance to identify problems 1.) in the design phase, 2.) before the product is shipped to the customer, 3.) just before the product is used or 4.) during product use. This process verifies that the product is ready for use and does not contain defects. Testability is a design characteristic that measures how quickly, effectively, and efficiently problems can be identified, isolated, and diagnosed. It is a major part of both producibility and reliability design. In this Chapter, testability represents design actions that improve effectiveness and efficiency of both test and inspection tasks.

Best Practices

- Comprehensive Test Plan
- Design for Effective, Easy and Efficient Inspection and Test
- Ergonomics
- Detailed Test Documentation
- Failure Information Used to Improve the Design
- Stress Related or Environmental Stress Screening
- Important Vendors are Part of the Process

TESTABILITY'S KEY PARAMETERS

One purpose of test is to verify that the product or service meets all requirements at the lowest cost. The goal is to minimize testing since it is a non-value added task. Testability is the process of designing a product so that it can be tested more efficiently and effectively. It also includes self-diagnostics and self-maintenance. The process is to identify the points or items to test and making the test process as effective and efficient as possible. Key design parameters for testability are accessibility, controllability, observability, and compatibility.

1. **Accessibility** is the physical, electrical or software ease of getting to (i.e., access) the area to be tested.
2. **Observability** is the ease of determining output results (i.e. verifying conformance to specification).
3. **Controllability** is the ease of producing a specified output when controlling the input.
4. **Compatibility** is to design the product to be compatible with existing user skills, test equipment, Internet, facilities and personnel.

The development of quick, reliable test and inspection processes poses a challenge to product development. Identifying which defects to focus on is a difficult decision since there are always more potential defects than could ever be tested and inspected. Other decisions include what data to record and display. Important manufacturing parameters that limit the number of defects to be evaluated and amount of data displayed/recorded include:

1.) Time to test and inspect that is available including human response time
2.) Initial cost of the test and inspection equipment
3.) Recurring cost of the test and inspection process itself

The current emphasis on quality, reliability, and the competitive state of the international market has resulted in both greater visibility and increased responsibility for design, software, test, reliability, manufacturing, and repair. This responsibility has resulted in an increased emphasis on the role of design for testability and inspection. A common trap is to assume that ingenuity in the design of manufacturing test equipment and test software can compensate for design and manufacturing deficiencies.

This chapter will cover product design techniques for inspection and testing.

15.1   IMPORTANT DEFINITIONS

Verification during production is the process of insuring that the manufactured product can meet all design requirements. The level of verification is related to the quality of manufacturing's test and inspection. It can be measured by how many parameters are checked and the level or depth that the parameters are tested.

Testing is defined as when some form of stimulus is applied to a product (i.e., hardware, software, or both) so that the product's functions can be measured and compared to design requirements. It actively exercises the product's (i.e., design) functions. For a product, testing wants to 1.) detect a problem, 2.) isolate the problem to a certain location, 3.) identify/diagnose what the problem is and 4.) respond to the problem (i.e. correct, stop, display, etc.).

Testability or Design for Test (DFT) is a design characteristic that measures how quickly, effectively, and cost efficiently problems can be identified, isolated, and diagnosed. The steps are to identify which defects to be tested and then design the product so the defects can be easily tested in product development, manufacturing, and in the field. Issues in testability include requirements, design, vendors, and manufacturing. For software, testability is the relative ease associated with discovering software faults.

Inspectability is a design characteristic that measures how easy an inspection task can be performed. Inspection is a process that identifies whether a process has been performed or makes status measurements to infer whether a process has been performed correctly.

15.2   BEST PRACTICES FOR TESTABILITY

Testing is the preferred method over inspection for ensuring quality for both hardware and software. Since inspection does not check part or product's functions, one can only infer from actively testing whether the process has been performed correctly and that the product performs according to requirements. For discussion purposes, testability will represent design actions that improve both test and inspection tasks. To be successful, testability should become an integral part of the design tradeoff process. The best practices for testability and inspectability are as follows:

- **Test plan** identifies in detail the requirements to test, methods to acquire data points, and their comparison to requirements.
- **Product is designed to be easily and efficiently inspected and tested** in test and evaluation, production, maintenance, and operational environments.
- **Ergonomics** develops a simple and effective human interface to improve human performance, reduce human errors and increase number of potential users

- **Design documentation** details test and inspection requirements and any special decision criteria to be used by quality control.
- **Failure information** explains the cause of each failure and which can be directly used by the self-maintenance system, test equipment, and repair personnel to correct the failure.
- **Stress related or environmental stress screening** transforms latent or hidden defects into functional failure modes that can be recognized in the testing process.
- **Important vendors** are included in testability efforts.

Nonconformance is defined as when a design, product, or part does not meet stated requirements. Nonconformance can include 1.) functional parameters that may result in a failure and 2.) aesthetic parameters that may cause customer dissatisfaction. A failure is an occurrence that renders a system unable to perform a normal function according to specification. Failures can be caused by the external environment or by some internal defect caused by design, vendors or manufacturing. An example of an aesthetic parameter is the quality of the paint color on an automobile. If the paint color does not meet company standards, the customer will be dissatisfied but it would not result in a failure. **A defect, fault, or nonconformance is defined as an imperfection in either the design or manufacture of the product that may or may not result in a failure or customer dissatisfaction.** Hardware commonly uses the term "defect" whereas software uses the terms "nonconformance" and "fault".

A latent defect is a nonconformance that has not yet been identified or caused a failure. A crack in a structure that has not broken and defective software code that has not yet been exercised are examples of latent defects. Since testing identifies failure, many tests are designed to stimulate the product at a high enough level to cause a latent defect to result in or change into a functional failure.

Important measures are the level of detection, isolation and diagnosis or explanation. The level of "detection" is the capability of detecting a specified minimal percentage of all product defects or failures that theoretically could occur. The level of "isolation" is the capability of defining where the detected defect/failure occurred using internal testing or diagnostic testing. Internal testing is called built-in-test. Equipment used in diagnostic testing for electrical circuits includes oscilloscopes, spectrum analyzers and modular measuring systems. The isolation requirement might be to correctly identify a single replaceable or repairable subfunction, such as a disk drive, circuit board, part, component, or lines of software code. The level of "explanation" is the capability for intelligently identifying parameters of the failure, statistically analyzing the data trends, and providing corrective actions to the equipment or operators.

Some contracts have specified the detection of 99 % of all faults and isolation of 95 % of all faults to three or fewer parts. Design requirements will continue toward more stringent testability requirements, with an ultimate goal of 100% detection and 100% isolation to a single component or line of software code. A new design technique is self-maintenance products. These products can diagnosis a problem and then repair it automatically.

Specific measures for testability and inspection for a notebook computer user self-test might include the following:

- Fault detection: 95% detection of all failures
- Fault isolation and diagnosis: 80% isolation to a replaceable part or line of software code.
- Fault self maintenance: 70% of all software failures can be self diagnosed and repaired automatically when connected to the company's diagnostic web-site.
- Mean time to test: 2 minutes (includes human response time but does not include web connection time)
- Software mean time to self test: 1 minute
- No special tools or equipment required
- Menu driven instructions for operator

Measures for factory test and factory repair might include:

- Mean time to test and repair (MTTR): 10 min. for 80% of all failures, 60 min. for all other 20% failures (includes human response time)
- Maximum time to test and repair: 2.0 hours
- Test equipment, ergonomics and procedures: compatible with existing maintenance concepts and skill levels
- Fault-tolerance requirements: maximum of 0.1% false alarms
- Test equipment: reliability of 99% with a maximum of 0.1% cannot duplicate failures
- Inspection: 100% measurement of all critical dimensions and identify presence of all manufacturing processes
- Total cost of test and inspection including amortized cost of test equipment: $5.00 per unit
- Mistake proof all hardware connections, and procedures
- Data recording capability and easy to read presentation

15.3   TEST PLAN

The test plan translates important design/product requirements into test requirements for each level of test and each vendor and environment. Planning for testability ensures that test capability is designed into the product and manufacturing processes. The objective is to develop an extensive but cost-effective testability approach. Testability is important since an incorrect design approach can make it difficult to properly test a product. The design team should identify technical advances that can reduce test times, test equipment cost, personnel skill requirements, and floor space. Efficiency (i.e. fast, low cost) and effectiveness (i.e. most possible defects tested) are the key goals.

The steps for testability and the test plan are to:

1. Identify all related requirements including test cost, cycle time, quality, false rejects/accepts, ergonomics and schedule
2. Determine what to test and what not to test for first pass tests and retesting of failed units
3. Develop a test flow chart and identify the best process and equipment for test and inspection
4. Design the product to be compatible with the selected test and inspection processes using testability guidelines
5. Qualify the test process

Major decisions are whether to test or not test at all, what needs to be tested, level of the test, and who should test it. If the quality of a part or software code is extremely high, it may be best to not test the item at all. When something is going to be tested, should the part be fully tested or would a simple go/no go test be adequate. Finally should the vendor test the item or should the item be part of a higher-level system test? Flow charts and simulation are used to ensure an effective but streamlined workflow.

The test plan and testability effort must consider the maintenance plan. For example, the notebook computer needs to provide enough information to decide whether the computer can repair itself or what the user should do. Should the owner call for technical support, connect to a web-based factory diagnostic center, send the unit back, repair it themselves, etc.

The earlier a failure is detected in product development, the less costly it is to repair. This is true for both hardware and software. The costs of detecting a failure during the different phases of product development are shown in Table 15.1.

Fault detection and identification are defined as the process of detecting defects and identifying the type of defect. Fault isolation is the process of identifying the exact location of the identified defect. Other considerations, such as the cost of test equipment and logistics requirements, should also be addressed in order to arrive at how much testing should be performed and at what level

Test and inspection are often classified according to the location where they are performed, such as test and evaluation, production, operational, and maintenance. The overall strategy for production testability is to design a system

TABLE 15.1 Failure Costs

Level of assembly	Historic nominal cost per failure ($)	Willoughby presentation 1985 ($)	Computer Product Marcoux, 1992 ($)
Component level	1	-	1
Circuit board level	10	395	100
Box level	100	495	200
System level	1000	1,100	300
User/Field Operational	2000-20,000	16,345	700

that can be effectively and efficiently tested at each appropriate stage from the lowest level component parts or software modules to the complete system. For example, the strategy of operational testability is to identify when systems have failed in the field. Maintenance testability is similar to production testability, in which the objective is to ensure that the system is operational after a maintenance action has been performed. For each level to be tested effectively and efficiently, testability goals must be traded off with the other design considerations.

Testability does not attempt to determine the seriousness caused by a particular failure. Testability can include defects with failure modes that may not critically affect system performance but still effect the usefulness of the product.

When inspection or test passes a faulty system or software it is probably caused by an incomplete test, incorrect parameters used in the test, failure modes exist that are not evaluated, poorly trained operators, or non-precise inspection procedures. Another problem in testing is to fail a system that is fault free (i.e., no fault exists), referred to as a false alarm. For very complex electronic systems, some have actually found that the number of false alarms has often exceeded the number of actual failures.

Problems occur when a failure is shown during one test but cannot be duplicated later when performing another test. This is often called a "cannot duplicate" failure and usually occurs when the test conditions (e.g., environment, people or test equipment) have different standards for identifying defects. Other problems include poor fault isolation and excessive test times. The effects of these problems are shown in Table 15.2.

15.4   EXAMPLES OF DEFECTS AND FAILURES

The following subsections discuss the various types of defects, faults, and failures that a testability and inspectability design effort should attempt to identify. These are:

* Incorrect and marginal designs

TABLE 15.2 Testability and Inspection Problems

Test and Inspection Problem	Effects
1. Pass faulty system	Decreased reliability, quality, and availability; increased warranty costs
2. False alarm (fails good system)and cannot duplicate failures	Increased test costs; decreased availability; increased number of systems required
3. Poor fault isolation	Increased time to test and repair; increased levels of repair skills and training
4. Excessive test and inspection times	Increased time to test and test cost
5. Test and inspection of non-critical parameters	Increased test cost and cycle time

- Production defects
- Test measurement errors
- Operational and maintenance induced failures

Incorrect and Marginal Designs

The test and evaluation phase attempts to identify design problems for the purpose of design improvement and verification. This is normally done with prototypes and validation lots to identify critical, major, and minor characteristics of the product. Due to complexities of most product designs, no better systematic approach exists other than "testing according to specification". An incorrect or marginal design problem is simply a case of the design being unable to perform a specified function. One cause of this type of failure could be a simple error in the drafting of a schematic, where two register data strobe lines were accidentally connected. A much more difficult case to identify is a marginal design. A marginal design problem is a design that cannot meet all specifications when subjected to unusual or worst-case conditions. An example of this type of problem occurs in circuit design when only nominal times are used to establish circuit setup times. If every device in the path is near its maximum or minimum delay time, problems occur. This condition of maximum delay is unlikely, but possible, and must be considered.

Production Defects

Production defects are flaws introduced during manufacturing, assembly, material handling, and test processes. It is hoped that these flaws are caused by exceptions to the normal process. Poorly defined specifications can also lead to mis-communication problems especially for purchased parts and test/inspection procedures and equipment. Inspection alone cannot identify all defects. This is especially true for non-visual parameters such as electrical parts (current, voltage, resistance), software (code execution), or after mechanical assembly when some areas cannot be seen. Sometimes assembly defects will satisfactorily function for a while until the environment creates sufficient corrosion, oxidation, jarring, or other normal environmental stress that in combination with the defect create a fault. An example of this is when cracks or voids are located in a solder joint. Over a period of time, environmental conditions will cause the cracks to slowly grow until a point where the joint will finally fail. A common misconception of designing for production test and inspection is that most failures are easily identified during static test conditions. Unfortunately, many defects caused by manufacturing are marginal and can only be identified in a dynamic test environment that simulates the environmental and operational conditions that the product will experience.

Test Process Measurement Errors

Test procedures, operators, and equipment can have errors just like any other manufacturing process. Some measurement errors with their common problem causes are shown in Table 15.3 as adapted from Kolts, 1997.

The process is to identify all error sources that could occur, quantify their effects, and design to eliminate, minimize, or compensate for the errors. A common method to eliminate these errors is to use more accurate test equipment or internal test methods.

TABLE 15.3 Test Measurement Errors (adapted from Kolts, 1997)

Requirement Errors	Mistakes in standards and requirements
	Insufficient documentation
	Incorrect procedures
Systematic Errors	Gain and offset
	Input/output impedance
	Cable and connector losses
Calibration errors	Non-linearity
	Environmental changes of temperature, humidity, etc.
	Uncertified equipment
Drift Errors	Changes of component values or instrument performance over time
Random Errors	Noise
	Thermal offsets
	Power line variations
	EMI and RFI
Operator Errors	Operator errors (omissions, incorrect, out of sequence, etc.)

Operational, Installation, and Maintenance Induced Failures

Operational and maintenance failures occur after a product has been tested and delivered. Some failures are immediate, but most occur later during the actual use of the product. Some of the common causes for these failures are installation, packaging, transportation, environment, aging, abuse, static discharge, and power fluctuations. Packaging and transportation can cause failures since boxes that are dropped in shipping can result in higher levels of shock and vibration than the equipment was designed for and would actually experience in operational use. Another type of operational failure is product abuse. Depending upon its definition, a great variety of things can constitute abuse of a system. Some common examples of abuse include dropping a product, operation in an overheated condition, use in an environment not intended for the product, not performing required maintenance tasks, improper storage, and using software products that were not intended for the particular use. An person banging on a computer keyboard out of frustration is one example.

Electronic components are occasionally subjected to environmental conditions that reduce their normal life span during operation. Some examples are high junction temperatures, radiation, contamination to nonhermetically sealed devices, chemical breakdown, and vibration that causes mechanical wear and stress fracturing of materials. During operation, heat is normally considered the major type of environmental stress for electronic products. In some cases, even room temperatures may be high enough to cause a product to overheat and fail. Thermal expansion of different materials can also produce mechanical stresses large enough to physically break the device.

Environmental stresses generally cause permanent failures, but can also cause intermittent failures. An example is in electronic circuits that operate under specified conditions and their outputs depend on the output impedance or resistance. An increase in temperature will increase the output impedance and therefore distorts the output

Most parts and components will eventually fail over long periods of time owing to the effects of wear and aging. These types of failures can often be minimized by preventive maintenance, such as printer ribbon replacement, mechanical adjustment, and tightening of assemblies that may come loose from vibration.

Static electricity is another condition capable of damaging electronic equipment. Methods are available to control the effects of static electricity not only on extremely sensitive electronic components, but also on all electronic components. Most static problems occur in the ordinary handling of printed wiring boards and components during repair or transportation.

Failures can be caused by power surges or transients on incoming power lines. Some of these are quite similar to the damage caused by static discharge. Unless precautions are designed into the product, many electronic components are susceptible to damage from high voltages. A power failure can cause other problems, particularly in products with forced cooling. During a power outage, the heat generated by the equipment is trapped in the components and could cause temperatures to exceed even the maximum storage ratings. Power transients can also induce temporary failures. Some examples of this include parity or error-correcting code (ECC) errors in electronic memory components, parity or cyclic redundancy code errors written to magnetic media devices, and interrupted data transmissions. All these examples require some sort of recovery process to resume normal system operation.

15.5 DESIGN FOR EFFECTIVE AND EFFICIENT TEST

The first step is identifying which defects to focus on. This is a difficult decision since there are more potential defects than could ever be tested and inspected. Defects that critically affect a product's performance, quality or safety are the highest priority. Those that are commonly found in the manufacturing and software processes being used are also usually checked. Failure modes and effects analysis (FMEA and PFMEA) or root cause analysis can provide defects to be evaluated. Important manufacturing parameters that limit the number of potential defects to be evaluated include the 1.) time to test and inspect that is available and the 2.) cost of the test and inspection process itself.

After successful identification of a defect, isolation to an area that permits cost-effective repair is the next design objective of testability. It is important that the test and inspection approach is effective, efficient and consistent to preclude ambiguous or inconclusive results. Consistency is especially important to ensure that failures caught by one test or inspection can be later duplicated when subsequent evaluations are performed. This reduces the number of faulty products to be returned because the failing condition could not be replicated. The last element of the test cycle is to support the product in the field. The maintenance concept is used as the baseline for the testing strategy. The levels of maintenance and their test requirements are determined using such factors as system architecture, operational environment, operational scenario, user expertise, and hardware complexity. A diagnostic and test strategy must be developed for each level of production and repair.

Many design techniques can significantly improve the testability of a product. Major trends in testing and inspection are:

1. Increasing levels of product complexity and number of new technologies to be tested
2. Increasing levels of automation including self testing and self maintenance
3. Higher levels of quality, inspection and testing required
4. Lower test cost and higher quality per tested and inspected function
5. More intelligent information on defects including statistical trends and recommendations

Other major areas include ergonomics, mistake proofing, documentation, and vendor participation that have been previously discussed in other sections.

15.6   DESIGN PROCESS FOR INSPECTABILITY

The purpose of inspection is to either:

- **Identify whether a process was performed or not performed** (e.g., is a hole drilled in the part? Are all parts inserted into the printed circuit board? Is there a software module in the program to compensate for time delays?)
- **Make static measurements or observations that can infer that the process was performed correctly** (i.e., measure the drilled hole size, measure the location or orientation of a part on a printed circuit board, count the lines of software code to estimate the time delay).

Inspection methods can include visual methods such as the human eye; optics aided methods such as microscopes; automated methods such as machine vision or software checking; and mechanical measurement methods such as gages, micrometers, and calipers. Some parameters that cannot be seen by the human eye can be inspected using X-ray, ultrasound and other radiation techniques. For instance, laser triangulation systems can measure the height and volume of solder paste. Cross-sectional X-ray systems, such as S-ray laminography, can measure solder joint parameters that include fillet height, solder volume, voids, and average solder thickness. Another example is disposable lighters where X-rays can measure the lighter fluid level inside the product. Inspection of software is the process of visually evaluating the quality of the program. Software inspection techniques include code inspections, documentation, and walk throughs.

The important design parameters that increase both manual and automated inspectability include:

- **Accessibility (physical and visual)**- ability to easily get into (i.e., access) the area to be inspected and easily visualize the area.
- **Product materials and parts** - select materials and parts whose parameters are easily identified (e.g. color contrast, size, etc.) and measured for the particular inspection process
- **Fixturing** - provide features which increase the ability to easily fixture the product for the inspection process
- **Lighting and other environmental parameters** - provide optimum environment and adequate space for the inspection task
- **Software methods** - use industry proven methods and techniques which allow easy inspection of software code

Manual inspection is still the most popular method in industry although there are problems with this approach such as:

- Operator boredom and fatigue resulting in human errors
- Performance and quality standard differences between individual inspectors
- Slow speed and high cost of performing manual inspection

15.6.1 Producibility for Machine Vision

The use of automated inspection systems such as machine vision is increasing due to improved technology and lower costs. Product design parameters can improve the overall efficiency of machine vision, including measures such as accuracy, reliability, and repeatability. It is important to recognize the important interaction between the product's design and the automated inspection process.

Producibility can improve machine vision in the acquisition of the image itself. Acquisition improvement can be made by product simplification techniques, such as reducing the number of components to be inspected or the number of different parts. Image acquisition is also sensitive to accessibility, lighting methods and color contrasts. Simple design decisions such as the color of the parts can make a significant difference. Attempts should be made to design for the particular vision system to be used and to incorporate common groups of fixtures. The less a vision machine has to correct or adapt to accomplish a particular task, the more efficient the process becomes (Casali, 1995).

Lighting is one of the most important aspects of any vision system. Product design can take into account a part's optical properties such as color, surface reflectance, and geometry. For example, where colors do not impact function, light versus dark colors can enhance the contrast associated with the task. This is especially important when using front lighting to inspect surface features such as solder line patterns or component location on a printed circuit board. Surface conditions must be carefully controlled in order to achieve repeatable results, because changing surfaces (textures, grains, etc.) produce different light intensities that complicate later image processing steps (Gyorki, 1994). Clear or semi-translucent materials may be needed in back lighting situations when the system will be inspecting part dimensions, different part outlines, or fluid level measurements. Side lighting equipment is beneficial for recognizing surface irregularities or material defects, but product designs should avoid the potential for confusing housing and fixture shadows and, in some cases, avoid highly reflective materials (Casali, 1995). Designs should consider possible direct effects by the vision system on a product's exposure to bright light or its heat-sensitivity.

Once the image has been acquired, the vision inspection system processes and analyzes significantly large amounts of data, making product design considerations important. The product design team must be aware of the particular threshold point of the particular vision system being used. Color intensity levels are evaluated to avoid inaccurate processing definitions. In theory, this will reduce the processing time by eliminating the decision logic routines associated with points within the threshold region (Casali, 1995).

Producibility can also affect the efficiency of the machine vision's interpretation. The system compares the processed image to a predefined or standard image stored in memory. "Indeed, the most significant challenge facing vision users today is deciding which key features to include in the reference image that defines the final product as having acceptable quality. Too few variables may not sufficiently define quality, and too many can needlessly bog the system down with invalid rejects" (Gyorki, 1994). The design specification of the product directly influences these decisions.

Technical difficulties can arise due to product alignment and orientation during image acquisition. Design practices for fixturing can include notches, guiding holes or pins, or slots to help alleviate this problem as well as to reference points or symbols that are stamped or printed directly on the product. These practices eliminate the need for sophisticated fixturing devices or sensors to locate or orient the part into position.

15.7   DESIGN PROCESS FOR TESTABILITY

Testability at only one level is not sufficient. The methodology must reach into all levels of the design. Today's functional units within a single design are as large as entire chips were just a few years ago. When all these units come together in a single product, a testability problem in any one of them can affect all of the others. There are four reasons for testability:

1.   **Higher quality:** Better test and inspection means better fault coverage so that fewer defective parts make it out of manufacturing and the test and evaluation design phase is more effective.
2.   **Easier and faster testing:** Reduces test times which reduces testing costs.
3.   **Less equipment and training cost:** compatibility with existing equipment and operator skills reduces cost.
4.   **Higher availability:** Built in testing gives the user constant feedback that the product is working and alerts the user when a problem occurs.

Some sample rules in a design for testability checklist for electronic equipment, developed by Bostak and others, (1993) are shown in Table 15.4.

TABLE 15.4 Excepts from a Design for Testability Checklist

- Is the product partitioned effectively, both functionally and physically, for layered testing so that faults can be detected and isolated independently and unambiguously?
- Are all shared signals and connectors capable of operating with the product and test equipment?
- Have the calibration procedures been included in the design of tests and built-in test?
- Does the instrument design allow for additional options without requiring access to previous subassemblies?
- Have all warm-up-related items been eliminated or minimized?
- Does the functionality of a module or assembly vary during the manufacturing process?

(adapted from Bostak et al, 1993)

15.8   SOFTWARE TESTABILITY GUIDELINES

There are basic design guidelines (i.e., heuristics) to make software easier to test. The guidelines encompass module size and complexity, scope of effect and control, number of fan-out/fan-in, and levels of nesting. The following sections are condensed from a report by Tracey Lackey (1996). For more information on software testability, the reader should review Byrne (1996), Pfleeger (1987), and Solvberg (1993).

Module Size and Complexity

Testing must focus on the critical or 'core' parts of the software. The core of the system is the complex portion of code that provides the power and functionality of the system. Consistent with the Pareto 80-20 rule, 20% of the code will have 80% of the errors. The core of the system must function reliably and be designed as a separate module or modules, so that the modules can be tested thoroughly and separately. Software can be logically partitioned into components that perform specific functions and sub-functions called modules.

Modules should average between 30 and 100 lines of code, with each module having low complexity. Keeping a module's complexity low is accomplished by reducing the number of decisions that occur within the module. Decisions include, for example, IF, WHILE, DO-WHILE, SWITCH, and FOR loops. These are control flow constructs, because they control the flow of data within the modules.

One tool for determining the complexity of a module is the McCabe's Cyclomatic Complexity that was discussed in the chapter on simplification. The McCabe's complexity is calculated by using a formula involving the number of decisions in the module to determine the module's complexity. The higher the complexity, the higher the number of defects should be expected. The McCabe's complexity, on average should be less than 10 (Byrne, 1996).

The modules should be independent of each other in order to make the modules easier to debug and test, more understandable, and easier to maintain. Independence between modules is called low coupling, where two modules communicate by parameters only. The modules have high cohesion, which means that all elements of a module are directed toward performing the same function (Pfleeger, 1987). All of the internal elements in a highly cohesive module are tightly bound with only one entry and exit point.

Scope of Effect and Scope of Control

"The scope of effect of a module should be limited to its scope of control" (Pfleeger, 1987). The scope of control of a module is the entire module's subordinates, all of the programs, or other modules that a particular module calls. The scope of effect of a module is every other module or program that is affected by decisions made within the first module (Solvberg, 1993). A module should not affect other modules that are not within its scope of control.

Span of Control (Fan-Out/Fan-In)

The fan-out of a module is the number of immediate subordinates of a module. The fan-out should be low, ideally less than seven. High fan-out indicates that few subprograms are ever used more than once in the system. Consequently, there is a higher potential for errors due to more code and logic.

Fan-in is the number of modules that call the given module. High fan-in is good because one piece of functionality is used many times, striving to create reusable code. Low fan-out and high fan-in helps to create modules that are easier to test with greater reliability (Solvberg, 1993).

Nesting

The control structures such as IF, WHILE, DO-WHILE, SWITCH, and FOR should not be too deeply nested. For example:

```
WHILE (counter < 10)
        FOR (x = 0, x > 10, x = x + 1)
            |
            FOR (y = 0, y > 10, y = y + 1)
                |
                FOR (z = 0, z > 10, z = z + 1)
                    |
                    |
```

This structure has a nesting level of three. A maximum level of five is recommended for ease of testing and debugging (Byrne, 1996)

In summary, a list of some design guidelines for software testability is as follows (adapted from Lackey, 1996):

- Average statement size <12
- Ratio of comments to statements > 20%
- Initialization is short (hopefully 1 vector) and easy
- A module (procedure, function, or subprogram) should be between 30 and 100 lines of code
- McCabe's level of complexity should be less than 10
- Modules should have a single purpose and be independent of each other
- Modules should have low coupling and high cohesion
- Modules should have only 1 entry and 1 exit point
- Fan-out should be low, less than 7, fan-in should be high
- Levels of nesting < 5
- Errors should be reported via flags with no "die" modes

15.9   TESTABILITY APPROACHES FOR ELECTRONIC SYSTEMS

The rest of this chapter illustrates various testability approaches used for electronic systems. These general approaches are also applicable for many other types of products. Testability design efforts for electronic systems can typically be broken into three major test approaches 1.) built in test, 2.) external diagnostic software, and 3.) external test equipment.

The testability techniques that will be discussed include:

- Built-in test (BIT) and Self-Maintenance
- Boundary scan testing
- Standardization
- Test points and control points
- Initialization
- Partitioning
- Feedback loops
- Redundant circuits
- Forcing error conditions
- Product, system and end-to-end testing

Built-in Test, Internal Test, Self-Diagnostics and Self-Maintenance

Future products will have more self-diagnostic and self-maintenance capabilities. Built-in test (BIT), built in self test (BIST) or internal test is when a product has the internal capability to initiate a test procedure, which electronically operates the product and identifies any failures. It is usually controlled by software, and can be initialized at power-up, invoked regularly by the operating software or hardware, or invoked upon request. Personal computers perform built-in tests during power-on initialization, for example virus scanning software. Regularly invoked BIT operation is usually controlled and scheduled by interrupt-driven software.

A future phase for testability is self-maintenance. Self-maintenance is where the product can maintain performance even when problems occur by identifying the problem and automatically implementing repairs. This requires the design to include sensors that can collect data, knowledge bases that can classify problems and identify solutions, and techniques that can automatically implement corrections.

Built-in test (BIT) started in military equipment and telephone systems. BIT's original mission was to reduce mean-time-to-repair (MTTR) in complex systems by locating a replaceable failed unit. The failed replaceable unit (e.g., a circuit board, a part, or software) would be replaced with a spare. Many manufacturers are turning to built in self-test to reduce the need for test equipment and because some devices can only be extensively tested by built-in-test capability.

The following key steps describes the approach typically used for implementation:

1. Identify key areas to be tested and the most likely cause and effects (i.e. test results) of various failure modes.
2. Identify how resident software and hardware can perform an internal check of the basic operation of the system. This includes microprocessors, associated memory, power supply, disk drives and the clock oscillator.
3. Implement built-in test into the hardware and software design.
4. Analyze the test results and determine the most likely cause and effect.
5. Provide test result information to the user through the display or test equipment, or access the repair center through an Internet connection.

The use of embedded microprocessors for built-in-test must be considered when creating any design. The embedded processor's capability and resident BIT software should be incorporated into the testing philosophy when appropriate. For example, a complex ASIC device may require several different BIT testers for different functional blocks. The testers may be networked to produce one overall pass or fail result for the entire ASIC. The BIT design can ensure that failures in the output or initiation devices themselves (i.e., the BIT hardware) are easily identifiable during troubleshooting. More than one scheme of BIT initiation and error display circuits can be used.

Boundary Scan

Boundary scan testing is a special type of in-circuit test and built-in-test for applications with extremely tight physical accessibility such as notebook computers and telephones. The main benefit of the boundary scan test strategy is the ability to individually set the input states of semiconductor devices such as an integrated circuit (IC) and to read the output states (Marcoux, 1992). This enables each IC to be individually tested within an assembly, without the use of probes in a bed-of-nails test fixture. Boundary scan places a scan register at the input/output (I/O) pins of an IC. The I/O pins are then connected into a sequential scan chain (Myers,. Boundary scan is useful for examining hard to access devices. The automated test equipment generates a stimulus for the input pins and then measures the response on the output pins using a serial data chain. Boundary scan allows the test to control and observe the state of the device without physical access.

For either boundary scan or built-in test strategies, the ability of a microprocessor to determine the operability of a product depends on the amount of observability and controllability of the hardware to be tested. Design team must interface with those engineers responsible for system-level diagnostics to ensure the testability of their function in the field.

Standardization

The number of different types of parts or connectors that interface with the test equipment should be held to a minimum. For example, the number of motherboard connector types for a system should be held to one type of connector (SO-pin, 90-pin, or whatever size is needed). All power, ground, and communication bus signals should be routed to the same pins. These two factors provide commonality, which assists test personnel and reduces the number of interface adapters needed to interface with the test equipment.

A simple method of standardization is to choose a standard pin on integrated circuits, transistors, diodes, and so on, and indicate that pin on the circuit board by an identifying mark. This can easily be done on both sides of the board using octagonal or hexagonal pads instead of the usual circular pads. Such marking assists in assembly and troubleshooting, especially when the board is viewed from the reverse side, and is particularly helpful on multilayer boards with a high density of components and feedthroughs.

Another technique for system or product level testability is the standardization of test-related equipment, such as connectors, controllers, communication modules, memory boards; front panel interface boards, and power supplies. The advantages of standardization are as follows:

- Spare part requirements are lowered.
- Slope of the learning curve(s) for new designers and technicians is increased.

- The reliability of a standard module produced in volume is usually greater than that of custom modules, and the cost is lower.
- The incorporation of standard indicator light configurations for power, status, and system failure cuts troubleshooting time and decreases maintenance costs.

Test Points and Control Points

The most commonly used method of designing for testability is the addition of test points and control points. A test and control point provides easy mechanical probe access for observing nodes or internal signals. It is not considered a part of the normal input or output signal group. Both test and control points can be added to a board at the unused pins on the motherboard connector or by adding a special test connector to the board. Test points and control points often use physically different connector pins; however, a single pin can sometimes be used as both a test point and a control point. Special attention must be given to the selection and use of test and control points to ensure that their presence does not adversely affect the board and its functions.

The reduction in board size and increase in design density and board complexity has greatly reduced the future effectiveness of this method. Many designs have test points on both sides of the board to gain complete test access. This technique requires testing with a double sided or "clamshell" fixture, which is not always desirable. Other problems occur due to tighter lead spacing, smaller test points, and non-standard dimensions. Testing becomes unrepeatable because the test pins do not have enough contact force to pierce the oxidation and solder mask that sometimes infringes on the test points (Laney and Loisate, 1992).

Initialization

Before a test starts it is important to know the current state and values of the system. A testability goal is to initialize all nodes on a board or system at the beginning of a test. Initialization refers to placing the signal nodes of a circuit board or system in a known and predetermined signal state. Since total initialization cannot always be accomplished, a good design practice is to initialize all memory devices. This can be accomplished by adding a control point or points that allow the previously uninitialized devices to be initialized.

Partitioning

Another method is to partition a design/product/module into subsystem functions that have functional independence. The objective of partitioning is to divide a complex system into smaller divisible units that can be easily tested. Figure 15.1 illustrates a complete system that has been partitioned at the top level. The initial partitioning of the design helps to sort the problem areas and also gives some insight into the grouping of functions. A good design practice is to separate the digital and analog portions. Functional partitioning allows the detected fault to be isolated to the lowest level possible and also allows replacement of the faulty box or board without affecting other areas of the system.

The next step is to partition the system into lower levels, such as a circuit board. Partitioning of a circuit board involves dividing circuit functions and logic families into separate areas. Partitioning by circuit function involves designing the circuit so that each function stands alone and can therefore be tested alone. If two or more logic types are used on the same board and interface with each other, the design team should design these functions in such a way that they can be tested alone as much as possible. Test and control points should be added at the interface to assist in fault detection and isolation.

Feedback Loops

Feedback loops are common in both analog and digital control circuitries. Testing of feedback loops is difficult because the output is dependent not only upon the input to the loop, but also upon past inputs, which have an effect upon the internal status of the feedback loop. To adequately test the feedback loop and to locate the faulty component(s), access to the feedback loop and/or control of the feedback loop must be provided.

At the board level, the most common method for providing testability for feedback loops is to use the motherboard connector to physically break the feedback loop. A second method of ensuring feedback loop testability is to provide an electrical means of breaking the feedback path. The two different schemes allow the circuit to be tested with no feedback or to be tested with external input data. In order for feedback loops to be tested properly at the box level, they must be designed in such a way that either the test equipment or software control can break the loop. When the loop is broken, the output becomes dependent only upon the input from the test equipment or the software controlling the feedback input.

Testing Redundant Circuits

Since redundant circuits are designed to be fault tolerant, they can be difficult to detect and difficult to isolate problems. Redundant circuits by their nature hide faults. They make the propagation of the fault almost impossible and fault isolation impossible. Redundant circuitry should be designed so that it can be separated into individual, nonredundant circuits. Highly sequential circuitry should be designed so that it can be divided into simple sequential circuits, as shown in Figure 15.2. This allows a fault to be detected and isolated.

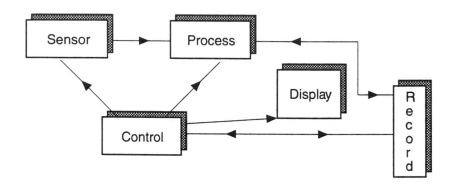

System partitioning

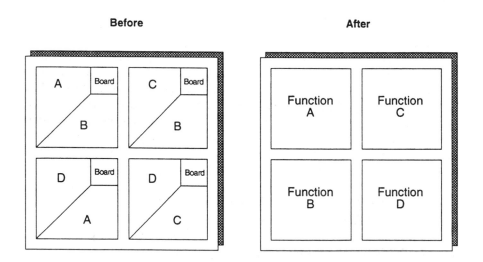

Subsystem partitioning

FIGURE 15.1 System-level partitioning.

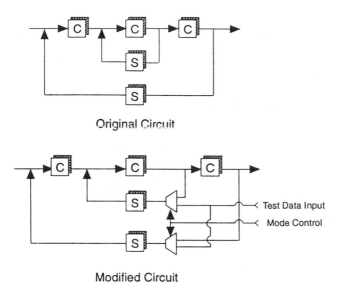

Original Circuit

Modified Circuit

FIGURE 15.2   Reconfiguring a circuit.

Forcing Error Conditions

To ensure the proper functioning of the built-in test and error correction capability, the ability to create error conditions is necessary. The capability should be provided to force error conditions in error-checking hardware. This applies to status signals, error interrupt signals, parity, and error correction logic.

This capability can be implemented via data to control. Test data patterns can be used to create errors. Control logic can also be used to force parity error and error interrupt bits. For error-correction logic in memory systems, hardware is needed that has the following capabilities:

- Read data memory with error-checking disabled.
- Read ECC memory onto a data bus.

Product, System, and End-to-End Testing

Product, system, and end-to-end testing are normally thought of as the final test of a system. This test is performed by generating known inputs and comparing the system's results to simulated results or previous results from the same system. System level testing, therefore, is required to identify system level failures such as subsystem integration problems or timing related failures that may go unchecked during lower level testing. A function of system level testing is to confirm the physical connectivity and dimensional integrity of the finished product. System test can be used as a simple go/no-go test, although this may not be recommended. It is often cost effective to run the test using operational software, in real time, and duplicating the operational environment of the system being tested. This can require large computers to verify results or predict results from the system. In addition, certain analog stimuli may have to be created to inject a problem condition into a system.

15.10   SUMMARY

Not much can be done to "add on" testability or inspectability after the product is designed. Design for testability and inspection must start at the very beginning of the design process. This will result in improved quality due to better fault detection and isolation, and lower manufacturing costs due to easier repair, and reduced time requirements for testing. The purpose of this chapter is to familiarize both experienced and inexperienced team members with key design principles of test and inspection by providing guidelines for their application.

15.11   REVIEW QUESTIONS

1.   What is the difference between inspection and test?
2.   What is the cost trade-off for increasing the level of testing?
3.   What are the roles of accessibility, controllability, and observability in testability?
4.   Describe the testing process as a design feedback process to improve quality.
5.   Why will built-in test and self-maintenance be so important in the future?

15.12   SUGGESTED READING

1.   P.P. Marcoux, Chapter 11 Testability, Fine Pitch Surface Mount Technology, Van Nostrand Reinhold, New York, 1992.

15.13   REFERENCES

1.   C.J. Bostak, C.S. Kolseth, and K.G. Smith, Concurrent Signal Generator Engineering and Manufacturing, Hewlett-Packard Journal, p. 30-37, April 1993.
2.   E. Byrne, Personal Interview with Tracey Lackey, April 12, 1996.
3.   R. Casali, Design for Machine Vision Inspection, Student Class Project, The University of Texas at Arlington, Spring 1995.
4.   J.R. Gyorki, Taking a Fresh Look at Machine Vision, In Machine Design, February, 1994.
5.   B.S. Kolts, Automated Precision Measurements, Evaluation Engineering, p. 136, May, 1997.

6. T. Lackey, Software Testability, Student Research Report, The University of Texas at Arlington, 1996.

7. S. Laney and S. Loisate, Circuit Density and Its Impact on Test, Circuits Assembly, p. 58-59, July, 1992.

8. P.P. Marcoux, Fine Pitch Surface Mount Technology, Van Nostrand Reinhold, New York, 1992.

9. M. Myers, Putting Boundary Scan to the Test, Circuits Assembly, p. 50-52, July, 1992.

10. S.L. Pfleeger, Software Engineering, New York: Macmillan Publishing Co., 1987.

11. A. Solvberg and D.C. Kung, Information Systems Engineering, New York: Springer-Verlag, 1993.

12. J. Willoughby, Jr., Certification of the Technical Disciplines in the Design and Manufacturing Process, Presentation to the Defense Science Board, Washington, D.C., 1985.

13. Brion Keller, Randy Kerr and Ron Walther, IBM Test Design Automation Group, Design for Testability, EE, March, 1999

INDEX